AF352137

PERGAMON MATERIALS SERIES
VOLUME 3

Wettability at High Temperatures

PERGAMON MATERIALS SERIES

Series Editor: Robert W. Cahn FRS
Department of Materials Science and Metallurgy, University of Cambridge, UK

Vol. 1 **CALPHAD (Calculation of Phase Diagrams): A Comprehensive Guide**
by N. Saunders and A. P. Miodownik
Vol. 2 **Non-equilibrium Processing of Materials** edited by C. Suryanarayana
Vol. 3 **Wettability at High Temperatures** by N. Eustathopoulos, M. G. Nicholas
and B. Drevet

A selection of forthcoming titles in this series:

Ostwald Ripening by S. Marsh
Structural Biological Materials edited by M. Elices
Phase Transformations in Titanium- and Zirconium-based Alloys
by S. Banerjee and P. Mukhopadhyay
Underneath the Bragg Peaks: Structural Analysis of Complex Materials
by T. Egami and S. J. L. Billinge
The Coming of Materials Science by R. W. Cahn
Nucleation by A. L. Greer and K. F. Kelton
Thermally Activated Mechanisms in Crystal Plasticity
by D. Caillard and J. L. Martin

PERGAMON MATERIALS SERIES

Wettability at High Temperatures

by

Nicolas Eustathopoulos

CNRS, Laboratoire de Thermodynamique et Physicochimie
Métallurgiques, ENSEEG, Institut National Polytechnique
de Grenoble, France

Michael G. Nicholas

Formerly at: Materials Development Division
Atomic Energy Research Establishment, Harwell, UK

Béatrice Drevet

Laboratoire de la Solidification et de ses Procédés
Centre d'Etudes et de Recherches sur les Matériaux
Commissariat à l'Energie Atomique – Grenoble, France

1999

PERGAMON

An Imprint of Elsevier Science

Amsterdam – Lausanne – New York – Oxford – Shannon – Singapore – Tokyo

ELSEVIER SCIENCE Ltd
The Boulevard, Langford Lane
Kidlington, Oxford OX5 1GB, UK

First edition 1999

Library of Congress Cataloging in Publication Data
Eustathopoulos, Nicolas.
   Wettability at high temperatures / by Nicolas Eustathopoulos, Michael G. Nicholas,
Béatrice Drevet.
     p. cm. – (Pergamon materials series ; v. 3)
   ISBN 0-08-042146-6
   1. Wetting. 2. Materials at high temperatures. I. Nicholas, Michael G. II. Drevet,
Béatrice. III. Title. IV. Series.

QD506.E78 1999
541.3′3–dc21   99-045829

British Library Cataloguing in Publication Data
A catalogue record from the British Library has been applied for.

ISBN: 0-08-042146-6

Printed in The Netherlands.

Contents

Series Preface

My editorial objective in this new series is to present to the scientific public a collection of texts that satisfies one of two criteria: the systematic presentation of a specialised but important topic within materials science or engineering that has not previously (or recently) been the subject of full-length treatment and is in rapid development; or the systematic account of a broad theme in materials science or engineering.

The books are not, in general, designed as undergraduate texts, but rather are intended for use at graduate level and by established research workers. However, teaching methods are in such rapid evolution that some of the books may well find use at an earlier stage in university education.

I have long editorial experience both in covering the whole of a huge field – physical metallurgy or materials science and technology – and in arranging for specialised subsidiary topics to be presented in monographs. My intention is to apply the lessons learned in 40 years of editing to the objectives stated above. Authors have been invited for their up-to-date expertise and also for their ability to see their subjects in a wider perspective.

I am grateful to Elsevier Science Ltd., who own the Pergamon Press imprint, and equally to my authors, for their confidence.

The third book in the Series, on a topic not previously treated in full-length book form, is presented herewith to our readership.

ROBERT W. CAHN, FRS
(Cambridge University, UK)

Preface

The purpose of this book is to bring together current scientific understanding of wetting behaviour that has been gained from theoretical models and quantitative experimental observations. The materials considered are liquid metals or inorganic glasses in contact with solid metals or ceramics at temperatures of 200-2000°C.

Wetting has been a significant scientific concern for the last two centuries and reference will be made to classical work by nineteenth century scientists such as Dupré, Laplace and Young that was validated by observations of the behaviour of chemically inert ambient-temperature systems. This classical work still provides the basic language for discussions of wetting behaviour and similar studies have continued in recent decades to provide understanding of the effects of factors such as the roughness or chemical heterogeneity of the solid surface. Due and proper notice is given both to such work and to studies dedicated to high-temperature systems. Although a limited number of wetting studies at high temperature were published before World War II, the first systematic studies, motivated mainly by the importance of wetting properties in powder metallurgy and steel-making, were performed in the fifties by Kingery and co-workers in the USA, Bailey and Watkins in the UK, Kozakevitch in France and Eremenko and Naidich in the Ukraine. A new impetus to wetting studies was given in the eighties by increased interest in the fabrication of metal-matrix composites by liquid routes and the joining of metals and particularly of ceramics by brazing and glazing. This intensification was accompanied by a marked increase in the number of research teams specializing in this field and it is significant that original papers have been published by teams from more than twenty different countries during the last five years.

Although many basic concepts established for chemically inert ambient temperature systems are also valid for high-temperature systems, these latter systems possess specific characteristics that exert a major influence on their wetting properties. First, the surface properties of nearly all metals and many ceramics at high temperatures are extremely sensitive to impurities, mainly oxygen, that are invariably contained in furnace atmospheres. For this reason, particular attention is paid throughout this book to interactions between oxygen and metallic or ceramic surfaces and to their effects on surface chemistry and wetting behaviour. Because only those wetting studies which were carried out during the last decade have benefited from high-resolution techniques for characterization of surfaces, recent experimental results are generally, but not exclusively, preferred to older

"Now, supposing the angle of the fluid to be obtuse, the whole superficial cohesion of the fluid being represented by the radius, the part which acts in the direction of the surface of the solid will be proportional to the cosine of the inclination ; and this force, added to the force of the solid, will be equal to the force of the common surface of the solid and fluid, or to the differences of their forces (…). And the same result follows when the angle of the fluid is acute."

The famous Young's equation, as described by Thomas Young himself in his original paper "An Essay on the Cohesion of Fluids" published in 1805 in Philosophical Transactions of the Royal Society of London (vol. 94, p. 65).

Chapter 1
Fundamental equations of wetting

This first Chapter provides a context and language for subsequent analysis of the many and varied important scientific and technical aspects of high temperature capillarity. It does so by first defining the thermodynamic and atomistic characteristics of the interfaces between different phases and then by developing the fundamental equations describing the wetting behaviour of *ideal* surfaces of chemically inert systems. Since a controversy regarding the validity of Young's equation periodically appears in the literature, particular attention is drawn to the effects of factors such as gravity, scale of observation, system size and time scale. Then, the wetting behaviour of *real* surfaces is presented taking into account the effects of roughness and chemical heterogeneity of the solid surface. The energetical quantities used to describe the various types of wetting occurring in practice (immersion, spreading, infiltration) are defined and discussed at the end of this Chapter.

1.1. SURFACE AND INTERFACIAL ENERGIES IN SOLID/LIQUID/VAPOUR SYSTEMS

The work needed for reversible creation of additional surface of a liquid L in contact with a vapour V, identified by the term σ_{LV}, was defined by Gibbs (1961) as :

$$\sigma_{LV} = \left(\frac{\partial F}{\partial \Omega} \right)_{T,v,n_i} \tag{1.1}$$

where F is the total free energy of the system, Ω the surface area, T temperature, v volume and n_i the number of moles of component i. The dimensions of σ_{LV} are energy per unit area. However this definition is not sufficient to describe the work needed for creation of a solid S surface because this can occur by two different processes.

First, the new surface can be created *at constant strain* by breaking bonds to increase the number of solid atoms (or molecules) which belong to the surface. A typical example of such a process is cleavage achieved without any elastic or plastic deformation of the solid (Figure 1.1.a). The reversible work done to create a surface of unit area by this process is simply σ_{SV} :

1

$$\sigma_{SV} = \left(\frac{\partial F}{\partial \Omega}\right)_{T,v,n_i} \tag{1.2}$$

Although σ_{SV} is a specific surface *free* energy quantity (it includes both energy and entropy), it is often called "surface energy" because of the greater simplicity of this term, and to avoid any confusion with the "specific surface free energy" defined by Gibbs which, in multicomponents systems, is different from σ_{SV} (Defay et al. 1966). In the following, σ_{SV}, as defined by equation (1.2), will be called the solid surface energy.

Second, the new surface can be created *without increasing the number of surface atoms* by purely elastic strain of the solid (Figure 1.1.b). The extra stress due to the surface, called by Gibbs "surface tension" or by other authors "surface stress" (Mullins 1963, Cahn 1989), is denoted as γ_{SV} and expressed as a force per unit length.

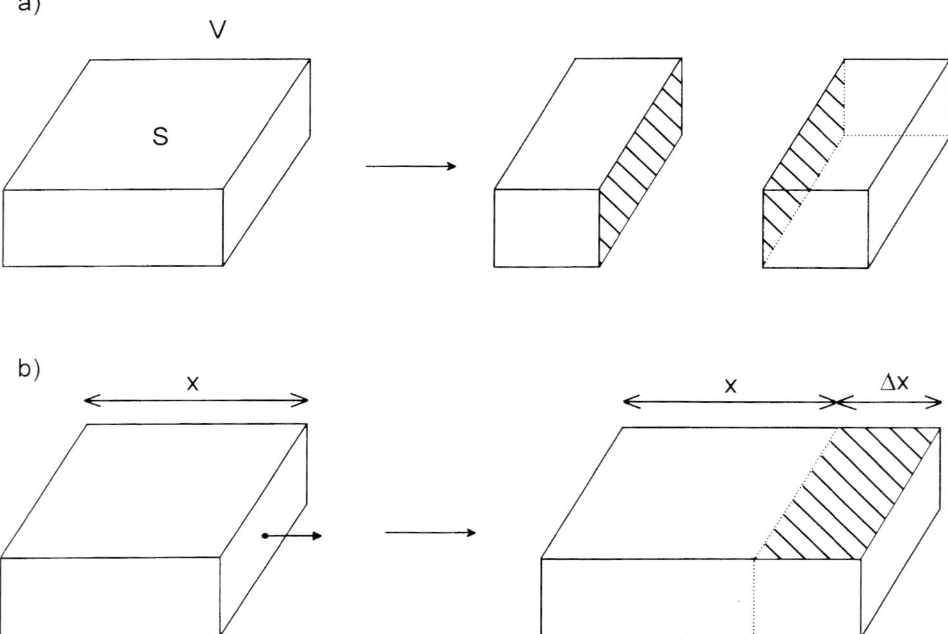

Figure 1.1. Creation of solid surfaces (shaded) by cleavage (a) and elastic deformation (b).

For liquids, σ_{LV} and γ_{LV} are equal because a reversible stretching of a liquid surface is identical to a reversible creation of new surface. In both cases, the liquid can increase its surface area only by the addition of new atoms to the surface. Note that, from a dimension point of view, an energy per unit area is equivalent to a force per unit length and the values are numerically equal when σ_{LV} is measured in J.m^{-2} (or mJ.m^{-2}) and γ_{LV} is measured in N.m^{-1} (or mN.m^{-1}).

For solids, σ_{SV} and γ_{SV} are different quantities. For instance, for each crystal face, there is a unique value of σ_{SV} (which is a scalar) while γ_{SV} depends also on the orientation along the face. Moreover, σ_{SV} is always a positive quantity (breaking bonds needs work to be done) while γ_{SV} can be either positive or negative (Nolfi and Johnson 1972). For high symmetry surfaces, the surface tension is related to the surface energy by the equation (Shuttleworth 1950) :

$$\gamma_{SV} = \sigma_{SV} + \frac{d\sigma_{SV}}{d\varepsilon} \tag{1.3}$$

where ε is a macroscopic elastic strain. The physical origin of the difference between γ_{SV} and σ_{SV} i.e., of the term $d\sigma_{SV}$ / $d\varepsilon$ in equation (1.3), can be explained taking into account the atomistic origin of σ_{SV}. As it will be seen below for monoatomic solids, σ_{SV} is proportional to the difference in potential energy between an atom of the surface and an atom of the bulk solid (equation (1.13)). When a new surface is created by stretching the solid (Figure 1.1.b), this difference does not remain constant. Indeed, because surface atoms are bonded weakly compared to those in the bulk, the work needed to stretch the surface is less than for the bulk material.

From now on, for S/V and S/L boundaries, only the surface and interfacial energies, σ_{SV} and σ_{SL}, will be considered. For L/V surfaces, both the surface tension γ_{LV} and surface energy σ_{LV} will be used interchangeably depending on the context.

Let us consider two bodies A and B which can be either two solids, or two liquids or a solid and a liquid, that have a unit cross-sectional area. A and B are surrounded by a vapour phase V at constant temperature. The free energy change corresponding to the reversible creation of two new surfaces of A and of B of unit area, by the process schematized on Figure 1.2, is :

$$\Delta F_{1\text{-}2} = 2(\sigma_{AV} + \sigma_{BV}) \tag{1.4}$$

Similarly, the free energy change corresponding to creation of two A/B interfaces by joining two surfaces of A and of B is equal to :

$$\Delta F_{2\text{-}3} = 2(\sigma_{AB} - \sigma_{AV} - \sigma_{BV}) \tag{1.5}$$

Finally, the transformation 1-3 corresponds to :

$$\Delta F_{1\text{-}3} = \Delta F_{1\text{-}2} + \Delta F_{2\text{-}3} = 2\sigma_{AB} \tag{1.6}$$

where σ_{AB} is the A/B interface energy.

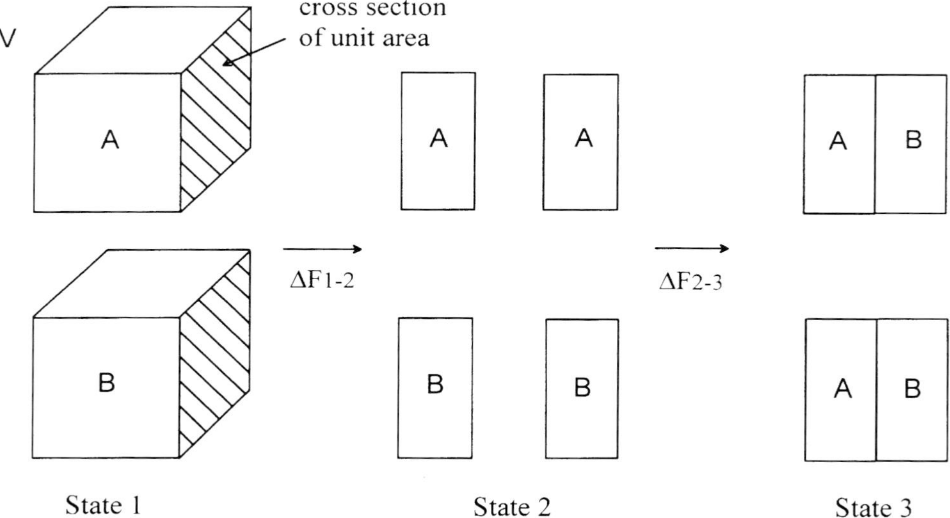

Figure 1.2. Formation of two A/B interfaces of unit area from pure A and B bodies.

For pure A or B, the quantity $2\sigma_{AV}$ or $2\sigma_{BV}$ defines the *work of cohesion* W_c of A or B :

$$W_c^A = 2\sigma_{AV} \tag{1.7.a}$$

$$W_c^B = 2\sigma_{BV} \tag{1.7.b}$$

In equation (1.5), the quantity $(\sigma_{AB} - \sigma_{AV} - \sigma_{BV})$ is equal but with an opposite sign to the *work of adhesion* defined by Dupré (1869) :

$$W_a = \sigma_{AV} + \sigma_{BV} - \sigma_{AB} \tag{1.8}$$

These macroscopic quantities can be related to the bond energies ε_{AA} and ε_{BB} in the bulk phases and ε_{AB} at the interface by a simple "nearest neighbour" interaction model. Assuming that :

– A and B are monoatomic solids that are chemically inert towards each other,
– A and B have the same crystal lattice and the same atomic volume and are perfectly matched at the interface,
– both cohesion and adhesion result from atomic interactions between nearest neighbours,
– entropy contributions to surface energies are negligible,

the three interfacial energies contained in equation (1.8) can be calculated simply from transformation 1-2 of Figure 1.2 :

$$\sigma_{AV} = -\frac{Zm_1}{\omega}\frac{\varepsilon_{AA}}{2} \tag{1.9}$$

$$\sigma_{BV} = -\frac{Zm_1}{\omega}\frac{\varepsilon_{BB}}{2} \tag{1.10}$$

and from transformation 1-3 :

$$\sigma_{AB} = \frac{Zm_1}{\omega}\left[\varepsilon_{AB} - \frac{\varepsilon_{AA} + \varepsilon_{BB}}{2}\right] \tag{1.11}$$

Using equation (1.8), one obtains :

$$W_a = -\frac{Zm_1}{\omega}\varepsilon_{AB} \tag{1.12}$$

In equations (1.9) to (1.12), ω is the surface area per atom, Z the number of nearest neighbours in the bulk crystal, m_1 the fraction of broken bonds at the surface of A or B per atom and ε_{AA}, ε_{BB} and ε_{AB} the bond pair energies, defined as negative quantities.

It is also interesting to derive the expression of the surface energy of a crystal, by equating this quantity to the difference in potential energy of an atom at the surface (E′) and a bulk atom (E) divided by the surface area per atom ω :

$$\sigma_{AV} = \frac{E' - E}{\omega} \tag{1.13}$$

In the framework of the nearest neighbour model, E′ and E equal the product of the number of atoms in nearest neighbour position and the pair energy per atom i.e., $E' = Z(m_1 + m_2)\varepsilon_{AA} / 2$ and $E = Z(2m_1 + m_2)\varepsilon_{AA} / 2$, where m_2 is the fraction of nearest neighbours located in the surface plane $(2m_1 + m_2 = 1)$ (Figure 1.3). Introducing the expressions for E′ and E into equation (1.13) leads to expression (1.9).

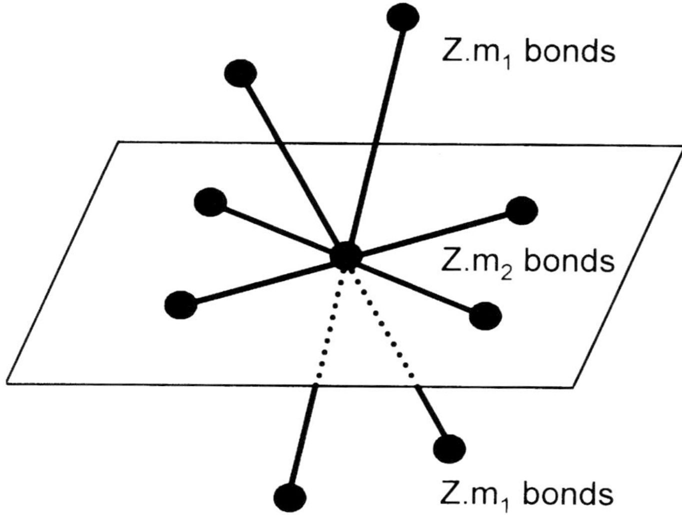

Figure 1.3. Definition of the fractions m_2 and $2m_1$ of nearest neighbours lying respectively in the same plane and in adjacent planes for a bulk atom lying in a symmetry plane of a crystal and having Z nearest neighbours.

The bond energies ε_{AA} and ε_{BB} are related to the energy of sublimation per atom L_s of the crystals A and B as follows :

$$L_s^A = -Z\frac{\varepsilon_{AA}}{2} \tag{1.14}$$

Combining equations (1.9) and (1.14) yields:

$$\sigma_{AV}\omega = m_1 L_s^A \qquad (1.15)$$

This equation shows that the surface energy *per atom* $\sigma_{AV}\omega$ of a moncatomic solid A is proportional to L_s^A. For a given family of compounds, a correlation between the surface energy per atom and the energy of sublimation (or evaporation) is expected provided m_1 is constant. Such correlations hold well for the solid and liquid surfaces of metallic bodies and also for the liquid surfaces of oxides and halides (see Figures 4.1, 4.9 and 4.10).

Equation (1.12) indicates that the magnitude of W_a directly reflects the intensity of interactions between A and B atoms across the common interface. Obviously, in real systems, the relation between W_a and bond energies is more complicated than equation (1.12) suggests. However, the physical meaning of W_a remains the same.

1.2. IDEAL SOLID SURFACES

Consider a flat, undeformable, perfectly smooth and chemically homogeneous solid surface in contact with a non-reactive liquid in the presence of a vapour phase. If the liquid does not completely cover the solid, the liquid surface will intersect the solid surface at a "contact angle" θ. The equilibrium value of θ, used to define the wetting behaviour of the liquid, obeys the classical equation of Young (1805):

$$\cos\theta_Y = \frac{\sigma_{SV} - \sigma_{SL}}{\sigma_{LV}} \qquad (1.16)$$

In this book, a contact angle of less than $90°$ will identify a wetting liquid, while a greater value will identify a non-wetting liquid. If the contact angle is zero, the liquid will be considered to be perfectly wetting (see also Table 1.1 in Section 1.4.1).

Equation (1.16) can be easily derived by calculating the variation of the *surface free energy* F_s of the system caused by a small displacement δz of the S/L/V *triple line*, usually hereafter referred to as TL. In Figure 1.4 the solid surface is vertical and TL is perpendicular to the plane of the figure and assumed to be a straight line, rendering the problem two-dimensional. Thus, the total length of TL is constant during its displacement, as in the case of a meniscus formed on a vertical plate. Moreover, the radius r of the TL region considered in this calculation (Figure 1.4) is much larger than the range of atomic (or molecular) interactions in the system.

For metallic and ionocovalent ceramics, this range is roughly 1–2 nm, while for bodies with long range van der Waals interactions it can attain 10 nm (de Gennes 1985). However, the radius r must be small compared to a characteristic dimension of the liquid, for instance the average drop base radius R in the sessile drop configuration (see Figure 1.6) or the maximum height of a meniscus formed on a vertical solid wall (see Figure 1.9) which are both typically in the millimetre range. Further, inside the region of radius r, the intersection of the L/V surface with the plane of Figure 1.4 is assumed to be essentially a straight line. With these assumptions, the variation of interfacial free energy per unit length of TL, resulting from a small linear displacement δz of TL, is:

$$F_s(z + \delta z) - F_s(z) = \delta F_s = (\sigma_{SL} - \sigma_{SV})\delta z + \cos(\theta)\sigma_{LV}\delta z \qquad (1.17)$$

and the equilibrium condition $d(\delta F_s)/d(\delta z) = 0$ leads to equation (1.16).

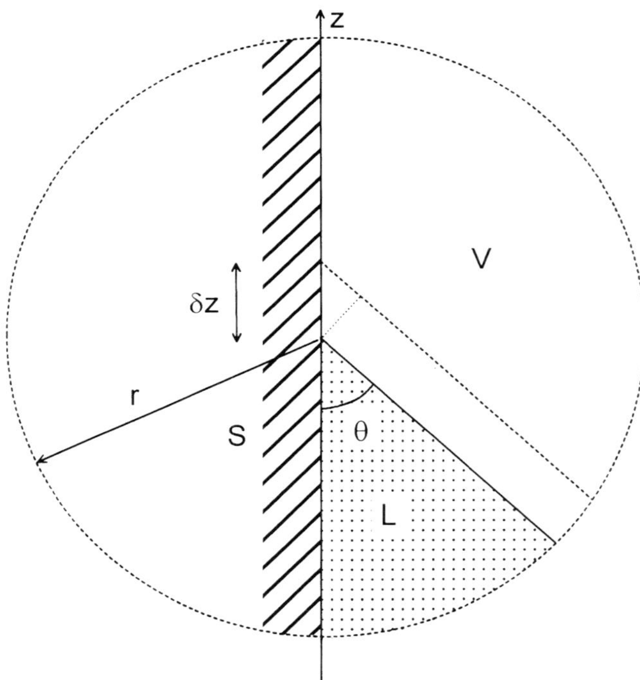

Figure 1.4. Displacement of a triple line around its equilibrium position that allows derivation of the Young equation. Only a small region close to the triple line is taken into account to neglect the curvature of the L/V surface.

1.2.1 Microscopic and macroscopic contact angles

In practice, the geometry in a *core* region around TL, i.e. inside a sphere of radius r_c which is of the order of the range of atomic interactions in the system, can be different from that shown in Figure 1.4. Due to long range interactions, the energy of an atom lying on a given interface inside this core region is different to the energy of an atom at the same interface far from TL. Thus, the three relevant interfacial energies close to and far from TL are different and this difference increases with the range of atomic interactions. Consequently, the local contact angle, as far as it can be defined at this nanometric scale, may be different to the macroscopic contact angle θ_M. An example is given schematically on Figure 1.5.

As pointed out by de Gennes (1985), the deviation of TL from its nominal position does not modify equation (1.17) provided this deviation does not change during the displacement δz. This is a reasonable assumption for small displacements around the equilibrium position.

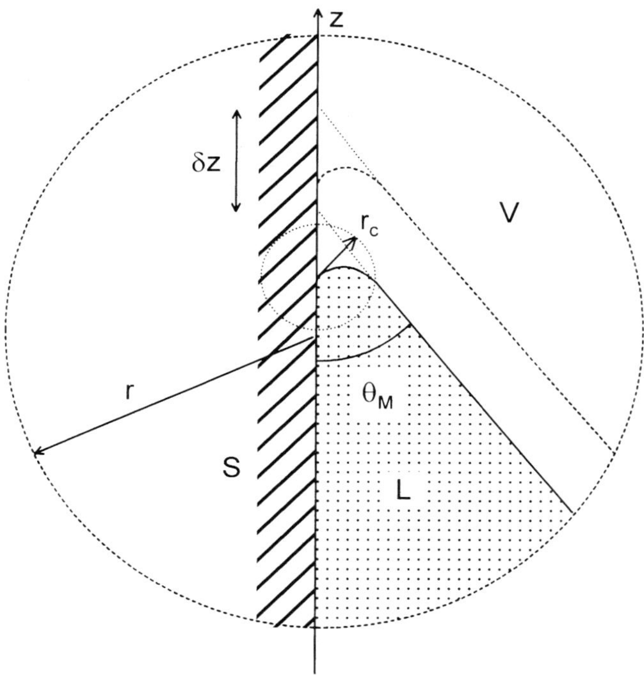

Figure 1.5. As Figure 1.4 but showing a microscopic contact angle different from the macroscopic θ_M.

1.2.2 Effect of system size

In the sessile drop configuration (Figure 1.6), when the drop size becomes of the order of magnitude of the radius r_c of the core region defined in Figure 1.5, the relevant contact angle is no longer the Young contact angle but the microscopic contact angle. The relation between these two angles has been discussed using the concept of line energy (or line tension). During wetting, the increase δR of the drop base radius leads to an increase of the TL length (Figure 1.6) and the free energy associated with the increase of the TL length must be taken into account in the interfacial energy calculation. The triple line can be treated phenomenologically as an equilibrium line defect with a specific *excess* free energy τ (per unit length). To some extent, it can be compared to the step energy on faceted crystals used to describe crystal growth or changes of crystal surface energy with crystallographic orientation in the case of vicinal surfaces. Then, the variation of interfacial free energy during wetting in the sessile drop configuration (Figure 1.6) is:

$$\delta F_s = 2\pi R(\sigma_{SL} - \sigma_{SV})\delta R + 2\pi R \cos(\theta)\sigma_{LV}\delta R + 2\pi\tau\delta R \tag{1.18}$$

Therefore, the equilibrium condition $d(\delta F_s)/d(\delta R) = 0$ leads to:

$$\cos\theta_{eq} = \cos\theta_Y - \frac{\tau}{R\sigma_{LV}} \tag{1.19}$$

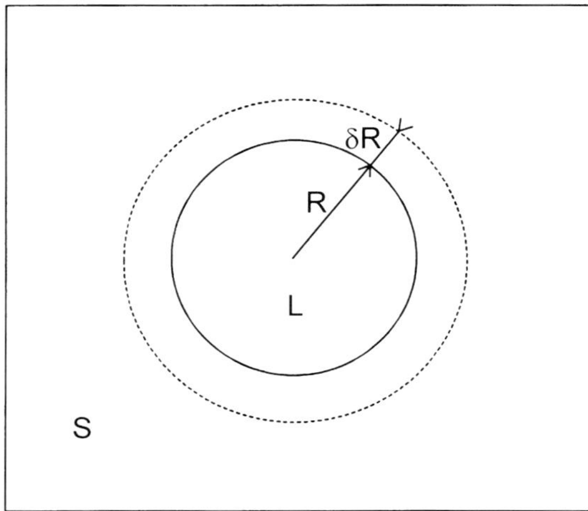

Figure 1.6. Top view of a sessile drop during spreading.

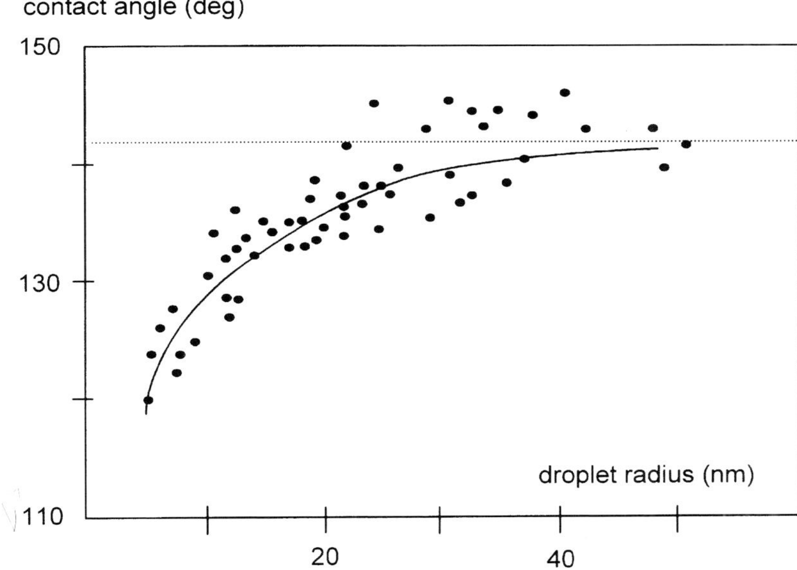

Figure 1.7. Dependence of contact angle on the metal droplet radius for Pb on vitreous carbon at a temperature close to the melting point of Pb. The horizontal line identifies the macroscopic contact angle. Data from work reported in (Chizhik et al. 1985).

The importance of the corrective term $\tau(R\sigma_{LV})^{-1}$ decreases when the drop size increases and it appears to be negligible for $R > 100$ nm (Shcherbakov et al. 1995). For instance, no significant differences in contact angle of pure Cu on monocrystalline Al_2O_3 were found between millimetre and micron size Cu droplets (Soper et al. 1996). However, the correction can be important for very small drops, for instance for those nucleated by condensation on to solid substrates, which can be as small as a few nanometers. For such small droplets, differences in contact angle of tens of degrees can exist, corresponding to differences in work of adhesion of 100%, as shown by the results in Figure 1.7 for pure Pb on vitreous carbon. Thus, care has to be taken in the use of contact angle data obtained with macroscopic droplets in small size systems because no information, either theoretical or experimental, about τ is available for high temperature materials.

Note that equation (1.19) has been established assuming the line energy τ to be a constant. However it was argued (Marmur 1997) that since this quantity results from the atomic or molecular interactions between the three phases in the vicinity of the triple line, it cannot be independent of the contact angle since the relative inclination of the phases affects the extent of these interactions. Calculations of τ

for liquid/solid systems bonded by van der Waals interactions confirmed this point and showed that, for this type of system, the sign of τ is positive for acute contact angles and negative for obtuse contact angles.

1.2.3 Effect of the curvature of the liquid/vapour surface

Until now the L/V surface in the vicinity of TL assumed to have a negligible curvature. Hereafter, this assumption will be removed by minimizing the free energy of the whole S/L/V system in the gravitational field, rather than the free energy of the region of radius r around TL. In this case, the curvature at each point Q of the L/V surface has to satisfy the equation of Laplace (1805) (see Appendix A):

$$P_L^Q - P_V^Q = \sigma_{LV}\left(\frac{1}{R_1} + \frac{1}{R_2}\right) \tag{1.20}$$

where P_L^Q and P_V^Q are the pressures on the liquid and vapour sides of the surface and R_1 and R_2 are the principal radii at point Q (Figure 1.8).

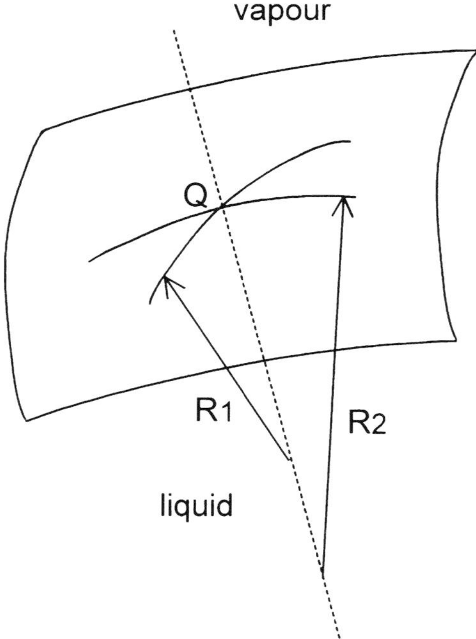

Figure 1.8. The principal radii of curvature R_1 and R_2 at a point Q on a curved liquid surface.

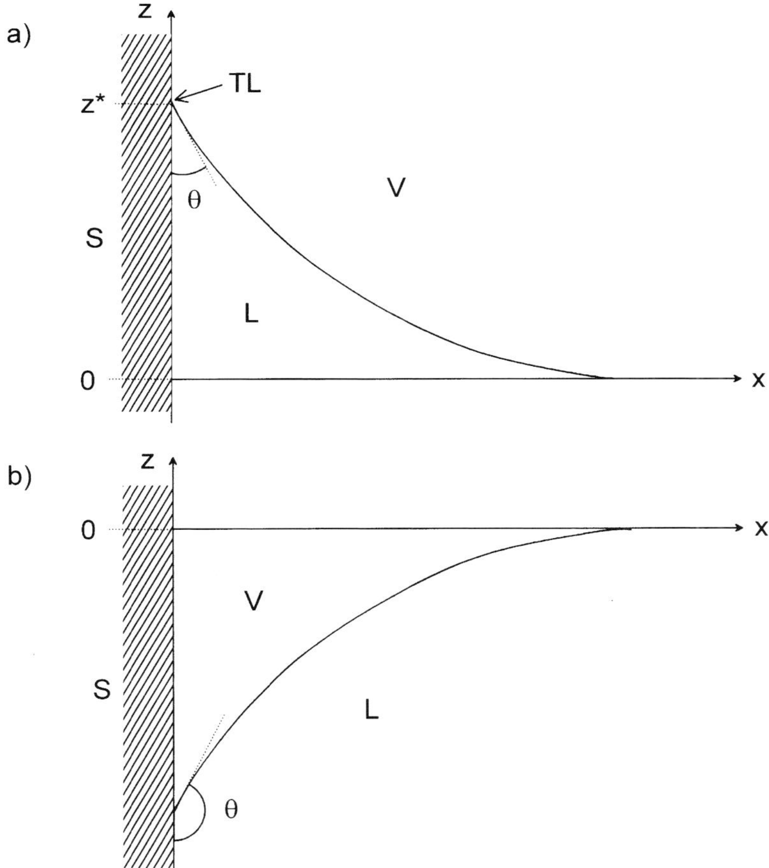

Figure 1.9. Meniscus rise on a vertical wall when $\theta < 90°$ (a) and depression when $\theta > 90°$ (b).

Because no analytical solution of the shape of the surface in the gravitational field is available in the sessile drop configuration (see Section 3.1.3.2), we prefer to consider a meniscus formed on a vertical plate (Figure 1.9) as in the papers of McNutt and Andes (1959) and Neumann and Good (1972). The calculations given below come from Neumann and Good (1972) but the conclusions are also valid for a tilted plate, as shown by Eick et al. (1975). The geometry used is given in Figure 1.9 with the vertical plate and TL perpendicular to the plane of the figure. The total free energy change ΔF can be calculated when a liquid surface initially in a horizontal position ($z^* = 0$, $\theta = 90°$) is raised (or depressed) to form a meniscus of height z^*, corresponding to a contact angle θ. Thus, the equilibrium contact angle can be determined by minimizing ΔF as a function of z^*, or equivalently of the contact angle θ. ΔF is composed of two terms: ΔF_b arising from changes in the potential energy of the bulk phase of the system, and ΔF_s resulting from changes in the surface and interfacial energies of the system. The last term takes into account both the difference between a S/V surface and a S/L interface ($\Delta F_{s,1}$) and the increase of the L/V area ($\Delta F_{s,2}$) during the meniscus formation. Calculations of these terms are given in Appendix B. For a triple line of unit length, the total free energy change ΔF is (Neumann and Good 1972):

$$\frac{\Delta F}{\sigma_{LV} l_c} = \mp A(1 - \sin \theta)^{1/2} + [2^{1/2} - (1 + \sin \theta)^{1/2}] + \tfrac{1}{3}[(2 - \sin \theta)(1 + \sin \theta)^{1/2} - 2^{1/2}]$$

$$(1.21)$$

where the minus sign before A corresponds to $0 \leq \theta \leq 90°$ and the plus sign to $90 \leq \theta \leq 180°$. In equation (1.21), A is a parameter that depends on the three interfacial energies of the system:

$$A = \frac{\sigma_{SV} - \sigma_{SL}}{\sigma_{LV}} \qquad (1.22)$$

For any rise z^* of the meniscus, z^* and θ are related by:

$$z^* = \pm \left(\frac{2\sigma_{LV}}{\rho g}\right)^{1/2} (1 - \sin \theta)^{1/2} = \pm l_c (1 - \sin \theta)^{1/2} \qquad (1.23)$$

where a positive value of z^* corresponds to $0 \leq \theta \leq 90°$ and a negative value to $90 \leq \theta \leq 180°$. The quantity $l_c = (2\sigma_{LV}/(\rho g))^{1/2}$, in which ρ is the liquid density and g the acceleration due to gravity, is often called "capillary length" and is simply the maximum rise of a liquid on a perfectly wetted vertical plate. (Note that in some papers the capillary length is defined without the factor $2^{1/2}$). The equilibrium contact angle of the system is obtained by setting $d\Delta F/d\theta = 0$, so that:

$$\pm A(1 - \sin \theta)^{-1/2} - (1 + \sin \theta)^{1/2} = 0 \qquad (1.24)$$

For $0 \leq \theta \leq 90°$, equation (1.24) yields only the solution:

$$\cos\theta_{eq} = A = \frac{\sigma_{SV} - \sigma_{SL}}{\sigma_{LV}} > 0 \tag{1.25}$$

which is identical to the Young equation (1.16), i.e. $\theta_{eq} = \theta_Y$. For $90 \le \theta \le 180°$, equation (1.24) again yields one solution:

$$\cos\theta_{eq} = A = \frac{\sigma_{SV} - \sigma_{SL}}{\sigma_{LV}} < 0 \tag{1.26}$$

The contact angle satisfying equation (1.25) or (1.26) corresponds to a minimum of the total free energy of the system $\Delta F(\theta)$. Examples are given on Figure 1.10 for $A = 0.5$ and $A = -0.5$ using equation (1.21) written with the minus sign for $0° \le \theta \le 90°$ and with the plus sign for $90° \le \theta \le 180°$. The minima of the curves are located at $\theta = 60°$ and $\theta = 120°$, values which correspond to arccosA.

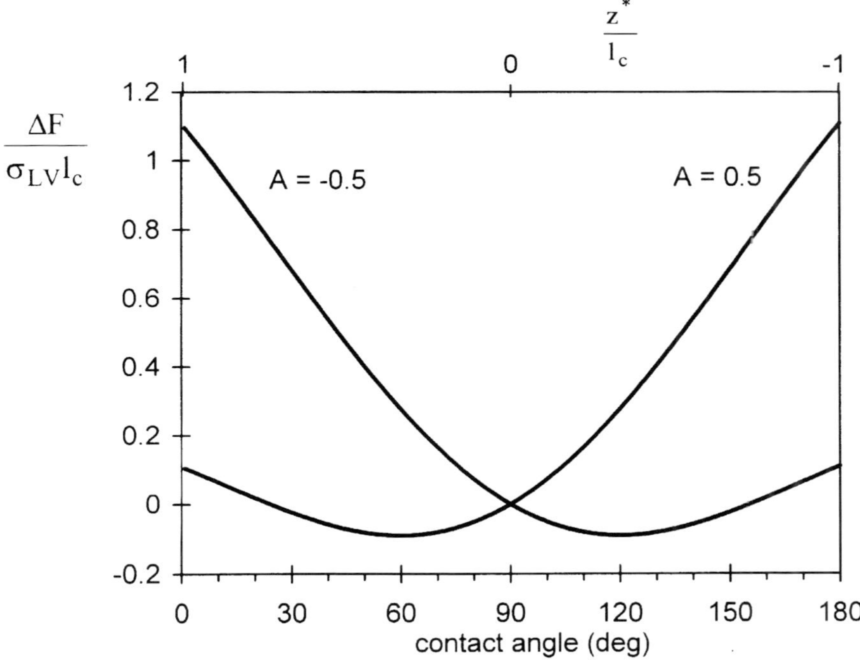

Figure 1.10. Reduced total free energy of a meniscus per unit length of triple line as a function of its height z^* or the corresponding contact angle for two values of the parameter $A = (\sigma_{SV} - \sigma_{SL})/\sigma_{LV}$. The minima of these curves correspond to the equilibrium contact angle of the system.

Figure 1.10 demonstrates that the equilibrium contact angle depends only on the value of A. Variations in the values of σ_{LV}, ρ or g (such as might be encountered in spacelab experiments) for a constant value of A will merely displace the $\Delta F(\theta)$ curve perpendicularly to the θ-axis, without changing the position of the minimum value of θ.

The above calculations show that the Young equation is valid in the presence of a gravitational field for the configuration of a vertical plate. This is also true for other configurations in which one curvature of the L/V surface is zero, as for a cylindrical "drop" on a horizontal plane (Marmur 1996). The validity of the Young equation in the classical configuration of an axisymmetric sessile drop has been discussed by various authors, for instance Johnson (1959), Collins and Cooke (1959) and more recently Garandet et al. (1998). In the last paper, a clear proof of the validity of the Young equation was achieved by minimizing the free energy of the system, including a gravitational term. Moreover, it was shown that the Young equation holds also when the fluid is subject to other kinds of body forces, e.g. dielectric or magnetic forces. Only localised "surface" forces such as those induced by Foucault's currents in high frequency heating can, in principle, modify the contact angle.

1.2.4 Metastable and stable equilibrium contact angles

The Young equation can be derived from minimization of the free energy of the system carried out by considering only displacements of the triple line parallel to an S/V surface assumed to be undeformable (Figure 1.11.b). Therefore, θ_Y corresponds to a metastable equilibrium configuration. Hereafter, deformation of the solid close to TL will be considered and the local stable equilibrium shown in Figure 1.11.c will be analysed. Equations describing the stable equilibrium in terms of three dihedral angles Φ_1, Φ_2 and Φ_3 can be obtained by regarding the displacement δh of TL (Figure 1.11.c) as two elementary displacements, one perpendicular to the intersection of the L/V surface in the figure plane (δh_1) and one perpendicular to the intersection of the S/L interface in the figure plane (δh_2). Assuming isotropic S/V and S/L surface and interfacial energies, the interfacial free energy change for the first displacement is (Figure 1.12):

$$\delta F_s = \sigma_{SV} \sin(\Phi_2)\delta h_1 - \sigma_{SL} \sin(\Phi_1)\delta h_1 \tag{1.27}$$

The equilibrium condition $d(\delta F_s)/d(\delta h_1) = 0$ leads to:

$$\sigma_{SV} \sin \Phi_2 = \sigma_{SL} \sin \Phi_1 \tag{1.28}$$

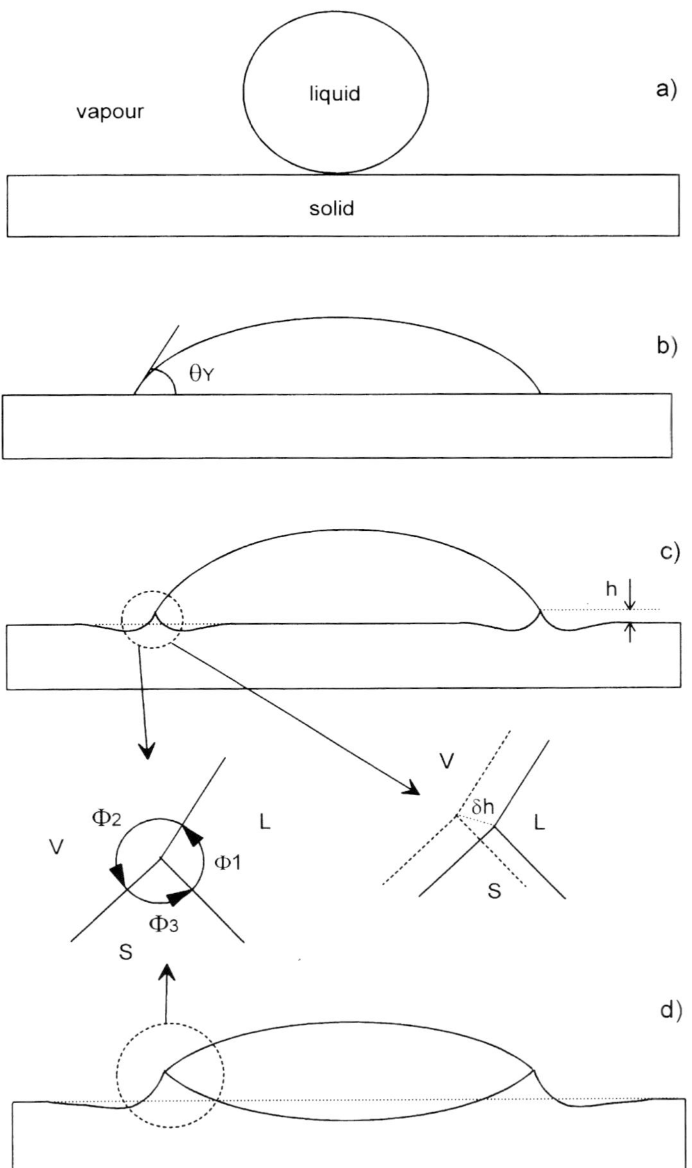

Figure 1.11. Metastable (b) and stable (c) equilibrium angles at a solid/liquid/vapour junction obtained after spreading of a liquid droplet (a). In configuration (d), both local equilibrium at the triple line and total equilibrium along the whole solid/liquid interface are attained.

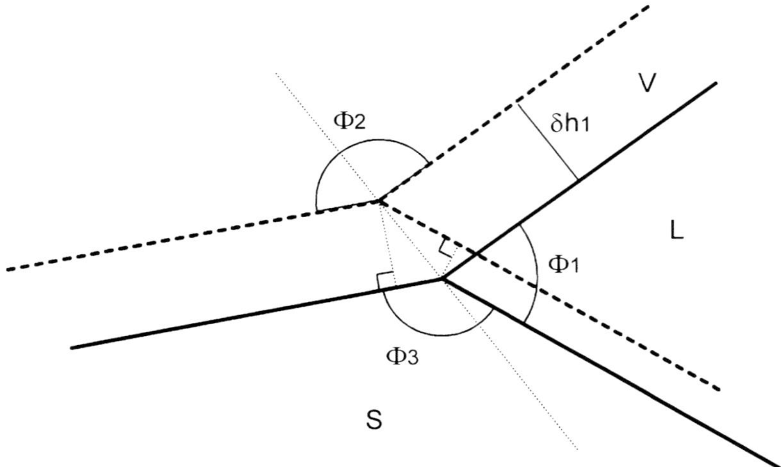

Figure 1.12. Displacement of the triple line around its equilibrium position when the solid is deformable. The displacement δh_1 is perpendicular to the intersection of the L/V surface in the figure plane and allows derivation of equation (1.28).

Similarly, $d(\delta F_s)/d(\delta h_2) = 0$ yields:

$$\sigma_{SV} \sin \Phi_3 = \sigma_{LV} \sin \Phi_1 \tag{1.29}$$

By combining equations (1.28) and (1.29), the relationship first given by Smith (1948) is obtained:

$$\frac{\sigma_{SV}}{\sin \Phi_1} = \frac{\sigma_{SL}}{\sin \Phi_2} = \frac{\sigma_{LV}}{\sin \Phi_3} \tag{1.30}$$

in which $\Phi_1 + \Phi_2 + \Phi_3 = 360°$. The configuration described by the Smith equation corresponds to local equilibrium. Total equilibrium requires an unchanging curvature at any point of the S/L interface (Figure 1.11.d), as in the case of a small liquid droplet on the surface of another immiscible liquid.

The actual TL configuration observed after a certain time of contact between the solid and liquid phases depends on the scale of observation and on the relative rates of two processes: (i) the movement of TL over large distances to satisfy the Young equation and (ii) the distortion of TL to satisfy locally the more general Smith equation. The kinetics of the two movements may be very different.

In non-reactive S/L couples and for liquids with a low viscosity, such as molten metals or certain oxide melts at high temperature, the lateral movement of TL is very fast, with spreading times for millimetre size droplets of 10^{-1} second or less (see Section 2.1.1). The movement of TL perpendicular to the initial S/V surface can occur first by elastic deformation of the solid. Remembering that the surface energy σ_{LV} of a liquid is equivalent to the surface tension γ_{LV}, the solid is strained at the triple line, the initial stress being $\gamma_{LV}\sin\theta$. The height h of the deformed region (Figure 1.13.a) is of the order of γ_{LV}/E, where E is the modulus of elasticity. Taking typical values for high temperature materials $\gamma_{LV} = 1$ N/m and $E = 10^{11}$ Pa, h is a negligible 10^{-2} nm (however, for "soft" or viscoelastic solids, h can attain several tens of nm (Carré and Shanahan 1995)).

The growth of the "wetting ridge", to easily measurable dimensions, can take place by mechanisms similar to those occurring during grain-boundary grooving, as described by Mullins (1957, 1960). When a grain boundary of energy σ_{gb} meets a S/L interface (or a S/V surface), a groove is formed with a dihedral angle Φ (Figure 1.13.b) which, at equilibrium, is given by:

$$\cos\left(\frac{\Phi}{2}\right) = \frac{\sigma_{gb}}{2\sigma_{SL}} \tag{1.31}$$

Once formed, the groove continues to grow, driven by differences in curvature κ at the S/L interface. These differences result in variations of the chemical potential of the solid, μ, according to the Gibbs–Thomson equation:

$$d\mu = v_m\sigma_{SL}\,d\kappa \tag{1.32}$$

where v_m is the molar volume of the solid. Thus, the S/L interfacial energy σ_{SL} provides a driving force for the transfer of solid atoms from regions of high curvature at the root of the groove to regions of small curvature far from the groove. As a result, although the shape of the groove and the angle Φ are constant with time, its linear dimensions, such as its depth h, increase continuously with time according to a power law:

$$h^n = K\,t \tag{1.33}$$

where n = 3 when the limiting process is diffusion in bulk solid or liquid (Mullins 1960), n = 4 for interface diffusion and n = 2 for dissolution-precipitation mechanism (or evaporation/condensation in the case of S/V surface) (Mullins 1957). For couples with low or negligible miscibility (for instance with an equilibrium molar fraction of solid component in the liquid of the order of 10^{-4} or less), the power law (equation (1.33)) leads to very small grooving rates and several hours or tens of hours are needed for groove depths to reach the micron scale (Eustathopoulos 1983).

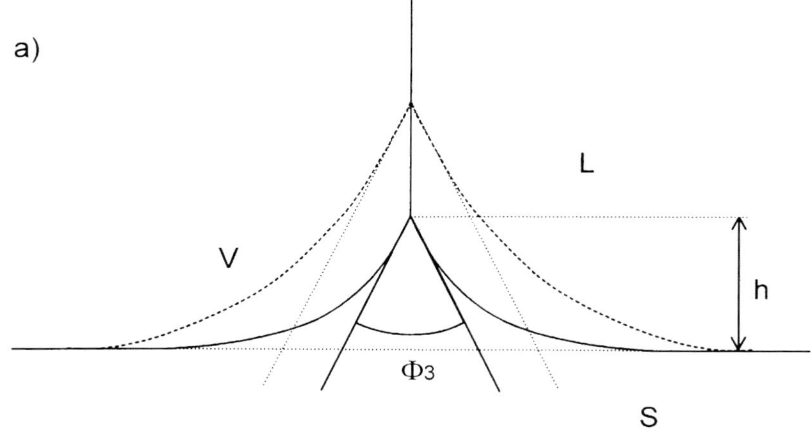

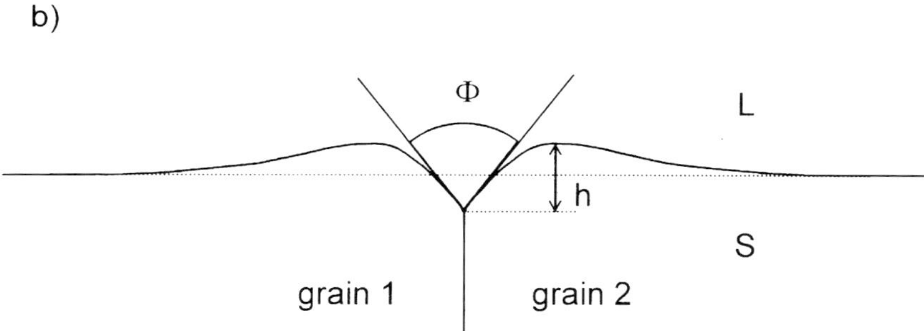

Figure 1.13. Formation of a wetting ridge at the triple line (a) by a mechanism similar to grain-boundary grooving (b).

In the vicinity of a triple line, a deformation of the solid surface can also occur by a mechanism in which the role of the grain boundary is played by the L/V surface (Figure 1.13.a). In this case, the mass transfer will occur from the S/L interface and S/V surface far from TL to regions close to TL where the curvature is relatively high but negative. Experimental evidence of the formation of wetting ridges, attributed to a mechanism similar to that occurring in grain-boundary grooving, was given by Saiz et al. (1998) for Cu and Ni droplets on Al_2O_3 surfaces in Ar. At the end of the spreading process, the droplets were maintained in contact with the substrate for several tens of minutes, allowing the ridge to grow to a size that permitted easy observation. For Cu, a wetting ridge with h = 10 nm was formed after 2 hours at 1150°C and for Ni h = 0.2 μm after 1 hour at 1500°C.

It is interesting to note that in the past, some authors used the stable TL configuration (Figure 1.11.c) to derive S/L and S/V interfacial and surface energies in nearly immiscible molten metal/solid metal systems by measuring the relevant dihedral angles, σ_{LV} values being measured independently by other techniques. To obtain deformed regions large enough to permit measurement of the angles, long duration experiments were needed. However, significant evaporation of the drop can occur during such experiments and cause some decrease in the drop volume resulting in instabilities in the position of the triple line and inaccurate measurements.

In liquid/solid systems with a high mutual solubility (as for many metallic systems), calculations show that the size of the deformed area close to the triple line can attain dimensions of about a micron in a minute or so. However, in this case, the equilibrium configuration at the triple line can be masked by the dissolution of the solid in the liquid (Warren et al. 1998) as discussed in Section 2.2.1.

From the above discussion, it appears that wetting of low viscosity liquid drops on solid substrates can occur in two stages. In the first rapid stage, the macroscopic contact angle approaches θ_Y (Figure 1.11.b), largely determining the area of the S/L interface and the L/V surface, followed by a much slower process occurring at the vicinity of TL to satisfy the requirements of a stable local equilibrium in accord with the Smith equation (Figure 1.11.c). Much longer times are needed to obtain a total equilibrium i.e., a constant curvature on the whole S/L interface (Figure 1.11.d). This description is also valid for the infiltration of liquids into porous media, as when sintering of metallic or ceramic powders in the presence of molten metals. On melting, infiltration of the liquid will occur rapidly by wetting non-deformed solid particle surfaces and the complete three-phase equilibrium will be developed at the solid particle/liquid/void contact line at a slower rate.

For vitreous solids, such as SiO_2, viscosity decreases strongly before reaching the melting point. In this case, the solid meniscus can be formed by viscous flow and the height h can reach easily measurable sizes in quite short times. For example, wetting of a Ni alloy droplet on a SiO_2 substrate at 1743 K is associated with the

The equilibrium condition $d(\delta F_s)/d(\delta z) = 0$ leads to the equation of Wenzel (1936):

$$\cos\theta_W = s_r \cos\theta_Y \qquad (1.35)$$

1.3.1.2 Effect of sharp edges. A sharp edge can pin the triple line at positions far from stable equilibrium, i.e., at contact angles markedly different from θ_Y. This effect is illustrated schematically in Figure 1.16.a where we consider a solid wall of a crucible consisting of a vertical surface S_1 and another surface S_2 inclined at an angle β. The initial equilibrium configuration of the liquid surface is identified 1 in Figure 1.16.a, corresponding to $MN = z^*_{max}$ and to a contact angle at point N on the S_1 surface $\theta_N(S_1) = \theta_Y$. Then, if the liquid volume is increased slowly enough for the liquid to retain capillary equilibrium, TL will advance on the S_1 surface and assume configuration 2 where $M'N' = z^*_{max}$ and $\theta_{N'}(S_1) = \theta_Y$. Thereafter, TL will be pinned at point N' and the macroscopic contact angle on S_1, $\theta_{N'}(S_1)$, will increase until the liquid surface assumes configuration 3 where $\theta_{N'}(S_1) = \theta_Y + \beta$ which corresponds to the establishment of the Young contact angle on the S_2 surface i.e., $\theta_{N'}(S_2) = \theta_Y$. Any further increase in the liquid volume will produce a rise of the liquid on the S_2 surface, for example to configuration 4 with $\theta_K(S_2) = \theta_Y$. Thus, the liquid can form an infinite number of *advancing* contact angles at point N' on the S_1 surface lying between θ_Y and $\theta_Y + \beta$. This last value defines the maximum advancing contact angle on S_1:

$$\theta_a(\text{max}) = \theta_Y + \beta \qquad (1.36.a)$$

Consider now that configuration 4 in Figure 1.16.a represents the initial liquid surface and that the liquid volume is slowly decreased. Using similar arguments for the retreat of a liquid on a solid wall, it can be shown that the liquid can form an infinite number of *receding* contact angles at point N' on the S_2 surface lying between θ_Y and $\theta_Y - \beta$. This last value defines the minimum receding contact angle on S_2:

$$\theta_r(\text{min}) = \theta_Y - \beta \qquad (1.36.b)$$

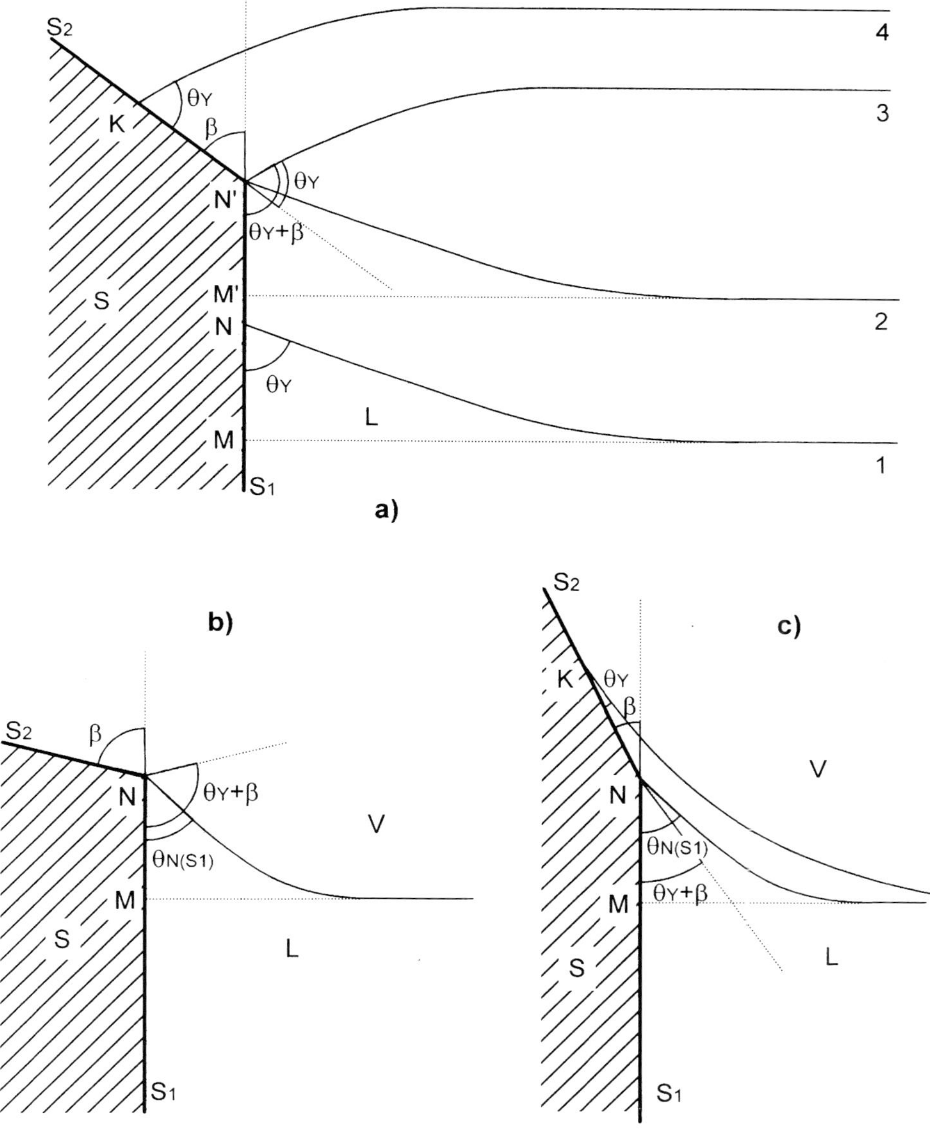

Figure 1.16. Effects of a sharp edge on the contact angle formed by a meniscus advancing on a solid surface. In (a) 1, 2, 3 and 4 denote the successive configurations of the liquid surface when the liquid volume is increased. With a constant volume of liquid, an advancing triple line is either pinned (b) or not (c) by the sharp edge.

Now consider the meniscus formation, on the same solid wall, of a liquid of *constant* volume initially in an horizontal position such that MN $\ll z^*_{max}$ i.e., the contact angle at point N on S_1, $\theta_N(S_1)$, is greater than θ_Y (Figures 1.16.b and c). Two cases can be distinguished depending on the value of β. If $\theta_N(S_1) < \theta_Y + \beta$ (Figure 1.16.b), TL will be blocked at point N. If $\theta_N(S_1) > \theta_Y + \beta$ (Figure 1.16.c), TL will be able to advance on the S_2 surface until capillary equilibrium is reached $(\theta_K(S_2) = \theta_Y)$.

Finally consider a vertical solid wall with one small groove (Figure 1.17) and a constant liquid volume. When TL reaches position 2, in principle it will be blocked by the groove if the inequality $\theta < \theta_Y + \beta$ is satisfied. In practice, whether TL will be pinned or pass over the groove when it reaches position 2 depends also on the other energies of the system, particularly the vibrational energy of the experimental device, as well as on the size of the groove.

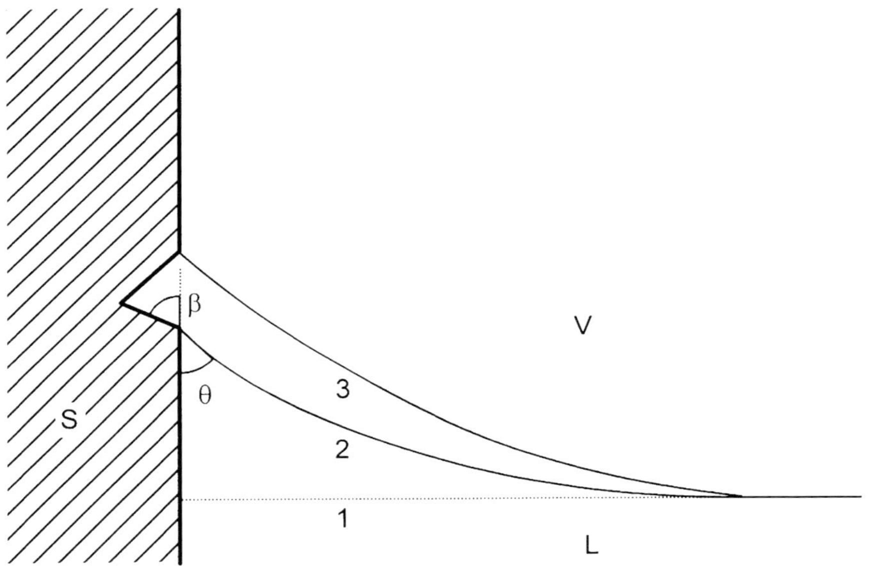

Figure 1.17. Meniscus rise on a vertical wall with a small groove.

1.3.1.3 Wetting on sawtooth surface. Now consider the system sketched on Figure 1.18. The solid surface is a series of facets of length L_f forming alternate angles β and $(180° - \beta)$ with the z-axis. In this geometry, the triple line is parallel to the grooves (Figure 1.19). The asperity wave-length ($2L_f\cos\beta$) is assumed to be small compared to the maximum height of the meniscus z^*_{max}. This assumption allows the definition, for each position z^* of the triple line, of a macroscopic contact angle θ_M between the vertical and the tangent to the L/V surface (Figure 1.18.a) and a microscopic contact angle θ (Figure 1.18.b). It should be noted that generally only θ_M can be measured experimentally. For the geometry of Figure 1.18, ratio s_r of the real surface to its projection on the z–y plane is $(\cos\beta)^{-1}$.

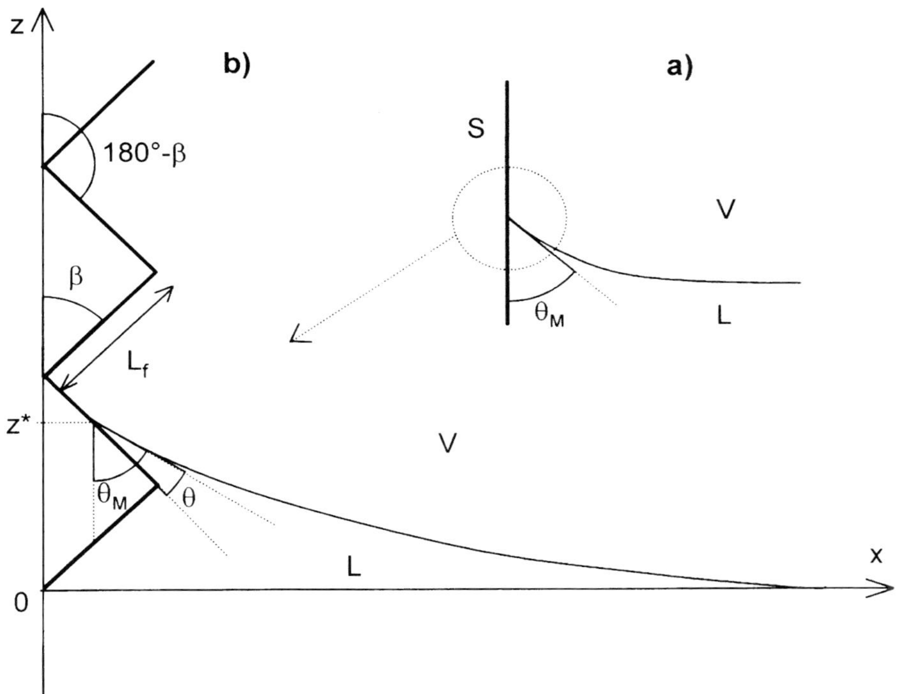

Figure 1.18. Model sawtooth surface used in calculations of the free energy of a meniscus. θ_M represents the macroscopic contact angle at each height z^* of the triple line.

As for the ideal smooth surface (Section 1.2.3 and Appendix B), the free energy change ΔF can be calculated as a function of θ_M, when a liquid surface initially in a horizontal position ($z^* = 0$, $\theta_M = 90°$) rises to form a meniscus of height z^* and a contact angle θ_M. The interfacial ($\Delta F_{s,1}$ and $\Delta F_{s,2}$) and potential (ΔF_b) energy contributions need to be modified to take into account the particular geometry of the rough surface. For instance the $\Delta F_{s,1}$ term, representing the change from a S/V surface to a S/L interface (equation (B.1) in Appendix B), is now equal to $s_r z^*(\sigma_{SL} - \sigma_{SV})$. Results of calculations are given in Figure 1.20 for $\beta = 10°$ and $L_f = 100\ \mu m$ (corresponding to a roughness parameter $s_r = 1.015$) and $\theta_Y = 40°$.

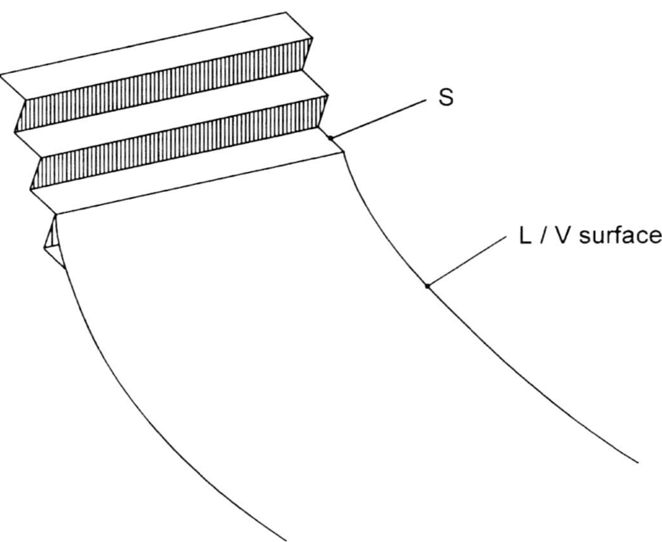

Figure 1.19. Sawtooth surface with grooves parallel to the triple line.

The modelling of Figure 1.20 leads to a number of conclusions:

(i) The contact angle at the minimum of the $\Delta F(\theta_M)$ curve, representing the equilibrium contact angle of a liquid on a rough and undeformable solid surface, is not equal to the Young contact angle θ_Y but to θ_W (equal to 39° in the calculations for Figure 1.20) as defined by the Wenzel equation (1.35). This equation was the first to take into account roughness effect on contact angles. However, it does not predict the existence of contact angle hysteresis and predicts that roughening the substrate improves wetting for wetting liquids ($\theta_W < \theta_Y < 90°$) but degrades wetting for non-wetting liquids ($\theta_W > \theta_Y > 90°$). In fact, experimental results show that the advancing contact angle increases with roughening not only for non-wetting liquids but also, in some cases, for wetting liquids (Figure 1.21).

(ii) Between the maximum value of the advancing contact angle and the minimum value of the receding contact angle, the $\Delta F(\theta_M)$ curve exhibits a succession of free energy minima, which correspond to metastable equilibria (Figure 1.20). The values of these extreme contact angles are given by equations (1.36.a) and (1.36.b), so that for $\beta = 10°$ and $\theta_Y = 40°$, the contact angle hysteresis domain lies between $30°$ and $50°$. In the case of a rough surface with a more realistic geometry and a variable β, for example sinusoidal surface shown in Figure 1.22, equations (1.36) become:

$$\theta_a(\text{max}) = \theta_Y + \beta_{\text{max}} \qquad (1.37.a)$$
$$\theta_r(\text{min}) = \theta_Y - \beta_{\text{max}} \qquad (1.37.b)$$

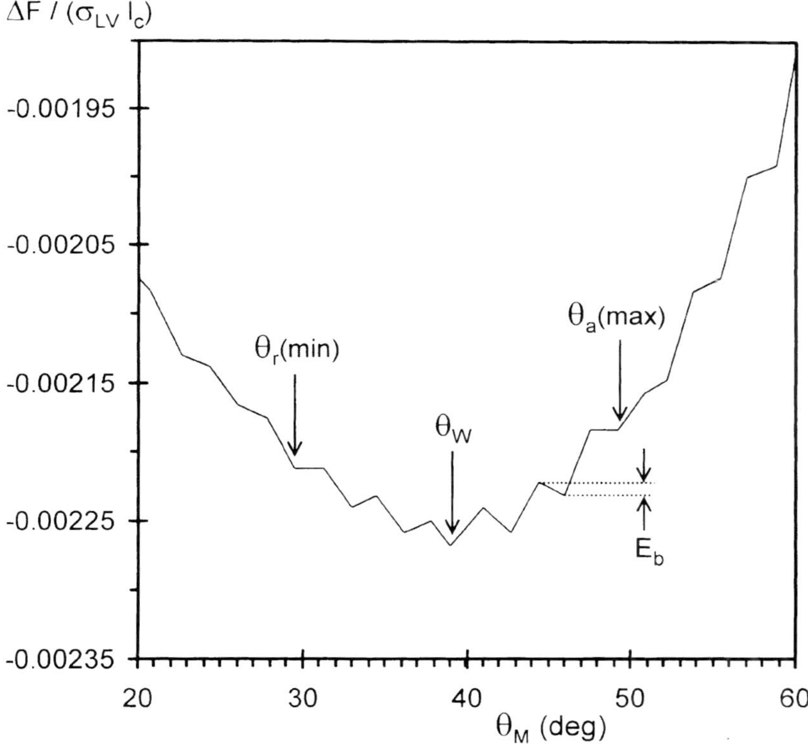

Figure 1.20. Reduced free energy per unit length of triple line of a meniscus on the solid surface shown in Figures 1.18 and 1.19 plotted as a function of the macroscopic contact angle. The physical parameters used in the calculation are $\rho = 7 \times 10^3\,\text{kg.m}^{-3}$, $\sigma_{LV} = 1\,\text{J.m}^{-2}$, $\theta_Y = 40°$, $g = 9.81\,\text{m.s}^{-2}$, $\beta = 10°$ and $L_f = 100\,\mu\text{m}$. Results from (Eustathopoulos and Chatain 1990) [1].

where β_{max} is the maximum slope of the real surface. Equations (1.37) have been used already by Shuttleworth and Bailey (1948) to interpret wetting hysteresis. These authors considered that surface asperities could pin the triple line in positions not predicted by the Wenzel equation as soon as the *microscopic* contact angle is equal to the Young contact angle, as expressed by equations (1.37) (Figure 1.22).

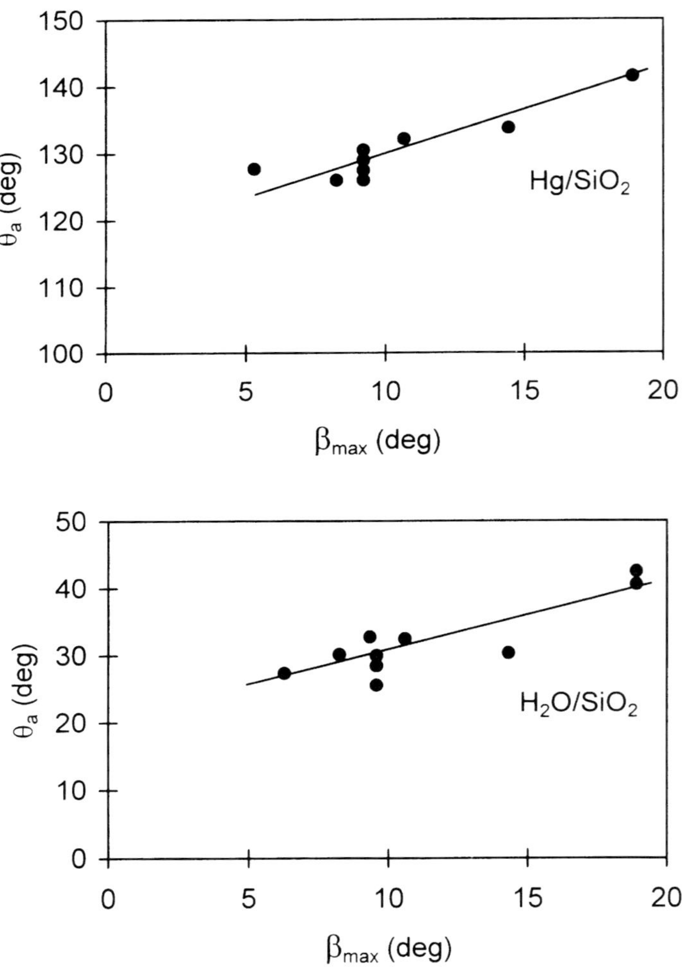

Figure 1.21. Advancing contact angles plotted as a function of the maximum slope of surface asperities for non-wetting (top) and wetting (bottom) systems (Hitchcock et al. 1981) [2].

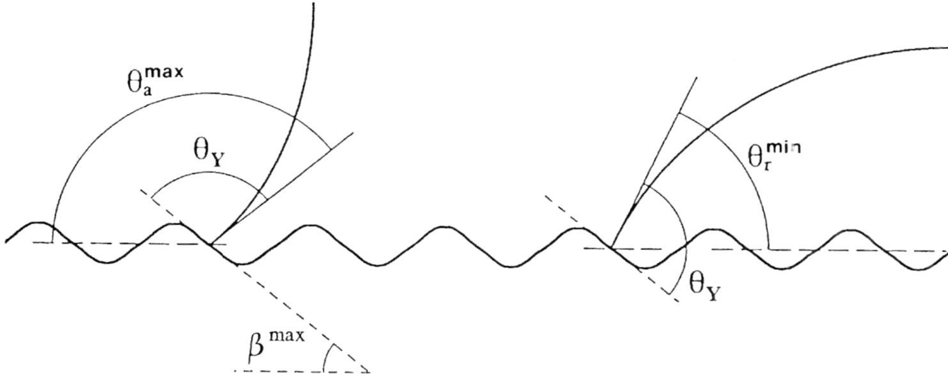

Figure 1.22. Identification of the maximum advancing contact angle and minimum receding contact angle on a rough surface.

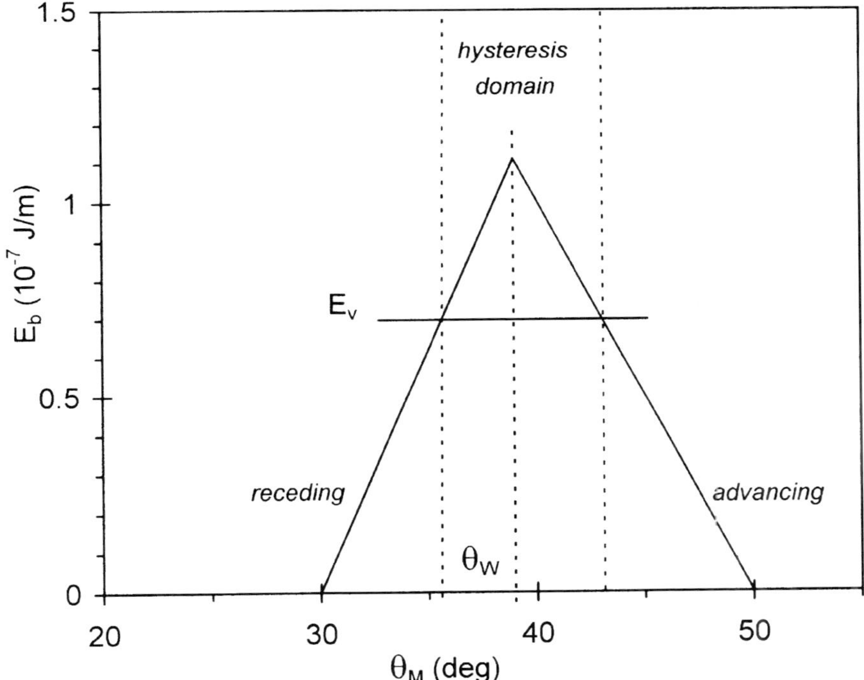

Figure 1.23. The energy barrier per unit length of triple line corresponding to the curve of Figure 1.20. From Eustathopoulos and Chatain (1990) [1].

(iii) As several states of metastable equilibrium exist, each of them being separated from neighbouring states by energy barriers E_b (Figure 1.20), it may be asked what is the metastable position chosen by the triple line in the course of an experiment. The variation of E_b in the metastable domain (Figure 1.23) shows that E_b is maximum when θ_M tends towards θ_W. According to Johnson and Dettre (1964), the triple line will be pinned in a position depending on the relative values of E_b and the "vibrational energy of the liquid" E_v (i.e., an energy due to vibrations which occur unavoidably in any experimental device). Thus, the values of advancing and receding contact angles would be determined by the condition $E_b = E_v$ (Figure 1.23). In the model of Johnson and Dettre, the whole triple line passes a defect in one move, and E_b is proportional to the length of TL and the physical size of the barrier. As pointed out by de Gennes (1985), the necessary energy would be very high and such a mechanism is thus not realistic. In fact, only a small part of the triple line needs to cross the defect : once this nucleation stage is realized, the remaining part of the triple line can advance by a lateral movement of the liquid along the valleys (Figure 1.24). In spite of this criticism, the vibrational energy concept of Johnson and Dettre was qualitatively confirmed by sessile drop experiments performed on substrates submitted to vibrations : the contact angle decreases irreversibly as a function of time (Figure 1.25).

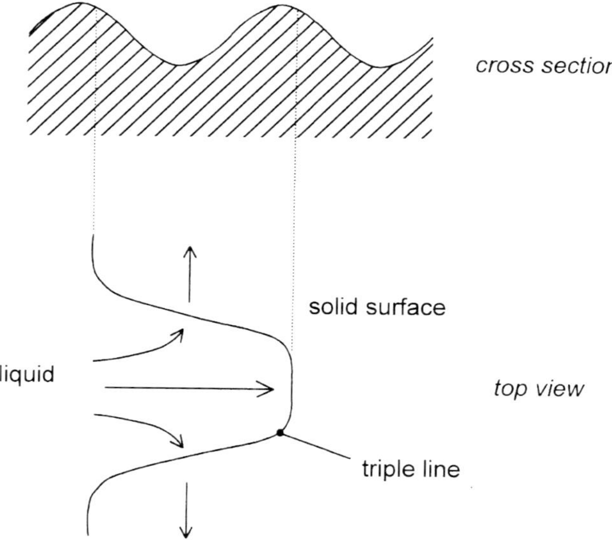

Figure 1.24. Movement of a triple line over a surface defect by a two step mechanism. According to de Gennes (1985).

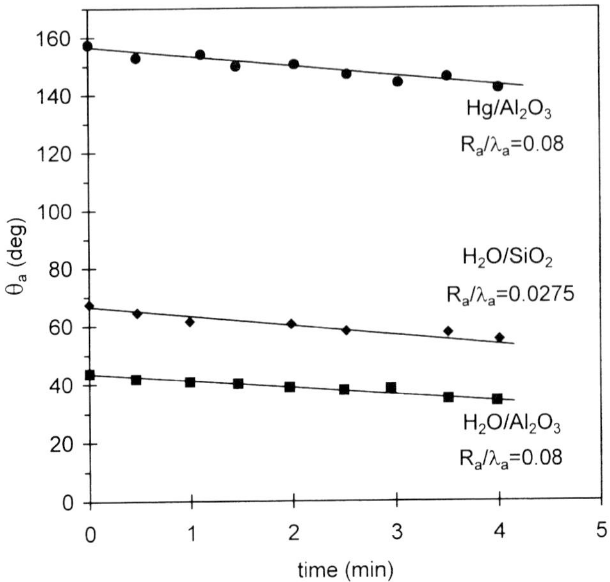

Figure 1.25. Effect of 50 kHz vibrations on the advancing contact angles of Hg and H_2O on rough substrates with different values of the ratio R_a/λ_a where R_a is the average roughness of the surface and λ_a is the average wavelength of the surface features. In practice $10R_a/\lambda_a \sim \tan\beta$. From (Hitchcock et al. 1981) [2].

An example of advancing and receding contact angles formed by liquid mercury on sapphire cylinders of variable roughness is given in Figure 1.26. Their roughness was characterized by the average displacement from the mean height of asperities R_a and the ratio of R_a to the average wavelength λ_a of asperities, as proposed by Hitchcock et al. (1981). R_a and λ_a are easily obtained by mechanical or optical surface profilometry techniques and ratio R_a/λ_a relates to the average slope of surface asperities. For very small roughness (a few nanometers), the hysteresis domain $(\theta_a - \theta_r)$ is only a few degrees, allowing the Young contact angle to be defined with reasonable precision as 150° for Hg/sapphire. With increasing roughness, the difference $(\theta_a - \theta_r)$ at first increases and then decreases, tending towards zero for very rough surfaces. This inversion produced at high roughnesses is due to the fact that a non-wetting liquid cannot infiltrate surface cavities, resulting in the formation of a *composite* interface, i.e. a mixed solid/liquid interface and solid/vapour surface (Figure 1.27.a). A characteristic of such an interface is the disappearance of the energy barriers E_b (Johnson and Dettre 1964). Indeed, the liquid surface contacts the solid only at a limited number

of points so that no pinning of TL can occur, explaining why the difference $(\theta_a - \theta_r)$ tends towards zero. Thus, mercury droplets on ceramic surfaces are very mobile. Indeed, owing to the poor wetting by mercury and the formation of a composite interface, droplets are never pinned and move easily.

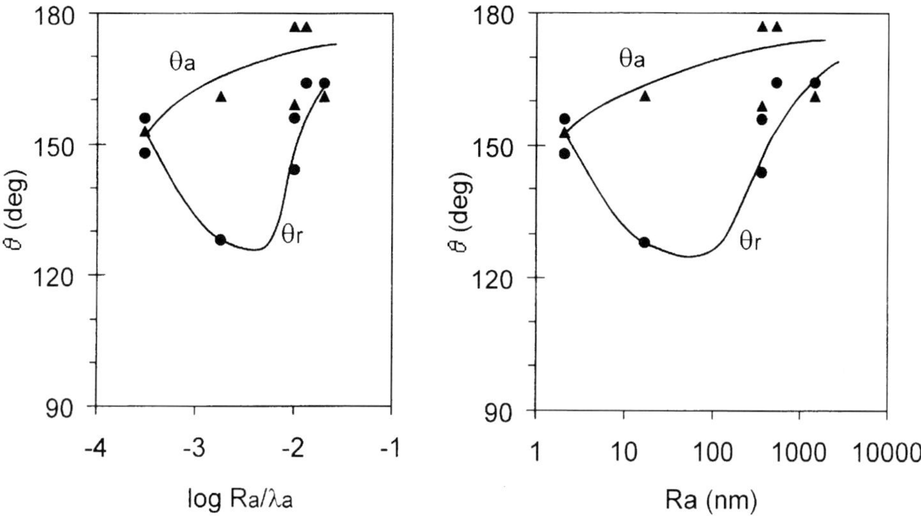

Figure 1.26. Advancing and receding contact angles in the Hg/Al$_2$O$_3$ system measured by the immersion-emersion technique as a function of substrate roughness. Data from work reported in (De Jonghe et al. 1990) [3].

To establish the condition for the transition from an S/L to composite interfaces, consider a liquid surface infiltrating a model triangular groove (Figure 1.27.b). The interfacial free energy change relative to a displacement δl along the groove wall is:

$$\delta F_s = -2\sigma_{LV}\cos(\beta)\delta l + 2(\sigma_{SL} - \sigma_{SV})\delta l \tag{1.38}$$

Setting $d(\delta F_s)/d(\delta l) = 0$ and recalling that $(\sigma_{SV} - \sigma_{SL})/\sigma_{LV} = \cos\theta_Y$, one gets $\cos\beta = -\cos\theta_Y$, that is:

$$\beta^* = 180° - \theta_Y \tag{1.39}$$

For any cavity with a β value higher than β^*, $d(\delta F_s)/d(\delta l)$ will be positive whatever the value of l, so the cavity will not be infiltrated by the liquid. Equation (1.39) shows that composite interfaces can be produced only for non-wetting liquids and very rough surfaces.

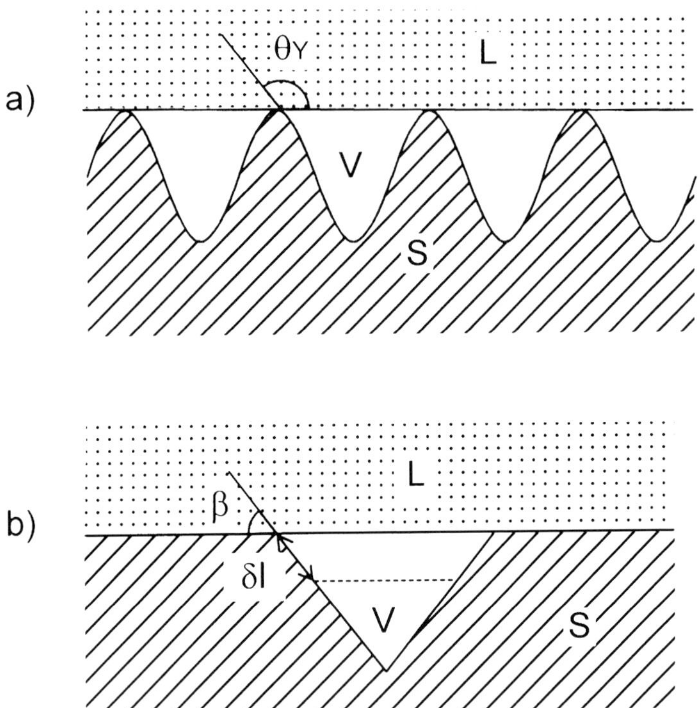

Figure 1.27. (a) Formation of a composite interface in the case of a non-wetting system. (b) Model triangular groove used in the calculation of the critical angle β^*.

Energy barriers to the movement of TL and contact angle hysteresis are predicted to exist only for grooves parallel to TL. When grooves are perpendicular to TL (Figure 1.28), no such barriers exist and the macroscopic contact angle is equal to the Wenzel contact angle given by equation (1.35) (Johnson and Dettre 1964). For wetting liquids, this equation predicts that roughness always promotes wetting and it is possible to obtain apparently perfect wetting ($\theta_W = 0°$) when the roughness parameter s_r is higher than $(\cos\theta_Y)^{-1}$. In this case, grooves act as open capillaries to reinforce wetting. This effect of surface texture is used for practical

purposes, for instance for the recycling of liquids in heat exchangers using liquid metals.

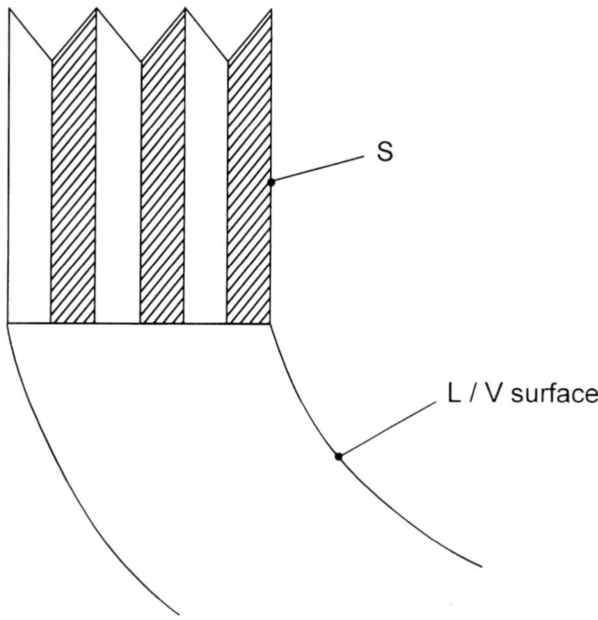

Figure 1.28. Sawtooth surface with grooves perpendicular to the triple line.

1.3.2 Heterogeneous surfaces

Many materials of practical interest are multi-phase solids with heterogeneous surfaces that can be either regular (orientated eutectics, unidirectional composites, etc) or random ("hard" alloys processed by liquid phase sintering, or alloys strengthened by precipitation, etc).

Consider a vertical solid wall of a crucible consisted of two macroscopic phases α and β separated by a plane intersecting the solid wall by an horizontal straight line (Figure 1.29.a). The intrinsic contact angles on the two phases are such that $\theta^\alpha > \theta^\beta$, and for the sake of simplicity $\theta^\alpha > 90°$ and $\theta^\beta < 90°$. Moreover, we assume that the level of liquid in the crucible far from the wall is such that $h_2 \ll z_\beta^*$, where z_β^* is the maximum meniscus height on the β phase i.e., z_β^* would be the meniscus rise on a homogeneous vertical plate of β (equation (1.23)). During

meniscus rise, when TL reaches the line of separation from β to α, it will be pinned at this position by the non-wetted α phase (configuration 2 in Figure 1.29.a). When the volume of the liquid in the crucible is increased and configuration 3 is assumed, TL does not move and the macroscopic contact angle θ_M takes a value such that $\theta^\beta < \theta_M < \theta^\alpha$. This value of θ_M does not satisfy the condition of capillary equilibrium but can be deduced from the geometry of the system and the quantity of liquid. A further increase in the liquid volume, to produce a surface contour shown as configuration 4 such that $h_4 = z_\alpha^*$ (where z_α^* is the capillary *depression* on the α phase), causes θ_M to equal the equilibrium contact angle θ^α, so that capillary equilibrium is established at TL on the α surface. If the liquid volume is increased again, TL will move up (configuration 5 in Figure 1.29.a) while θ_M will remain equal to θ^α.

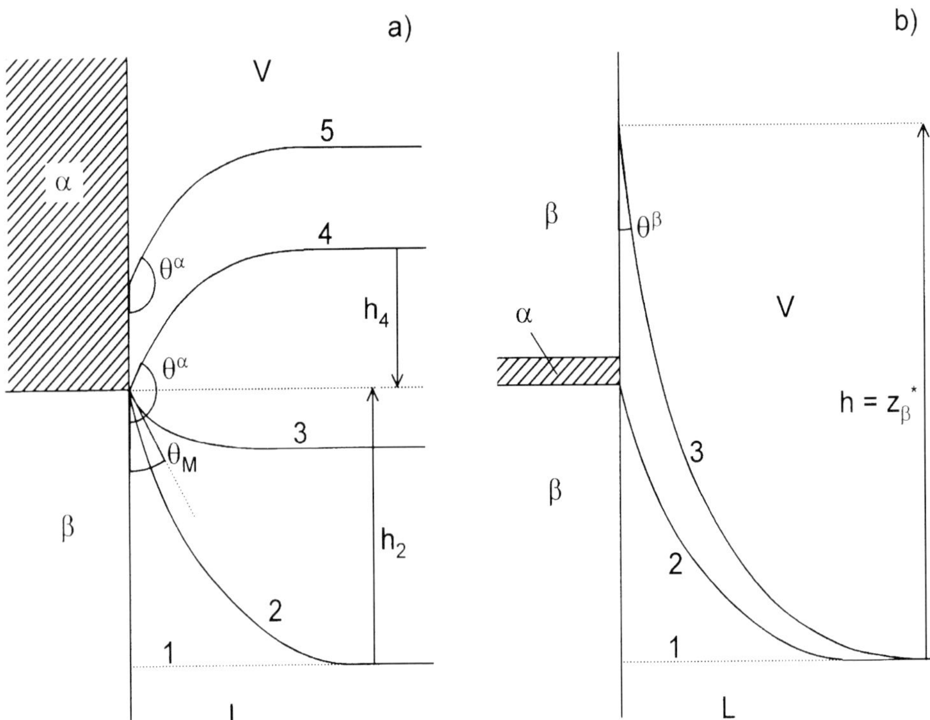

Figure 1.29. (a) Successive configurations of a meniscus on a vertical solid surface consisting of two macroscopic phases when the liquid volume is increased. (b) Effect of a small chemical heterogeneity on the advance of such a meniscus.

The photograph in Figure 1.31 of a tin drop on a striped surface shows that anisotropic wetting occurs preferentially parallel to the strips. For a strip width of 100 μm, the experimental contact angle measured in the direction where TL is perpendicular to the strips was 80°, in good agreement with the 79° predicted by the Cassie equation (1.41). For the direction where TL is parallel to the strips, the experimental advancing contact angle is 120°, close to the maximum value of 125° given by equation (1.42.a). This result can be explained by the large strip width but testifies also to a low vibration level in the experimental device.

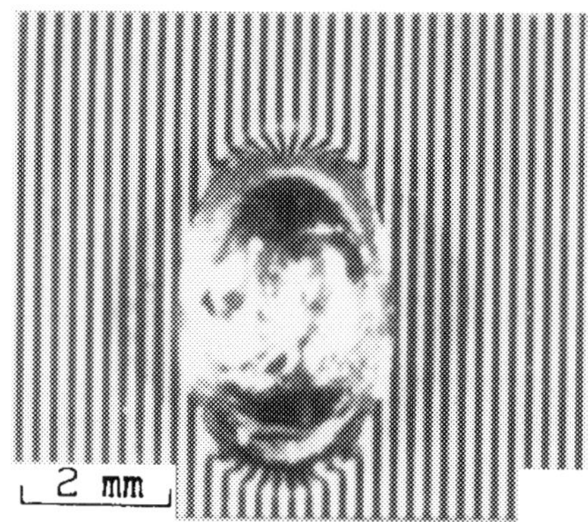

Figure 1.31. Tin drop on a striped surface consisted of non-wetted (oxide glass) and wetted (molybdenum) strips. Reprinted from (Voitovich 1992) with kind permission of the author.

As a general rule, metastable states are predicted for patched surfaces. For such surfaces, Horsthemke and Schröder (1981) proposed the following expressions for advancing and receding contact angles:

$$\cos \theta_a = f_{max}^{\alpha} \cos \theta^{\alpha} + (1 - f_{max}^{\alpha}) \cos \theta^{\beta} \qquad (1.43.a)$$

$$\cos \theta_r = f_{min}^{\alpha} \cos \theta^{\alpha} + (1 - f_{min}^{\alpha}) \cos \theta^{\beta} \qquad (1.43.b)$$

where f_{max}^{α} and f_{min}^{α} are the maximum and minimum surface fractions of the least wetted α component in regions adjacent to TL, which is assumed to be a straight line. In the extreme case of a strip surface, equations (1.43) reduce to the Cassie equation (1.41) for TL perpendicular to the strips (Figure 1.30.a), and to equations (1.42) for TL parallel to the strips (Figure 1.30.b).

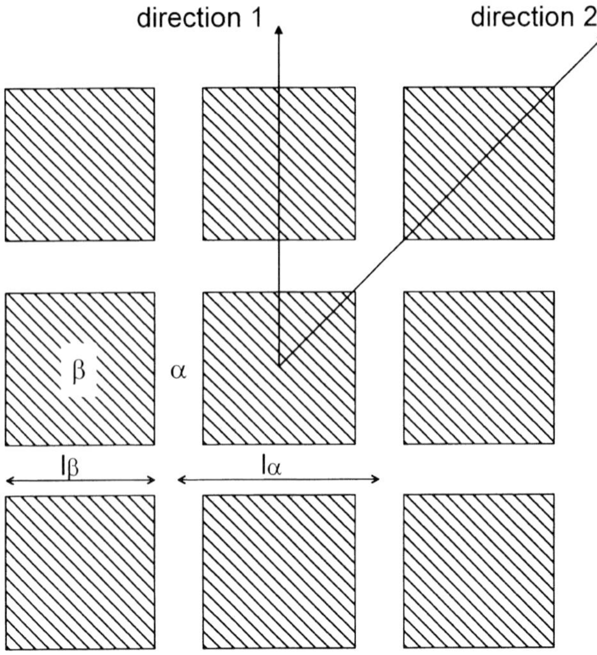

Figure 1.32. Patched surface with β surface fraction $f^\beta = 0.5625$ ($l_\beta/l_\alpha = 0.75$). In the experiments of Naidich et al. (1995) (see results on Figure 1.33), the continuous α matrix was an oxide glass ceramic ($\theta^\alpha = 125°$) and the β squares were Molybdenum ($\theta^\beta = 18°$).

Experimental determinations of wetting hysteresis on patched surfaces have been made by De Jonghe and Chatain (1995) and Naidich et al. (1995). These last authors performed wetting experiments with liquid tin on surfaces composed of wettable β square patches (molybdenum) dispersed in a continuous α matrix (oxide glass ceramic). Such a surface is shown in Figure 1.32 with a β surface fraction, f^β, of 0.562. The experimental contact angle hysteresis is determined as a function of f^β for direction 1 (Figure 1.33.a) parallel to the sides of the patches, and direction 2 (Figure 1.33.b) diagonally across the patches. Theoretical values of advancing and receding contact angles for $f^\beta = 0.25$ and $f^\beta = 0.562$, as calculated by equations (1.43), are also presented on Figure 1.33 that are in good agreement with the experimental data. It can be seen that for $f^\beta = 0.25$, the contact agle hysteresis is identical in directions 1 and 2, as these two directions have identical values of f^α_{max} and f^α_{min} (equations (1.43)). For $f^\beta = 0.562$, the hysteresis domains are different, so one gets $f^\alpha_{max} = 1$, $f^\alpha_{min} = 0.25$ in direction 1 and $f^\alpha_{max} = 0.5$, $f^\alpha_{min} = 0.25$ in direction 2 (see Figure 1.32).

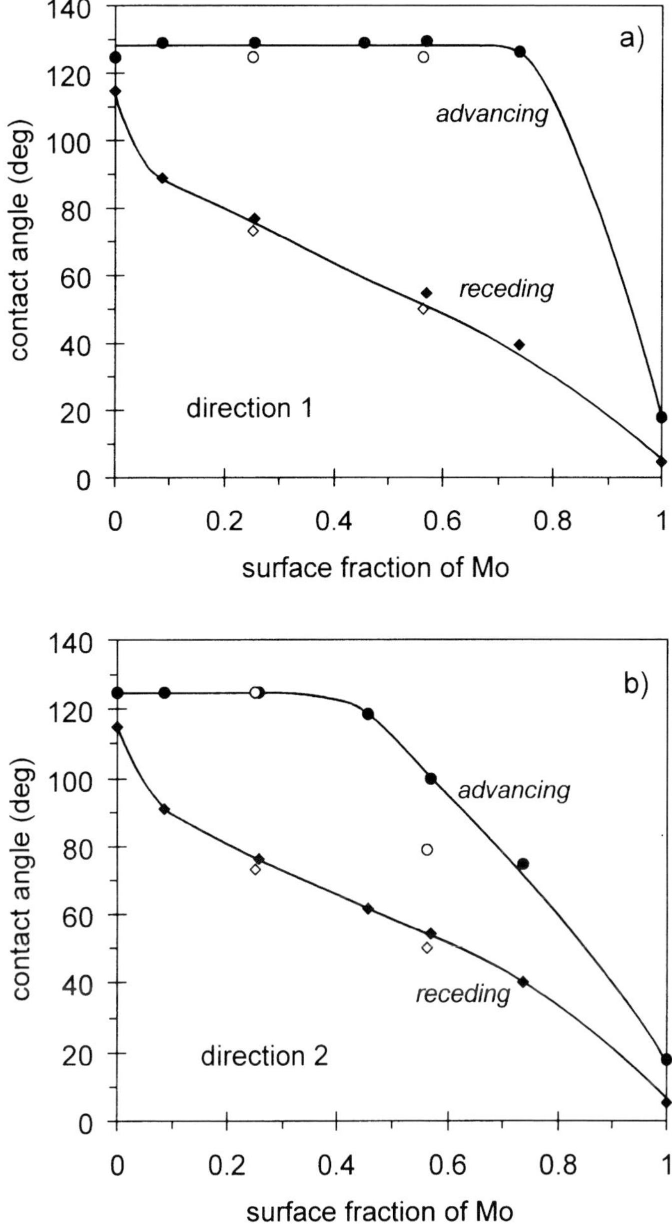

Figure 1.33. Advancing and receding contact angles vs surface fraction of Mo, f^β, for a triple line moving along the two directions of the composite surface shown on Figure 1.32. Open symbols : calculated data, full symbols: experimental data from work reported in (Naidich et al. 1995).

To sum up, the effects on static contact angles of the departures from ideality of solid surfaces are qualitatively well understood and some of these effects are used in practice to improve or reduce wettability. Moreover, for simple geometries, a semi-quantitative agreement is obtained between experimental results and theoretical predictions. For surfaces with random roughness, predictions of wetting hysteresis present a great difficulty because the relevant size of defects is not yet well-established.

1.4. DIFFERENT TYPES OF WETTING

In the past, various thermodynamic quantities, resulting usually from linear combinations of interfacial energies, were introduced to characterize the different types of wetting.

1.4.1 Adhesive wetting – The Young-Dupré equation

Consider a system consisting of a solid with a cross-section of unit area and a liquid of a different species, immiscible with the solid, held at constant temperature. Both the solid and liquid surfaces are assumed free of any adsorbed species. When the solid surface is lowered towards the liquid until contact is established (Figure 1.34.a), the change of interfacial free energy of the system, or the work gained, is given by:

$$-W_a^0 = \sigma_{SL} - (\sigma_{SV} + \sigma_{LV}) \tag{1.44}$$

where W_a^0 is the work of adhesion of the liquid on the solid. Combining equations (1.44) and (1.16), the following fundamental equation of wetting, known as the Young-Dupré equation (Dupré 1869), is obtained:

$$\cos\theta = \frac{W_a^0}{\sigma_{LV}} - 1 \tag{1.45}$$

Introducing into equation (1.45) the expressions (1.9) and (1.12) for σ_{LV} and W_a^0 derived from the simple nearest-neighbour model described in Section 1.1, this equation can be re-written as a function of S/L and L/L pair energies:

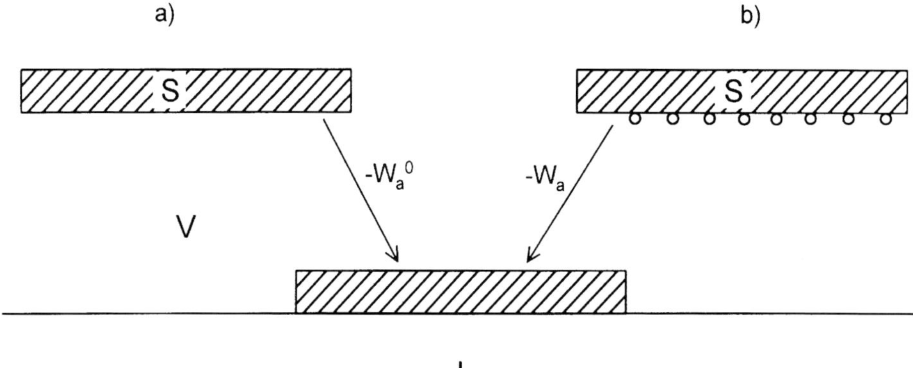

Figure 1.34. Work of adhesion in absence (a) and in the presence (b) of adsorbed vapour.

$$\cos\theta = \frac{\varepsilon_{SL}}{\varepsilon_{LL}/2} - 1 \qquad (1.46)$$

Equation (1.46) shows that the contact angle results from the competition of two types of forces: cohesion forces responsible for σ_{LV} ($= W_c/2$) and adhesion forces responsible for W_a^0. Depending on the strength of S/L and L/L interactions, different contact angles can be obtained (Table 1.1).

Table 1.1. Degrees of wetting according to the strength of S/L and L/L interactions.

S/L interactions →	weak	strong
↓ L/L interactions		
weak	wetting ($0 < \theta < 90°$)	perfect wetting ($\theta = 0$)
strong	no wetting ($90 \le \theta \le 180°$)	wetting ($0 < \theta < 90°$)

As shown in the Table, wetting can be obtained not only in systems featuring strong L/S interactions (ionic, covalent, metallic or some mixture of them), but also in systems featuring weak L/S interactions (for instance van der Waals interactions) provided that L/L interactions are weak too. The nearly immiscible molten Cu/W system ($\theta = 10°$, Table 5.2) is a typical example of strong L/L and

S/L interactions, which are both metallic. No wetting is observed in the Cu/Al_2O_3 system ($\theta = 128°$, Table 6.1) which features strong L/L but weak S/L interactions. However, many organic liquids, whose cohesion is due to weak, physical, interactions, wet silica well despite the fact that their S/L interactions are rather weak (Johnson and Dettre 1993).

From the above, it is obvious that good wetting does not necessarily mean a thermodynamically strong interface: good wetting means that the interfacial bond is energetically nearly as strong as the cohesion bond of the liquid itself.

1.4.2 Equilibrium and non-equilibrium work of adhesion – Work of spreading

Adsorption of vapour of a pure liquid on the solid surface was neglected in the previous Section. Although it does not change the general tendencies revealed by Table 1.1, adsorption must be taken into account when considering experimental results of wetting for a particular system, especially when these results are compared with theoretical predictions.

Consider now that some adsorption of liquid vapours on the solid surface occurs (Figure 1.34.b), leading to a reduction of its surface energy by a quantity $\Delta\sigma_{SV}$ (we will continue to neglect any adsorption of vapours of the solid on the liquid surface). For the sake of clarity, we denote by σ_{SV}^0 and W_a^0 the solid surface energy and the work of adhesion *in the absence of adsorption*, while σ_{SV} and W_a refer to these quantities when adsorption has occurred. Because in a pure solid/pure liquid/vapour system held at constant temperature, the solid surface is in equilibrium with a saturated vapour of the liquid at a partial pressure of P_{sat}, the equilibrium values of σ_{SV} and W_a are denoted $\sigma_{SV}(P_{sat})$ and $W_a(P_{sat})$. According to these definitions:

$$\sigma_{SV}^0 = \sigma_{SV}(P_{sat}) + \Delta\sigma_{SV}(P_{sat}) \tag{1.47}$$

$$W_a^0 = W_a(P_{sat}) + \Delta\sigma_{SV}(P_{sat}) = \sigma_{LV}(1 + \cos\theta) + \Delta\sigma_{SV}(P_{sat}) \tag{1.48}$$

with $\sigma_{SV}(P_{sat}) > 0$. A typical wetting experiment usually allows measurement of σ_{LV} and θ, so that it provides only the value of the *equilibrium* work of adhesion $W_a(P_{sat}) = \sigma_{LV}(1 + \cos\theta)$. Note that only the non-equilibrium work of adhesion W_a^0 reflects the interaction energy between the solid and the liquid phases and this can be calculated from theoretical models. For instance, in the nearest-neighbour model, the quantity $-Zm_1\varepsilon_{SL}/\omega$ in equation (1.12) equals W_a^0 and not W_a.

At this point, it is useful to introduce the *work of spreading* W_s, sometimes called the spreading coefficient, which is defined by:

$$W_s(P_{sat}) = \sigma_{LV} + \sigma_{SL} - \sigma_{SV}(P_{sat}) = 2\sigma_{LV} - W_a(P_{sat}) \qquad (1.49)$$

Thus, $W_s(P_{sat})$ represents the change in the interfacial energy of the system resulting from the formation on a solid surface of a continuous liquid film with a macroscopic thickness. Taking into account the expressions (1.9) and (1.12) for σ_{LV} and W_a, it appears that the conditions for stability of such a film can be discussed using parameters representative of the cohesion energy of the liquid (ε_{LL}) and the adhesion energy of the liquid on the solid (ε_{SL}). Depending on their values, two very different situations can arise (Table 1.2):

(1) If the adhesion energy is higher (in absolute value) than the cohesion energy of the liquid ($|\varepsilon_{SL}| > |\varepsilon_{LL}|$), significant adsorption of liquid atoms on to the solid can occur as a monolayer or multilayer depending on the ratio P/P_{sat} and the type of S-L interactions even for $P \ll P_{sat}$ (Figure 1.35). Increasing P causes the growth of a macroscopic film at a negligible supersaturation i.e., without any significant nucleation barrier. This corresponds to the 2D or Frank-van der Merwe growth of thin films (Chernov 1984). In a wetting experiment, the partial pressure of a pure liquid close to the triple line is fixed at P_{sat}, implying perfect wetting as expressed by the conditions $\theta = 0$ and $\Delta\sigma_{SV}(P_{sat}) > 0$. Although there exists some experimental evidence that these conditions hold for certain metal/metal couples (Chernov 1984), quantitative measurements of $\Delta\sigma_{SV}(P_{sat})$ have been made only for low cohesion energy liquids on various solid substrates. These experiments consisted of determining the adsorption isotherms of a species on a solid substrate for $0 \leq P \leq P_{sat}$, and the energy gained by adsorption (i.e. the quantity $\Delta\sigma_{SV}(P_{sat})$) was then deduced by integrating the Gibbs adsorption equation (Defay et al. 1966):

$$\Delta\sigma_{SV}(P_{sat}) = RT \int_{0}^{P_{sat}} \Gamma d(\ln P) \qquad (1.50)$$

between $P = 0$ and $P = P_{sat}$. In equation (1.50), Γ is the surface excess of the condensate on the solid which, neglecting the density of liquid atoms in the vapour, reduces to the number of moles of adsorbed species per unit surface area (see also Section 6.4.1). Results show that the term $\Delta\sigma_{SV}(P_{sat})$, also called the *spreading pressure*, can be high and in some cases even higher than the cohesion energy of the liquid, $2\sigma_{LV}$. For instance for water on clean silica, $\theta = 0$, $W_a = 2\sigma_{LV} = 145.6$ mJ/m^2 and $\Delta\sigma_{SV}(P_{sat}) = 316.4$ mJ/m^2 corresponding to $W_a^0 = 462$ mJ/m^2. Thus, the non-equilibrium adhesion energy for this couple is three times higher than the cohesion energy of water (Johnson and Dettre 1993).

Table 1.2. Different types of solid surface in equilibrium with a saturated vapour of a liquid.

Energy conditions	Wetting	Adsorption on the solid	Examples of adsorption isotherms of Figure 1.35	Example
$|\varepsilon_{SL}| > |\varepsilon_{LL}|$	$\theta = 0°$	$\Delta\sigma_{SV}(P_{sat}) > 0$	1	H_2O/SiO_2 $(\theta = 0°)$[a]
$|\varepsilon_{SL}| \ll |\varepsilon_{LL}|$	$\theta \gg 0°$	$\Delta\sigma_{SV}(P_{sat}) = 0$	2	Pb/graphite $(\theta = 117°)$[b]

[a](Johnson and Dettre 1993), [b](Gangopadhyay and Wynblatt 1994)

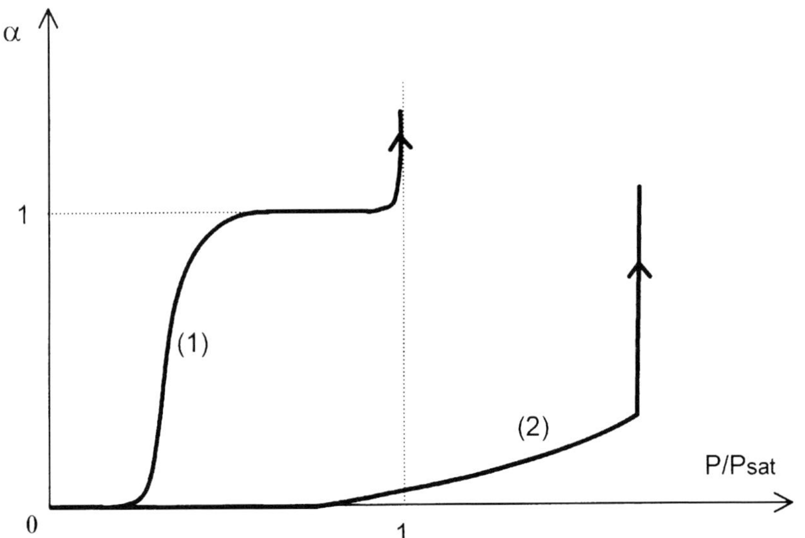

Figure 1.35. Schematic monolayer adsorption isotherms (surface coverage α versus P/P_{sat}) for the two cases in Table 1.2. At $P/P_{sat} = 1$, the solid surface is "wet" in case (1) and nearly "dry" in case (2).

(2) When the adhesion energy is much lower than the cohesion energy of the liquid ($|\varepsilon_{SL}| \ll |\varepsilon_{LL}|$), adsorption is negligible or weak, even at $P = P_{sat}$ (Figure 1.35). Growth of a macroscopic film on the substrate from the vapour is possible only if the supersaturation is sufficient to overcome nucleation barriers. This corresponds to 3D nucleation or Volmer-Weber mode of growth of thin films. In wetting experiments, $P = P_{sat}$ and the spreading coefficient W_s is positive so that a

macroscopic film is not stable and will collapse to form droplets (i.e. $\theta > 0°$). Typical examples are non-reactive metal/ionocovalent oxide and metal/graphite couples, in which adhesion is due to physical interactions that are weak compared to the cohesion energy of the metal. Experimental $\Delta\sigma_{SV}(P_{sat})$ data exist only for three metal/oxide couples (Section 4.2.2) : Sn/Al_2O_3 (T = 1200°C, $\theta = 121°$), Co/Al_2O_3 (T = 1505°C, $\theta = 113°$) and Cu/UO_2 (T = 1400-1600°C, $\theta = 90\text{-}100°$ (Hodkin and Nicholas 1973)). They were obtained using the multiphase equilibria technique (see Section 4.1.3) by measuring the surface energy of the oxide in vacuum or in a neutral atmosphere and in a saturated vapour of the metal. No significant difference between σ_{SV}^0 and σ_{SV} was found in any case. A negligible adsorption of metal atoms on the substrate surface was also revealed by direct Auger electron spectroscopy analysis of the Pb-Ni/graphite samples that had been heated at temperatures close to but slightly lower than the melting point of Pb (Gangopadhyay and Wynblatt 1994). In that system, the contact angles were in the range 100-120°, depending on the molar fraction of Ni in Pb.

To sum up, fragmentary results show that solid surfaces in equilibrium with a saturated vapour of a species exhibit a transition from a "wet" surface ($\Delta\sigma_{SV}(P_{sat}) \gg 0$) to a "dry" surface ($\Delta\sigma_{SV}(P_{sat}) \cong 0$) when the contact angle increases from zero to about 90°. According to Johnson and Dettre (1993), for organic liquids on low cohesion energy solids, if $\theta > 0$ then $\Delta\sigma_{SV}(P_{sat}) \ll W_a(P_{sat})$ and $W_a(P_{sat}) \cong W_a^0$. However, there is experimental evidence that in some nearly immiscible couples consisting of a low-melting-point metal on a high-melting-point metallic substrate, for which very low but finite contact angles are observed ($0 < \theta \leq 20°$), adsorption is significant (see Table 4.4 and Section 5.2). On the other hand, a negligible effect of saturated vapour of U on the surface energy of uranium carbide was found in the range 1300–1600°C, the contact angles being close to 60° (Hodkin et al. 1971). While interesting, these results are not sufficient to define accurately the range of θ corresponding to a transition between a "wet" and a "dry" solid surface.

1.4.3 Immersion

Consider now a liquid in a crucible that is unwetted by the liquid and surrounded by a vapour phase (Figure 1.36). The triple line between the liquid surface, the crucible wall and the vapour, is at a depth z_{max}^* below the general level of the liquid. For a crucible radius much larger than the capillary length l_c, z_{max}^* is given by (see equation (1.23)):

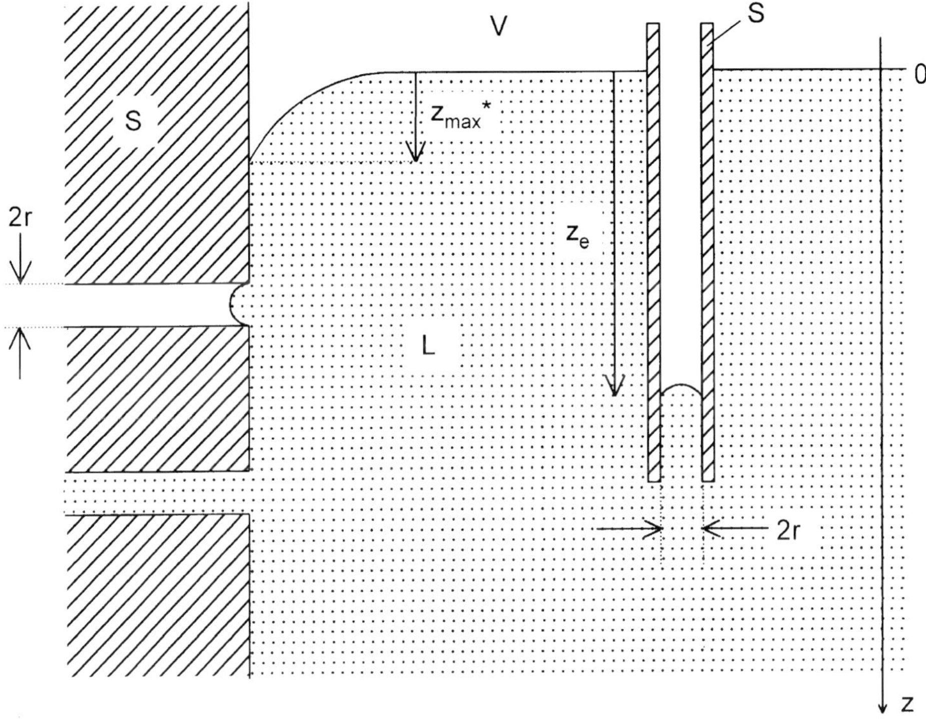

Figure 1.36. Formation of menisci of height z^*_{max} on a crucible wall and of height z_e in a capillary of the same material.

$$z^*_{max} = l_c(1 - \sin\theta)^{1/2} \tag{1.51}$$

where θ is the equilibrium contact angle. Note that z is defined here as being positive when directed downwards. We introduce into the liquid a capillary of internal radius r made of the same material as the crucible (Figure 1.36). The depression z_e of the liquid in the capillary can be calculated by minimizing the total free energy change of the system as a function of depth z. This free energy change is:

$$\Delta F = F(z) - F(z = 0) = (\sigma_{SV} - \sigma_{SL})2\pi r z + (\pi r^2 z)\frac{z}{2}\rho g \tag{1.52}$$

Setting $d(\Delta F)/dz = 0$, the equilibrium depression z_e is:

$$z_e = \frac{2(\sigma_{SL} - \sigma_{SV})}{r\rho g} \qquad (1.53)$$

The quantity (σ_{SL}–σ_{SV}) is called the *work of immersion* W_i because it defines the surface energy change when a S/V surface of unit area is replaced by a S/L interface of equal area by immersion of a solid in a liquid. In both the immersion and capillary rise processes, S/V and S/L areas change but the total L/V surface remains constant. Using the Young equation (1.16), W_i and z_e can be written:

$$W_i = \sigma_{SL} - \sigma_{SV} = -\sigma_{LV}\cos\theta \qquad (1.54)$$

$$z_e = -\frac{2\sigma_{LV}\cos\theta}{r\rho g} \qquad (1.55)$$

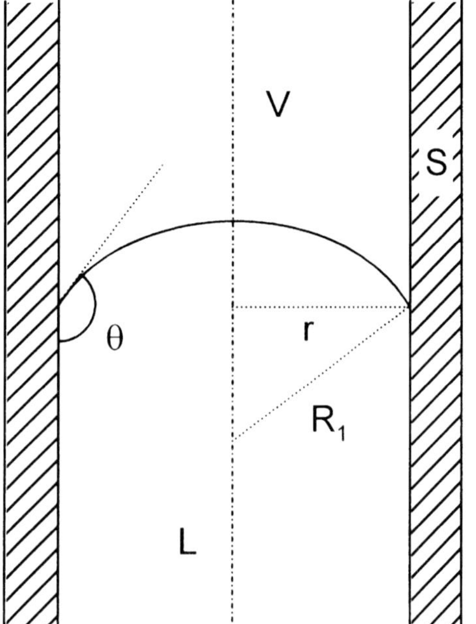

Figure 1.37. Configuration of the surface of a meniscus in a capillary.

Note that for $\theta > 90°$, z_e is positive i.e., it corresponds to a depression; for $\theta < 90°$, z_e is negative and corresponds to a capillary rise. Equation (1.55) can also be derived by a mechanical approach, considering the hydrostatic pressure $\Delta P_h = \rho g z$ and the capillary pressure ΔP_c. Applying the Laplace equation (1.20) to the capillary configuration with $R_1 = R_2 = -r/\cos\theta$ (see Figure 1.37), ΔP_c is:

$$\Delta P_c = -\frac{2\sigma_{LV}\cos\theta}{r} \tag{1.56}$$

By equating ΔP_h and ΔP_c, one finds again equation (1.55).

The work of immersion W_i is a thermodynamic quantity that describes any process of infiltration of liquids into porous media, for instance fabrication of composites by liquid routes, liquid state sintering or infiltration of refractories by molten metals or salts. In the example of Figure 1.36, at a depth z, any porosity (assumed cylindrical and open) of radius r larger than $(-2\sigma_{LV}\cos\theta)/(\rho g z)$ will be infiltrated by the non-wetting liquid, while for smaller porosities no infiltration will occur.

All the energetic quantities used to describe the different types of wetting (W_a, W_s, W_i) can be determined experimentally by measuring the contact angle and the L/V surface energy. When the measurement of θ is performed to quantify atomic or molecular S/L interactions, great care must be taken to ensure that the measurement is of the Young contact angle and that adsorption of vapours of the liquid on the solid surface can be safely neglected.

REFERENCES FOR CHAPTER 1

Cahn, J. W. (1989) *Acta Met.*, **37**, 773

Carré, A. and Shanahan, M. E. R. (1995) *Langmuir*, **11**, 3572

Cassie, A. B. D. (1948) *Discuss. Faraday Soc.*, **3**, 11

Chernov, A. A. (1984) *Modern Crystallography III: Crystal Growth*, ed. M. Cardona, P. Fulde and H. J. Queisser, Springer-Verlag, Berlin

Chizhik, S. P., Gladkikh, N. T., Larin, V. I., Grigor'eva, L. K., Dukarov, S. V. and Stepanova, S. V. (1985) *Poverkhnost*, **12**, 111 (in Russian)

Collins, R. E. and Cooke, C. E. (1959) *Trans. Faraday Soc.*, **55**, 1602

Defay, R., Prigogine, I., Bellemans, A. and Everett, D. H. (1966) *Surface Tension and Adsorption*, Longmans, London

de Gennes, P. G. (1985) *Reviews of Modern Physics*, **57**, 827

De Jonghe, V., Chatain, D., Rivollet, I. and Eustathopoulos, N. (1990) *J. Chim. Phys.*, **87**, 1623

De Jonghe, V. and Chatain, D. (1995) *Acta Met. Mater.*, **43**, 1505

Dupré, A. (1869) *Théorie Mécanique de la Chaleur*, Chapter IX, Actions moléculaires (suite), published by Gauthier-Villars, Paris

Eick, J. D., Good, R. J. and Neumann, A. W. (1975) *J. Colloid and Interface Science*, **53**, 235

Eustathopoulos, N. (1983) *Int. Metals Rev.*, **28**, 189

Eustathopoulos, N. and Chatain, D. (1990) *Entropie*, **26**, 40

Gangopadhyay, U. and Wynblatt, P. (1994) *Metall. Mater. Trans. A*, **25**, 607

Garandet, J. P., Drevet, B. and Eustathopoulos, N. (1998) *Scripta Mater.*, **38**, 1391

Gibbs, J. W. (1961) *The Scientific Papers of J. Willard Gibbs*, Dover Publications, New York

Hitchcock, S. J., Carroll, N. T. and Nicholas, M. G. (1981) *J. Mater. Sci.*, **16**, 714

Hodkin, E. N., Mortimer, D. A., Nicholas, M. G. and Poole, D. M. (1971) *J. Nucl. Mater.*, **39**, 59

Hodkin, E. N. and Nicholas, M. G. (1973) *J. Nucl. Mater.*, **47**, 23

Horsthemke, A. and Schröder, J. (1981) *J. Chem. Ingen. Technic.*, **53**, 62

Johnson, R. E. (1959) *J. Phys. Chem.*, **63**, 1655

Johnson, R. E. and Dettre, R. H. (1964) *Adv. Chem.*, **43**, 112

Johnson, R. E. and Dettre, R. H. (1969) *Surface Colloid Sci.*, **2**, 82

Johnson, R. E. and Dettre, R. H. (1993) in *Wettability*, Surfactant Science Series, vol. 49, ed. J. C. Berg, Marcel Dekker, Inc., New York, p. 1

Laplace (Marquis), P. S. (1805) *Traité de Mécanique Céleste*, fourth volume, first section (théorie de l'action capillaire) of the supplement to Book 10 (sur divers points relatifs au système du monde), published by Chez Courier, Paris

Marmur, A. (1996) *Langmuir*, **12**, 5704

Marmur, A. (1997) *J. Colloid and Interface Science*, **186**, 462

McNutt, J. E. and Andes, G. M. (1959) *J. Chem. Phys.*, **30**, 1300

Merlin, V. (1992) Ph.D. Thesis, INP Grenoble, France

Mullins, W. W. (1957) *J. Appl. Phys.*, **28**, 33

Mullins, W. W. (1960) *Trans. Met. Soc. AIME*, **218**, 354

Mullins, W. W. (1963) *Metal Surfaces: Structure, Energetics and Kinetics*, ASM

Naidich, Y. V., Voitovich, R. P. and Zabuga, V. V. (1995) *J. Colloid Interface Sci.*, **174**, 104

Neumann, A. W. and Good, R. J. (1972) *J. Colloid and Interface Science*, **38**, 341

Nolfi, F. V. and Johnson, C. A. (1972) *Acta Met.*, **20**, 769

Saiz, E., Tomsia, A. P. and Cannon, R. M. (1998) *Acta Mater.*, **46**, 2349

Shcherbakov, L. M., Samsonov, V. M. and Shilovskaya, N. A. (1995) in *Proc. Int. Conf. High Temperature Capillarity*, Smolenice Castle, May 1994, ed. N. Eustathopoulos (Reproprint, Bratislava), p. 33

Shuttleworth, R. and Bailey, G. L. J. (1948) *Discuss. Faraday Soc.*, **3**, 16

Shuttleworth, R. (1950) *Proc. Phys. Soc.*, **A63**, 444

Smith, C. S. (1948) *Trans. AIME*, **175**, 15

Soper, A., Gilles, B. and Eustathopoulos, N. (1996) *Materials Science Forum*, Transtec Publications, Switzerland, **207-209**, 433

Voitovich, R. (1992) Ph.D. Thesis, Institute for Problems of Materials Science, Kiev (in Russian)

Warren, J. A., Boettinger, W. J. and Roosen, A. R. (1998) *Acta Mater.*, **46**, 3247

Wenzel, R. N. (1936) *Ind. Eng. Chem.*, **28**, 988

Young, T. (1805) *Phil. Trans. Roy. Soc. Lond.*, **94**, 65

Chapter 2
Dynamics of wetting by metals and glasses

In Section 2.1 spreading kinetics will be presented for non-reactive liquid/solid systems i.e., for systems in which the mass transfer through the interfaces is very limited and has a negligible effect on the interfacial energies. Thus these quantities can be taken constant with time. The discussion of the results for molten metals and glasses will be based on models developed mainly for low cohesion energy viscous liquids at room temperature. The dynamics of wetting in reactive systems are developed in Section 2.2 taking into account the substantial progress made in the 90's in understanding the driving force and spreading kinetics for this important type of wetting.

2.1. NON-REACTIVE WETTING

2.1.1 Experimental facts

In general, two types of dynamic wetting can be distinguished. In *spontaneous wetting*, a drop is placed on a solid substrate and the rate at which it tends towards its equilibrium shape is measured (Figure 2.1.a). In *forced wetting*, a relative velocity of one phase (liquid or solid) with respect to the other one is imposed and the contact angle is measured as a function of the velocity. An example of forced wetting is the steady immersion (or withdrawal) of a plate in a liquid bath (Figures 2.1.b and 3.16). In the following, we will focus on spontaneous wetting which is the most studied of the high temperature systems.

A typical spreading experiment performed using the sessile drop configuration consists of measuring the rate of advance of the triple line over the solid surface when the contact angle decreases towards its final i.e., equilibrium, value θ_F. Achieving reproducible results that can be interpreted in terms of scientific fundamentals needs (i) isothermal conditions, (ii) clean liquid and solid surfaces (otherwise spreading may be controlled for instance by deoxidation kinetics), (iii) negligible or at least controlled roughness of solid substrate, and (iv) negligible external forces (although the study of the effect of a well-quantified external force on spreading can also be of great interest).

There is only a limited number of studies which satisfy all these criteria and many of them were devoted to complex metal/metal, metal/ceramic or slag/oxide systems of interest for applications in steel-making or in composite processing (this is true also for reactive systems). The major part of these studies have been performed by Russian and Ukrainian teams, whose results were reviewed by Popel (1994).

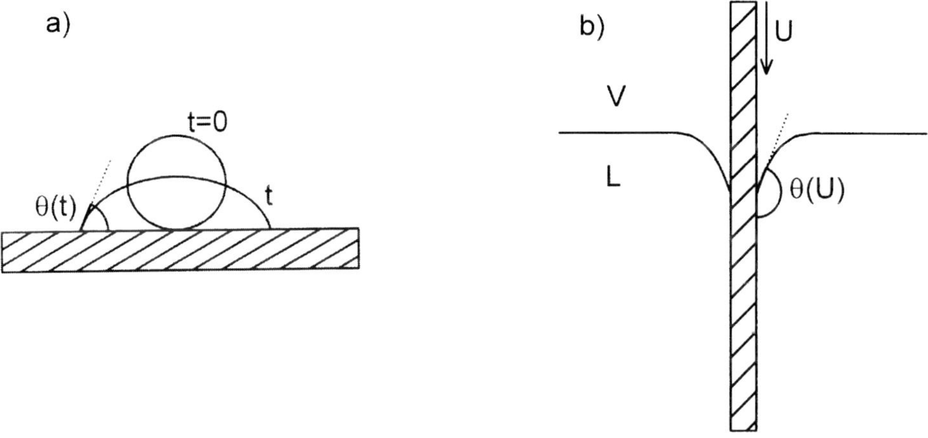

Figure 2.1. Examples of spontaneous (a) and forced (b) wetting.

For low-viscosity liquids like metals or certain oxide melts at high temperature, the spreading time for millimetre size droplets is as brief as a few tens of milliseconds. In these systems, it is impossible to carry out truly isothermal spreading experiments by the conventional sessile drop technique (configurations a and b of Figure 3.7) in which the droplet is formed by the melting of a small solid piece. Indeed, in this case, spreading is strongly influenced by kinetics of the melting process. This drawback is avoided using arrangements c and d of Figure 3.7, in which the drop and the substrate are heated separately before they are brought in contact at a fixed temperature. These techniques allow clean liquid and solid surfaces to be prepared before contact, for instance by an appropriate heat treatment. However, for both of these arrangements, the spreading process may be perturbed by external forces. Thus, in the dispensed drop method (Figure 2.2.a), the kinetic energy of a falling drop transiently can increase the driving force for wetting and lead to oscillations of the triple line around the final drop base radius R_F. Indeed, the mechanical energy ($m_d gz$) of a drop of mass m_d and radius r_d falling from a height z of a few millimeters is of the same order of magnitude as its surface energy $4\pi r_d^2 \sigma_{LV}$. As for the transferred drop variant (Figure 2.2.b), when the liquid bridge is detached from the lower substrate, it moves towards the upper substrate with an acceleration which can again lead to oscillations of the triple line. Both in the dispensed and transferred drop techniques, using a small drop mass diminishes these perturbations. Finally, the effect of substrate roughness on spreading kinetics of high temperature melts has been little studied and most experiments were carried out on what were described as "well polished" substrates. In the following, spreading kinetics will be reviewed for selected systems. Particular attention will be paid to the distinction between features due to the experimental technique and features characteristic of the intrinsic behaviour of the system.

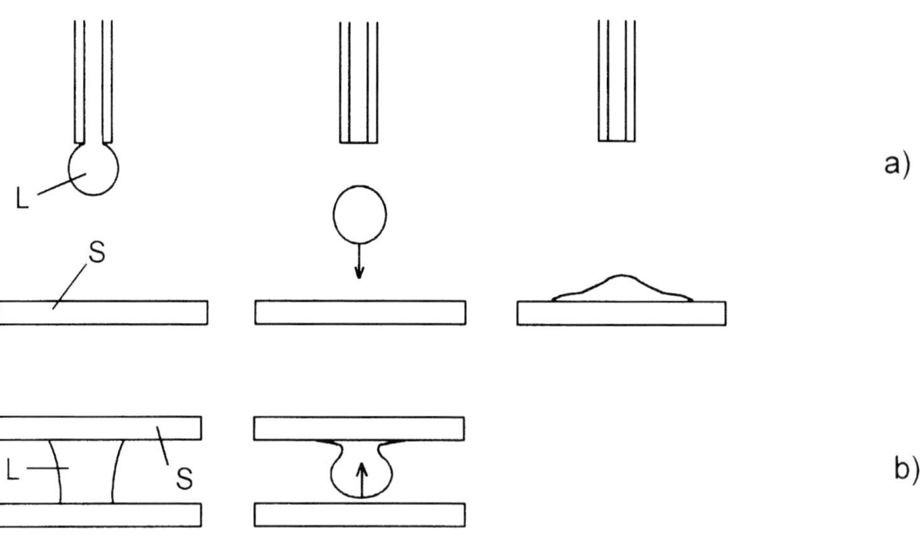

Figure 2.2. Shape of a drop when it contacts the substrate in the dispensed drop method (a) or when the liquid bridge is detached from the lower substrate in the transferred drop method (b).

Molten Pb on solid α-Fe. The maximum solubility of α-Fe in molten Pb at a temperature close to 400°C is less than 1 ppm. Spreading kinetics in this inert system have been studied by two different teams (Popel et al. 1974, Ebrill 1999) who used dry H_2 atmospheres to reduce Fe before contact with Pb. Droplets of about 200 mg were dropped on high purity Fe substrates from 5 to 10 mm height. Smooth surface substrates were used ($R_a = 5$ nm in the experiments of Ebrill). Good wetting is expected in this system and, indeed, the final contact angle θ_F was found to be close to 40–45° by both teams. Figure 2.3 shows the changes with time of the drop base radius R and of the instantaneous contact angle measured by the tangent method.

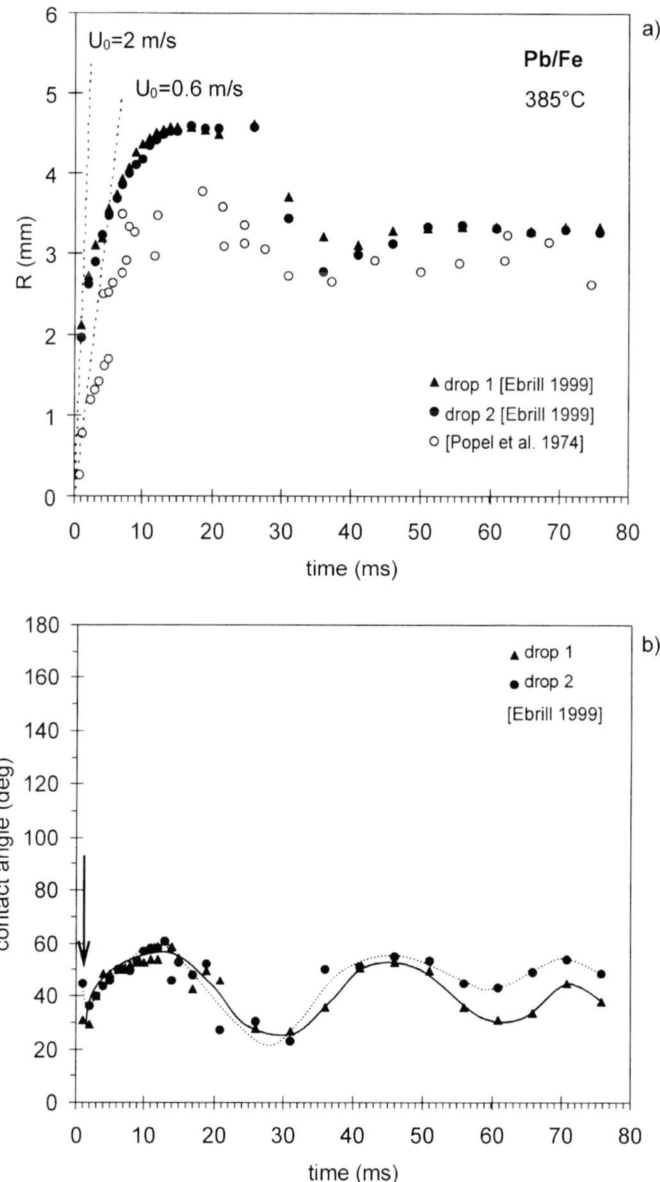

Figure 2.3. Variation of drop base radius (a) and contact angle (b) with time for the Pb/Fe system at 385°C using the dispensed drop method. The frametime of the video technique used in the experiments did not allow following of the initial decrease of contact angle from 180° to about 40°. Data from works reported in (Popel et al. 1974, Ebrill 1999).

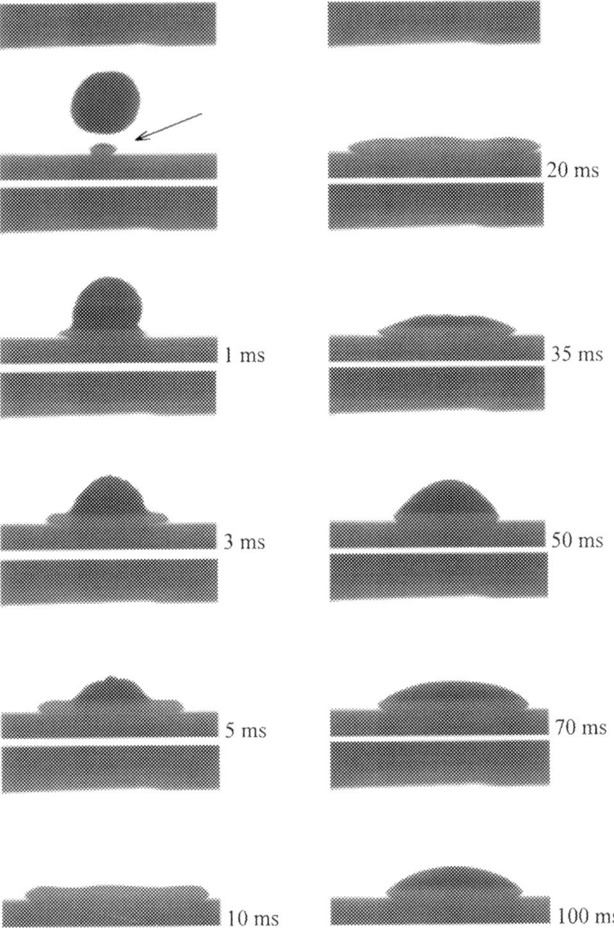

Figure 2.4. Successive images taken with a high speed camera of a Pb drop spreading on a Fe substrate at 370°C. The first image represents the falling drop before contact. The arrow shows the shadow of the drop on the solid substrate. Reprinted from (Ebrill 1999) with kind permission of the author.

In all the experiments, oscillations of the triple line around the final position (corresponding to $\theta_F \cong 40\text{--}45°$ and $R_F \cong 3\text{--}3.5$ mm) are observed. Curves in Figure 2.3 show that the first contact angle, measured after about 1 ms, is close to the final value θ_F, but at this moment the drop is clearly non-spherical (Figure 2.4). Thereafter, oscillations around R_F and θ_F are observed which decay in about 10^2 ms. At this time, the droplet becomes a nearly spherical cap (Figure 2.4). Because external forces act at the moment of impact of the drop on the solid, it is not clear if this behaviour is an intrinsic feature of the system or if it is only due to external forces. For the same reason, the initial spreading rate $U_0 = dR/dt$ which is of the order of 1 m/s (Figure 2.3.a) may be overestimated. However, the order of magnitude of U_0 in these experiments seems to be right. Indeed, the same order of magnitude has been found for other metal/metal couples using the transferred drop configuration (see below).

Molten Sn on solid Mo. This system, in which the mutual solubility is negligible, was studied by Sabuga (1990) and Naidich et al. (1992) using the transferred drop technique in high vacuum. Before contact, both the liquid and the solid were cleaned *in situ* by heating at a temperature higher than that used in the subsequent experiment. The volume of the drop placed on a lower graphite substrate was 0.025 ml. Changes of R and θ with time, illustrated by photographs, are given in Figure 2.5 for an experiment performed at 1190°C. Similar results were obtained at lower temperatures (down to 760°C) but at such temperatures, the phase diagram indicated the presence of intermetallics.

The complete transfer of the drop on to the Mo substrate occurs after 15 to 21 ms (Figure 2.5.b). Then, the acceleration of the drop is so high that a satellite drop is formed which also moves towards the upper substrate and coalesces with the main drop. As a result of the transfer process, the triple line oscillates around its equilibrium position. Thus, an unforced spreading takes place for 15 ms, with an initial rate U_0 close to 0.75 m/s (because there are no experimental points for $t < 1$ ms, this value may be too low). The contact angle decreases towards 40° in about 5 ms and thereafter changes comparatively slowly with time. At $t \geq 5$ ms, the contact radius R increases significantly at a nearly constant rate. During this stage, spreading is due to fresh liquid coming from the lower part of the bridge, while the configuration of the triple line i.e., the θ value, changes by comparison, only slightly.

Molten Sn on solid Ge. In this system, the phase diagram does not exhibit intermetallics and there is no solubility of Sn in solid Ge. The solubility of Ge in Sn is small at temperatures much lower than the melting point of Ge. Moreover, the changes in the configuration of the triple line due to dissolution occurs at comparatively longer times (see Section 2.2.1.1) and, as a result, there is no significant difference in spreading kinetics between pure Sn and Sn saturated in Ge on solid Ge for $t < 50$ ms (Naidich et al. 1972). Figure 2.6 shows results obtained using the transferred drop technique in high vacuum. As in the Sn/Mo system, the initial spreading rate is about 0.5–1 m/s and the contact angle reaches a nearly constant value in only 5 ms while the drop base radius changes significantly for about 20 ms.

oxides and other ionocovalent liquid compounds can vary by several orders of magnitude.

Molten mixtures of CaF_2-Al_2O_3 rich in CaF_2 have a viscosity of the order of a few mPa.s, comparable to that of liquid metals. Results of Sorokin et al. (1968), reported by Popel (1994), showed that the initial spreading rate of these mixtures on Al_2O_3 substrate is close to 1 m/s, as for liquid metals on solids, and decreases strongly with time, particularly when the instantaneous contact angle becomes very low.

The viscosity of the ternary silicate glass SiO_2-Na_2O-TiO_2 is in the range 10^2–10^3 Pa.s i.e., five orders of magnitude higher than that of CaF_2-Al_2O_3 molten mixtures. Results obtained by the conventional sessile drop technique for millimetre size droplets on polished Pt are given in Figure 2.7 for two different compositions of glasses (Hocking and Rivers 1982). In these experiments, the drop surfaces maintained a nearly spherical shape during the whole spreading process. Both R and θ vary significantly up to about 500 seconds. Thereafter, spreading continues very slowly, specially when θ tends towards the equilibrium value ($\theta_F = 5°$–$15°$). Dramatically slow spreading has been observed also for viscous SiO_2-Al_2O_3-CaO-MgO glass on Al_2O_3 when the contact angle tends towards the equilibrium value close to $10°$ (Ownby et al. 1995).

To sum up, in spontaneous spreading, i.e., in the absence of any external force, the spreading time for millimetre-size droplets of high viscosity ($\eta > 10$ Pa.s) is greater than 10^2 seconds and drops maintain a spherical shape during the whole process, indicating equilibrium of pressures inside the droplet at any instant. For the low viscosity liquid metals ($\eta \approx 1$ mPa.s), the spreading time is 10^{-2}–10^{-1} second and a lower limit of the initial spreading rate is 1 m/s. However, owing to experimental difficulties relating to such high spreading rates, no reliable information exist on the drop shape during spreading.

2.1.2 Modelling

Modelling of non-reactive spreading has been performed mainly for the configuration of a small droplet spreading on a horizontal and perfectly smooth solid substrate so that gravitational forces are negligible. The effect of gravity will be briefly described at the end of this Section.

In the absence of any external forces, including gravity, two shapes may be observed during spreading depending on the relative importance of viscous and inertial forces. In the configuration of Figure 2.8.a, the shape of the drop is not spherical. The instantaneous macroscopic contact angle rapidly approaches the final equilibrium value θ_F and spreading consists of an increase of R while the configuration of the triple line remains nearly unchanged. The spreading rate will

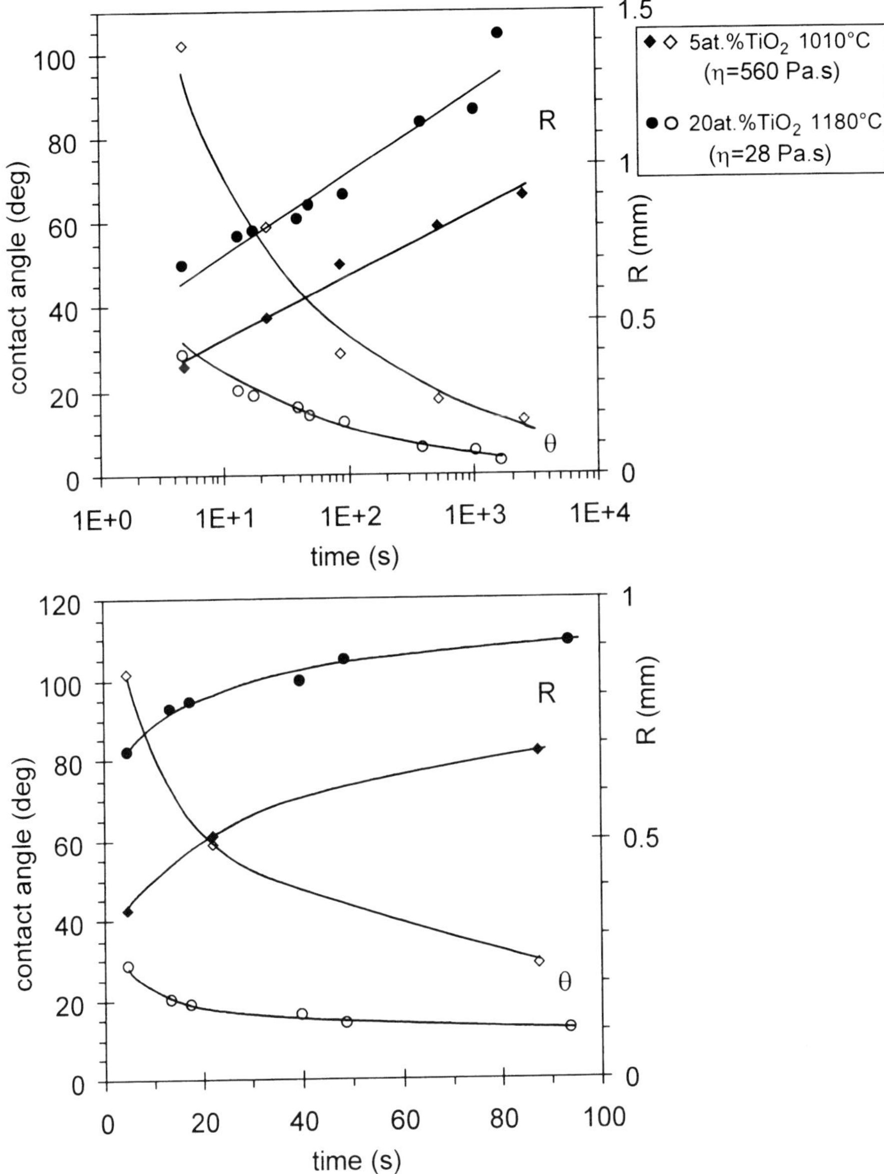

Figure 2.7. Variations of contact angle with time for ternary silicate glasses containing SiO_2 and Na_2O in a 4:1 mole ratio with additions of TiO_2 on Pt substrates (Hocking and Rivers 1982). The R(t) curves are calculated from contact angles values assuming a spherical shape for the drop and taking a drop volume of 0.52 mm^3.

depend on the rate at which the matter moves from the bulk drop to the triple line, driven by the differences in capillary pressure given by the Laplace equation (1.20). In the configuration b of Figure 2.8, both θ and R change with time while the shape of the drop is a nearly spherical cap.

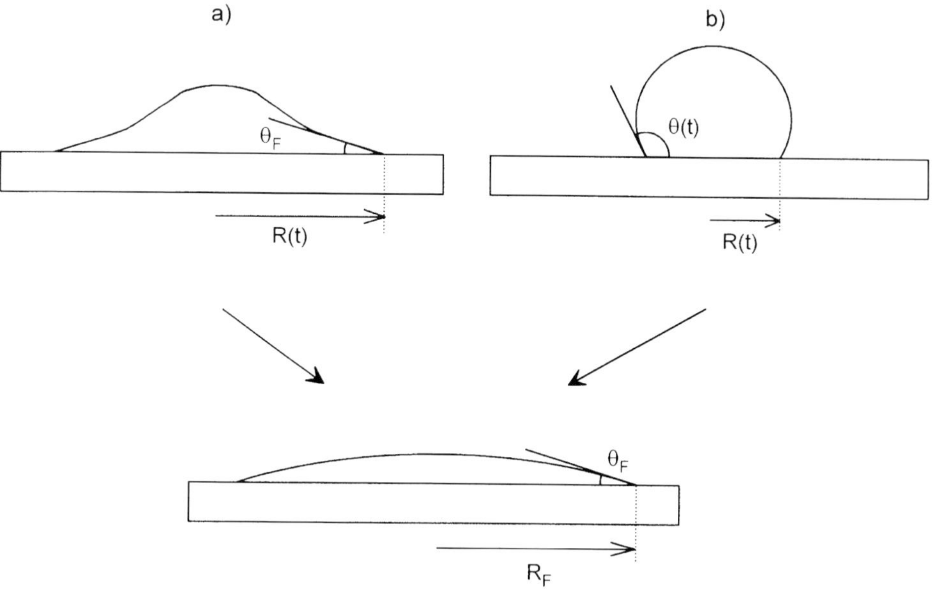

Figure 2.8. Inertial (a) and viscous (b) spreading of a sessile drop in a non-reactive liquid/solid system.

The first case is expected to apply for liquids with a very low viscosity which permit high triple-line velocities and in which, once the contact angle approaches its equilibrium value, spreading is controlled by inertial forces responsible for the non-spherical shape of the drop at t $\ll$ t_F, where t_F is the time needed to reach a constant drop base radius. The second case corresponds to a spreading controlled by viscous resistance. Because, as we will see below, in viscous spreading the dissipation of energy occurs mainly in the liquid close to the triple line, the remaining liquid can move more easily, maintaining a macroscopic spherical shape i.e., a constant capillary pressure. In this case, the driving force for wetting per unit length of triple line, f_d, is given by the change in the surface and interfacial energy F_s of the system resulting from a lateral displacement of the triple line:

$$f_d = -\frac{dF_s}{dx} = \sigma_{SV} - \sigma_{SL} - \sigma_{LV}\cos\theta = \sigma_{LV}[\cos\theta_F - \cos\theta] \qquad (2.1)$$

where x is the abscissa of the triple line and θ is the instantaneous contact angle. When the triple line advances, the work done by surface forces, *per unit length of triple line and unit time*, is equal to:

$$W_d = U.f_d \tag{2.2}$$

where U is the triple line velocity ($= dR/dt$).

The viscous spreading of a sessile drop has been modelled by several authors (Tanner 1979, Hocking and Rivers 1982, de Gennes 1985, Summ et al. 1987). Although several differences exist between these treatments, they all regarded the drop as divided into two regions : (i) the bulk in which the motion of the fluid is easy and energy dissipation by viscous flow negligible and (ii) a macroscopic wedge near the triple line of typical width 0.1 mm. The motion of the fluid in this region has been observed by Dussan and Davis (1974) who marked the upper surface of a viscous liquid with small spots of dye and watched their motion. They found a characteristic rolling motion (Figure 2.9), similar to that of a caterpillar-tracked vehicle, which gives rise to viscous friction. Therefore, the work W_d is mainly dissipated by viscous friction in the wedge.

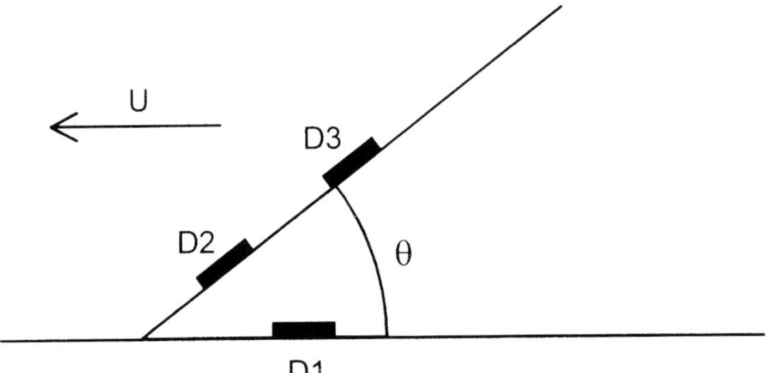

Figure 2.9. Motion of the liquid in the vicinity of the triple line in viscous spreading. In the Dussan and Davies experiment, spots of a dye D1, D2, D3 are laid on the free surface of an advancing edge. The spots slide down and then get stuck to the solid. From (de Gennes 1985) [4].

De Gennes (1985) considered also a precursor film ahead of the triple line, formed when the ratio $(\sigma_{SV} - \sigma_{SL})/\sigma_{LV}$ is greater than unity, which corresponds to the case of more than perfect wetting. Because more than perfect wetting is not

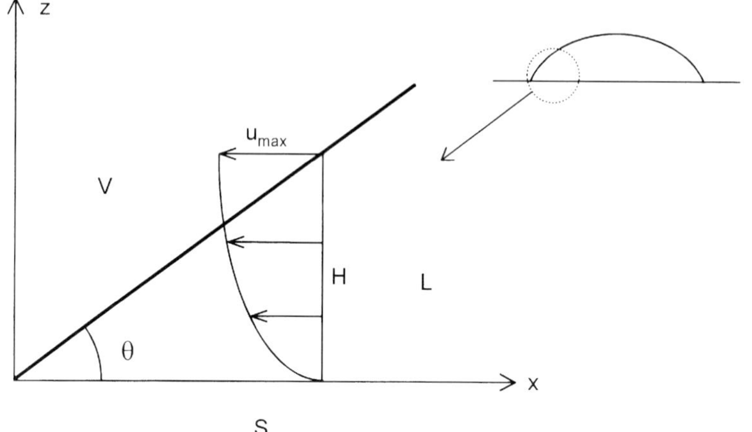

Figure 2.10. Viscous spreading : fluid velocity profile in the liquid wedge close to the triple line. According to (de Gennes 1985) [4].

very common in high temperature systems, we consider (below) the case where $\theta_F > 0°$. Viscous spreading will be discussed in the framework of the de Gennes model because it has the advantage of being analytical.

The fluid velocity in the wedge region is calculated by a "lubrication" approximation in which the fluid velocity u parallel to the interface varies parabolically between $u = 0$ at $z = 0$ and $u = u_{max}$ at the drop surface, as shown in Figure 2.10. (Note that this description is not valid near the drop center where the velocity field is much more complex (Denesuk et al. 1993); however, viscous dissipation in this region is deliberately neglected in the de Gennes model). The average value of u at a fixed distance x from the triple line is approximately equal to the triple line velocity U. At a fixed x, the viscous dissipation integrated over the liquid column of height H is $3\eta U^2/H$ and the total dissipation in the wedge calculated between $x = x_{min}$ and $x = x_{max}$ is

$$E_{diss} = \frac{3\eta U^2}{\tan \theta} \ln \left| \frac{x_{max}}{x_{min}} \right| \tag{2.3}$$

x_{max} is a distance related to the macroscopic size of the droplet ($x_{max} \approx R$) while x_{min} must be different from zero otherwise dissipation would diverge. Although all authors (Hocking and Rivers 1982, Marmur 1983, de Gennes 1985) agree on the necessity to remove this divergence by allowing slippage of the liquid layer in contact with the solid over a distance x_{min}, neither the value of x_{min} nor the

capillary pressure is nearly zero). By considering that the capillary energy of a droplet of initial radius r_d is tranformed into kinetic energy, Joanny (1985), cited by de Gennes (1985), found a typical value of the spreading velocity given by:

$$U \cong \left(\frac{\sigma_{LV}}{\rho r_d}\right)^{1/2} \tag{2.10}$$

For a typical metal and taking $r_d = 10^{-3}$ m, this expression leads to $U \cong 0.3$ m/s, much closer to the experimental data than values calculated from the viscous model.

Wetting ridge regime. As shown in Section 1.2.4, complete equilibrium at the triple line is obtained only with a displacement of the triple line in the direction perpendicular to the solid surface. For "soft", viscoelastic, substrates, this local deformation, or "wetting ridge", is measurable (several tens of nm) and the lateral displacement of the ridge may control spreading (Carré and Shanahan 1995). For "hard", high cohesion-energy, solids, the elastic deformation at the triple line is negligible (of the order of 10^{-2} nm, see Section 1.2.4) and cannot affect the spreading process. However, at high temperature, atomic mobility can contribute to the growth of the ridge, as discussed in Section 1.2.4. The possibility that this ridge plays an important role in the spreading of metallic liquids on "hard" solids was discussed in (Saiz et al. 1998). The starting point of this work is the study of Mullins (1958), who calculated the effect of thermal grooving on grain-boundary motion.

When a grain boundary forms an angle α with the direction perpendicular to the interface, there is a critical value α_c above which the grain boundary can move, driven by the difference in interface curvature across the groove root (Figure 2.13). When atomic movement in the groove is assumed to occur by surface diffusion, Mullins (1958) found a stationary solution for $\alpha \geq \alpha_c$ corresponding to a constant grain boundary velocity and a constant groove depth h. For $\alpha < \alpha_c$, the grain boundary is pinned by the groove (i.e., the groove acts as an asperity pinning the triple line in wetting) and the groove grows according to the power law $h^n = Kt$ (equation (1.33) with n = 4).

Saiz et al. (1998) considered that in the case of a triple line, the L/V surface can play the role of a grain boundary and the "wetting ridge" can move either by bulk or surface (or interface) diffusion of solid atoms (Figure 2.14). They treated the case of surface diffusion with n = 4, taking into account the difference of diffusivities at the S/V surfaces and S/L interfaces. In their experiments with Cu and Ni droplets on Al_2O_3 surfaces (see Section 1.2.4), Saiz et al. maintained the droplets in contact with the substrate for several tens of minutes at the end of the spreading process, allowing the ridge to grow to a size that permitted easy observation. For Cu, a wetting ridge with h = 10 nm was formed after 2 hours at 1150°C and for Ni h = 0.2 μm after 1 hour at 1500°C. Much greater ridge heights (several tens of μm) were observed for silicate glass droplets on Co at 1440°C, i.e., at a temperature very close to the melting point of the substrate.

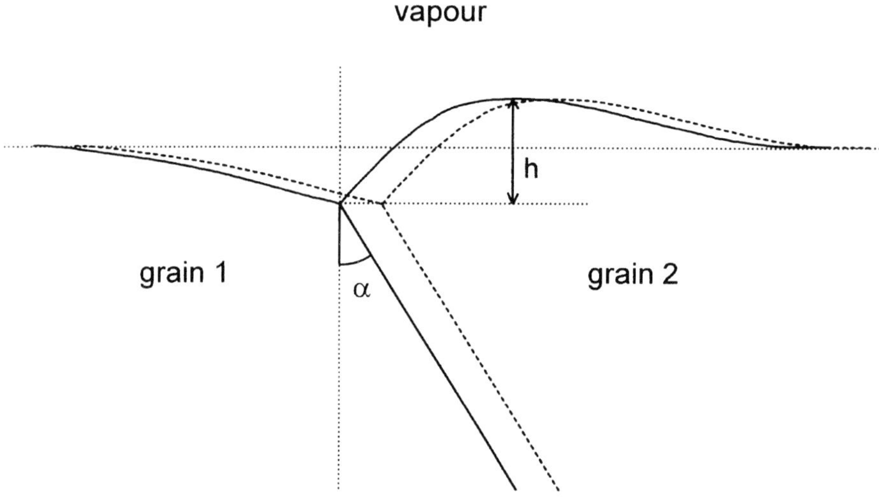

Figure 2.13. Traveling grain-boundary groove according to Mullins (1958).

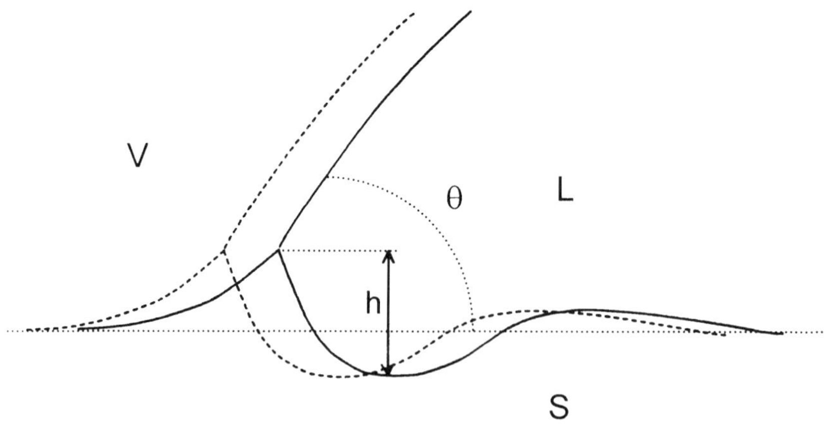

Figure 2.14. Wetting ridge at the triple line formed during spreading according to (Saiz et al. 1998).

A wetting ridge can affect spreading kinetics only if its size *during spreading* is significant with respect to the size of surface roughness of the substrate. Spreading kinetics studies are usually performed on polished surfaces with a typical average roughness of 10 nm. Accordingly, wetting ridges must have $h > 10$ nm in order to affect spreading rates. For Ni, the time t_r needed for the ridge to attain $h = 10$ nm can be extrapolated from data obtained for $t = 1$ h using equation (1.33) with $n = 3$, which leads to $t_r \cong 0.5$ s. Although the conclusions of this calculation would be the same with $n = 4$, it appears more plausible, at least on the liquid side of the ridge, for the ridge growth to be governed by diffusion through the liquid for which $n = 3$ rather than by interface diffusion for which $n = 4$. If we consider a triple line with a lateral velocity U not yet affected by the ridge, the dwell time of the triple line on the area of a growing ridge of width w (w being of the order of h (Mullins 1960)) is $t_w \cong h/U$. For the two processes of ridge growth and drop spreading to interfere, the time t_r for ridge formation must be close to or lower than t_w, which will occur only for $U < U^* = h/t_r$. In the example of Ni, U^* is 20 nm/s. Therefore non-reactive spreading of Ni, for which U is expected to be of the order of 10^2–10^3 mm/s, cannot be affected by wetting ridges, except at times t approaching t_F i.e., when U tends *asymptotically* towards zero. The same conclusion is *a fortiori* valid for Cu on Al_2O_3.

Another example is the molten Pb/solid Fe system for which the depth of a groove formed at the L/S interface is 10 μm after 150 h at 1100°C (at this temperature the molar fraction of Fe dissolved in Pb in equilibrium with γ-Fe is 8×10^{-4}) (Eustathopoulos 1983). The value of U^* calculated from these data is 20 μm/s, which is again negligible compared with the expected spreading rate of non-reactive metals, except at t very close to t_F. Note that values of the spreading rate of the order of 1–10^2 μm/s are observed for several cases of reactive wetting and thus wetting ridges can, in principle, influence spreading kinetics in these cases.

Effects of gravity. Until now, spreading kinetics have been discussed for small droplets for which gravity can be neglected. For large droplets, gravity can modify spreading by two different effects : (i) by increasing the value of the final drop base radius R_F compared to the zero gravity situation (Figure 2.15) and (ii) by increasing the driving force of wetting f_d (equation (2.1)) which now will be given by:

$$f_d = -\frac{dF_s}{dx} - \frac{dE_p}{dx} \tag{2.11}$$

where the second term of the right hand side takes into account the decrease in potential energy E_p of the spreading drop. The first effect tends to increase the spreading time for a fixed value of the drop mass while the second effect acts in an

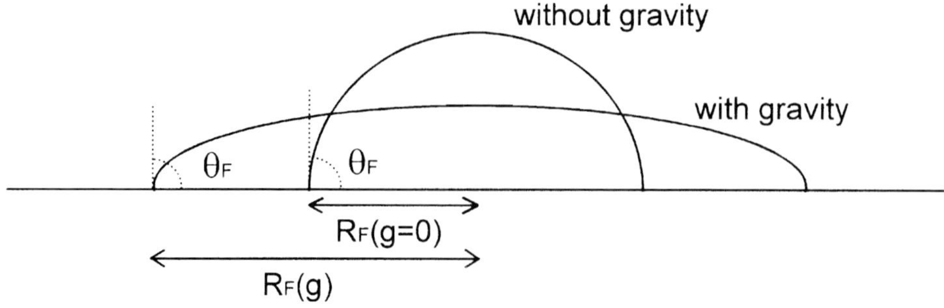

Figure 2.15. Equilibrium shape of a sessile drop with and without gravity.

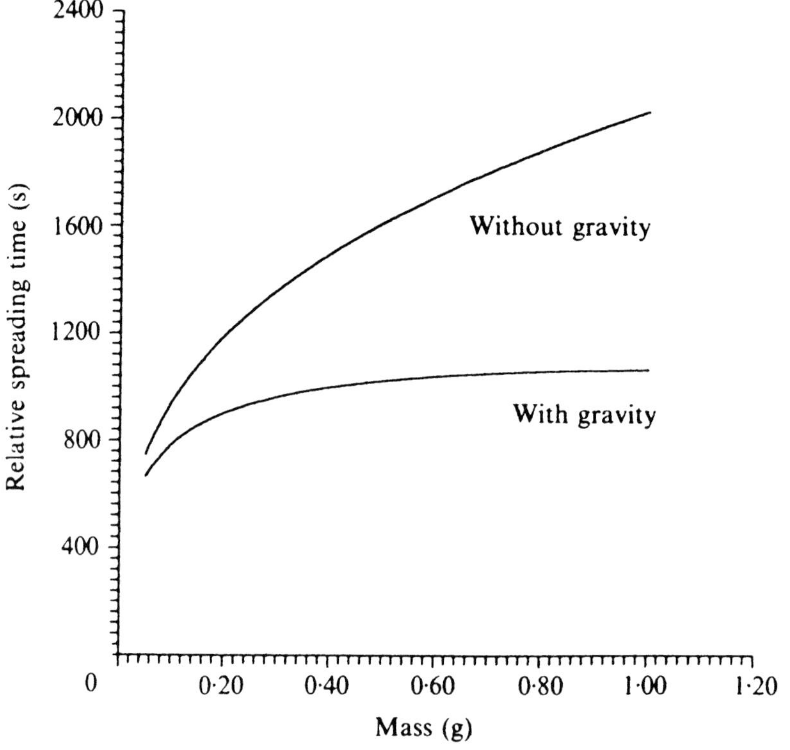

Figure 2.16. The predicted time, with and without gravity, required for a droplet to change from a contact angle of 100° to 10° as a function of the total droplet mass ($\theta_F = 5°$, $\sigma_{LV} = 0.2$ J.m^{-2}, $\eta = 5 \times 10^3$ Pa.s, $\rho = 5 \times 10^3$ kg.m^{-3}). Reprinted from (Denesuk et al. 1993) [6] with kind permission of the authors.

opposite manner. Figure 2.16 shows values of spreading times calculated for a viscous liquid as a function of drop mass m_d with and without gravity (Denesuk et al. 1993). With gravity, the curve shows that the spreading time first increases with m_d and then varies only weakly indicating that the two effects are partially compensated. At any m_d, the action of gravity leads to a decrease in spreading time but this effect is significant only for $m_d > 0.1$ g.

2.1.3 Concluding remarks

For viscous liquids such as most glasses ($\eta > 10$ Pa.s), the spreading time for millimetre size droplets is longer than 10^2 seconds. For these liquids, spreading rates at $\theta < 90°$ are described satisfactorily by models in which capillary energy is dissipated only by viscous friction, which occurs mainly in a wedge close to the triple line. For low viscosity liquid metals ($\eta \sim 10^{-3}$ Pa.s), much shorter spreading times are observed (10^{-1}–10^{-2} second) and the lower limit of the initial spreading rate is 1 m/s. For these liquids, models of viscous spreading cannot explain existing data, except probably in the region of very small contact angles. For higher θ values, other processes may be limiting, specially inertia.

2.2. REACTIVE WETTING

We will first discuss the case of simple dissolution of the solid in the liquid (Section 2.2.1) and then the case of formation at the interface of a 3D compound by reaction between the solid and the liquid (Section 2.2.2).

2.2.1 Dissolutive wetting

Two extreme cases will be considered. In the first one, dissolution of the solid into the liquid is assumed not to change significantly the surface and interfacial energies σ_{LV} and σ_{SL}. Therefore, dissolution modifies only the geometry at the triple line (Figure 2.17.a). In the second case, the interfacial energies are modified due to dissolution of small quantities of tensio-active species of the solid but the solid/liquid interface is assumed to remain nearly flat i.e., the equilibrium contact angle is still given by the Young equation (Figure 2.17.b).

2.2.1.1 Dissolution-insensitive σ_{LV} and σ_{SL} values. This case was studied by Warren et al. (1998) and Yost and O'Toole (1998) and modelled by the first authors for a liquid metal B/solid metal A system in which, at the experimental temperature, there is no formation of intermetallics and the solubility of B in A is negligible.

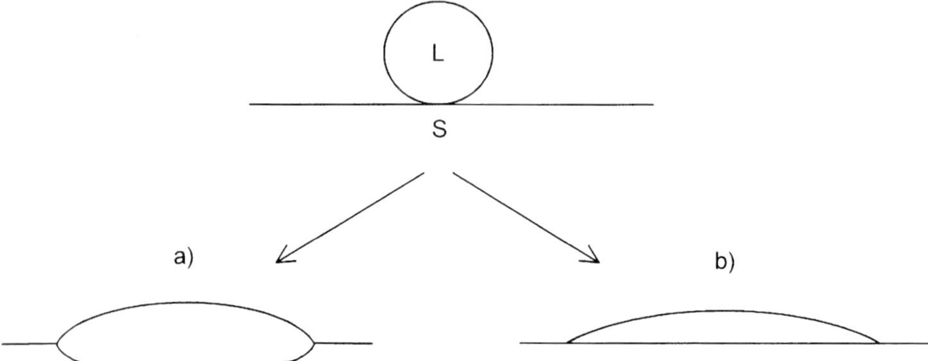

Figure 2.17. Two extreme cases of dissolutive wetting. (a) Dissolution of the solid modifies the geometry at the triple line. (b) A slight dissolution is enough to modify the surface energies of the system while the S/L interface remains macroscopically planar.

The treatment of Warren et al. provides an insight into the parameters of importance for this type of wetting. According to these authors, when a droplet of pure B, or B undersaturated in A, is placed on pure solid A, the approach to equilibrium occurs in three stages having different time scales (see Figure 2.18 for meaning of symbols used in the following):

(i) Non-reactive spreading takes place in a short time t_1. During this stage, the effect of dissolution on the macroscopic morphology is negligible and the contact angle at $t = t_1$ is nearly equal to the Young contact angle of the system (Figure 2.18.b). We saw in Section 2.1.1 that $t_1 \approx 10^{-2}$ s.

(ii) Then, dissolution of the solid in the liquid occurs (Figure 2.18.c) and affects the macroscopic contact angle after a certain time (in the experiments of Warren et al. on molten Sn/solid Bi couple, the S/L interface remained macroscopically flat even at $t = 5$ s and was curved close to the triple line at $t = 90$ s, see Figure 2.20). This is followed by diffusion, driven by differences of concentration, C, between the S/L interface, where $C \cong C_i$, and the drop bulk, where $C = C_\infty$ (with $C_\infty(t = 0) = C_0$). This process increases the total volume of the liquid phase and the final volume v_t depends on the initial volume v_0, the initial concentration of A in B and the phase diagram. The changes of morphology with time were calculated by Warren et al. (1998) by considering that the *microscopic* dihedral angle at the triple

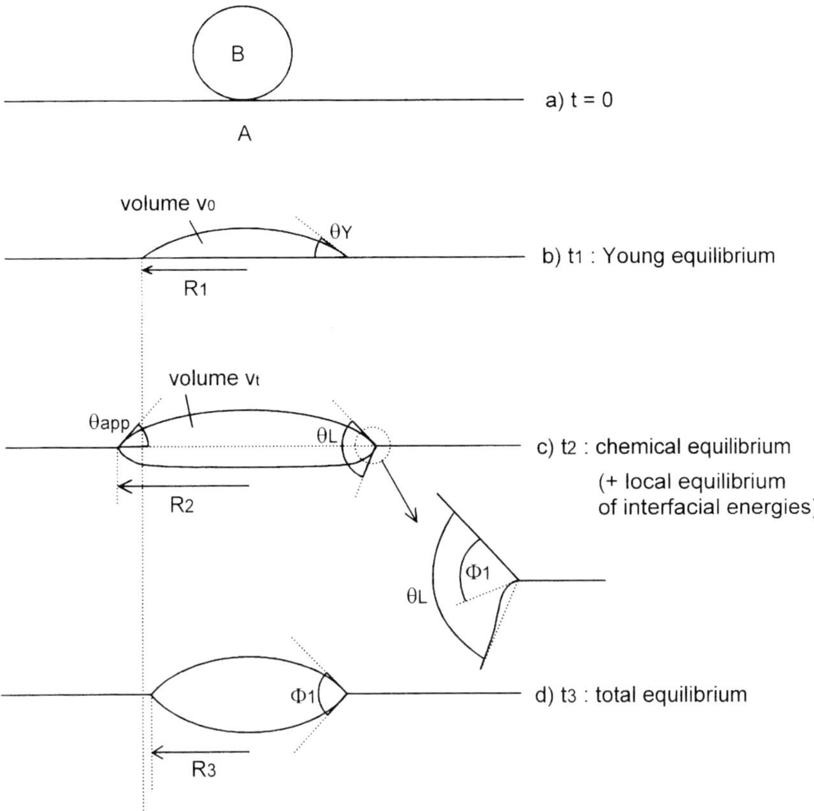

Figure 2.18. Dissolutive wetting: the approach of local (c) and total (d) equilibrium in the sessile drop configuration. In liquid metal/solid metal systems, the characteristic times of the different stages are $t_1 \sim 10^{-2}$ s, $t_2 \sim 10^2$ s and $t_3 \sim 10^8$ s. According to the calculations of Warren et al. (1998).

line is given by the Smith equation (1.30) and assuming that the triple line remains on the plane of the substrate. Moreover, assuming that the transfer of atoms through the interface is rapid compared to diffusion in the liquid alloy, it follows that the solid and liquid phases are in thermodynamic equilibrium at the interface. This equilibrium is described by the Gibbs-Thomson equation (1.32) which, expressing the chemical potential μ as a function of the concentration in the liquid C, allows, after integration, the equilibrium concentration C_i in the liquid at an interface of curvature κ to be related to the equilibrium concentration C_{eq} for a flat interface:

$$C_i = C_{eq}(T)[1 + \Lambda\kappa] \tag{2.12}$$

In this equation, Λ is the Gibbs-Thomson parameter defined by $\sigma_{SL}v_m/(RT)$ (v_m being the molar volume of the solid). Then, the dissolution rate is derived by calculating the diffusion flux from the moving interface to the drop bulk. Such calculations were performed by Warren et al. (1998) for a Bi-Sn drop, of radius $R_1 = 1.9$ mm at t_1 and initial composition $C_0 = 30.5$ at.% Sn, on a Bi substrate at 245°C. At this temperature, the equilibrium concentration C_{eq} is 16.4 at.% Sn, so that equilibration doubles the liquid volume. During this process, the drop base radius R increases moderately, the maximum value of $(R_2 - R_1)/R_1$ being 7%. Most of this increase occurs during the first 10^2 seconds, which, for millimetre size droplets, can be regarded as the order of magnitude of the time t_2 needed to approach chemical equilibrium in the liquid drop/solid system. During this time, the depth of the crater in the solid is greatest close to the triple line where the comparatively fresh liquid in contact with the solid causes a more rapid dissolution (Figure 2.19). At any instant, the apparent contact angle, θ_{app}, is lower than the Young contact angle $\theta_Y \cong \theta(t_1)$. Finally, it should be noted that the macroscopic dihedral angle θ_L is different from the true contact angle Φ_1 defined by the Smith equation (Figure 2.18.c). This is because at these relatively short times, the geometry of the system close to the triple line i.e., the value of θ_L, is controlled by the diffusion process and not by interfacial energies. Indeed, as argued by Warren et al., based solely on diffusion and fluid flow, the liquid angle θ_L "wants" to be near 90° since the liquid surface admits no-flux of solute and the liquid/solid interface is essentially an isoconcentrate. This is verified experimentally as shown by micrographs of the triple region for Bi-Sn/Bi (Figure 2.20), Al/Si (Nakae and Goto 1998) and Fe/C (Hara et al. 1995).

(iii) At much longer times, the shape of the solid/liquid interface will change to assume a uniform curvature. During this stage, θ_L tends asymptotically towards Φ_1 while the contact radius R decreases towards a value R_3 of less than R_1 (Figure 2.18.d). However, this decrease in R starts only at $t = 60$ h and R_3 is approached at a time t_3 estimated to be a few years!! This is because the shape change is driven by concentration differences along the S/L interface induced by curvature variations (equation (2.12)). For liquid metals, the Gibbs-Thomson parameter Λ is small (typically 2×10^{-10} m) which, combined with a small curvature ($\kappa \approx (r_{drop})^{-1} \approx 10^3 \, m^{-1}$), leads to extremely small concentration gradients along the interface ($C_i - C_{eq} \approx 2 \times 10^{-7} C_{eq}$).

Calculations by Warren et al. for the second stage are in semi-quantitative agreement with experimental results. For instance, the calculated value of R_2/R_1 at $t = 300$ s is 1.07, lower but of the same order of magnitude as the experimental value of 1.26. To sum up, dissolution in metal/metal systems increases the contact area between the liquid and the solid, mainly by increasing the available liquid (Yost and O'Toole 1998). In the absence of tensio-active effects, this increase of contact area is rather limited.

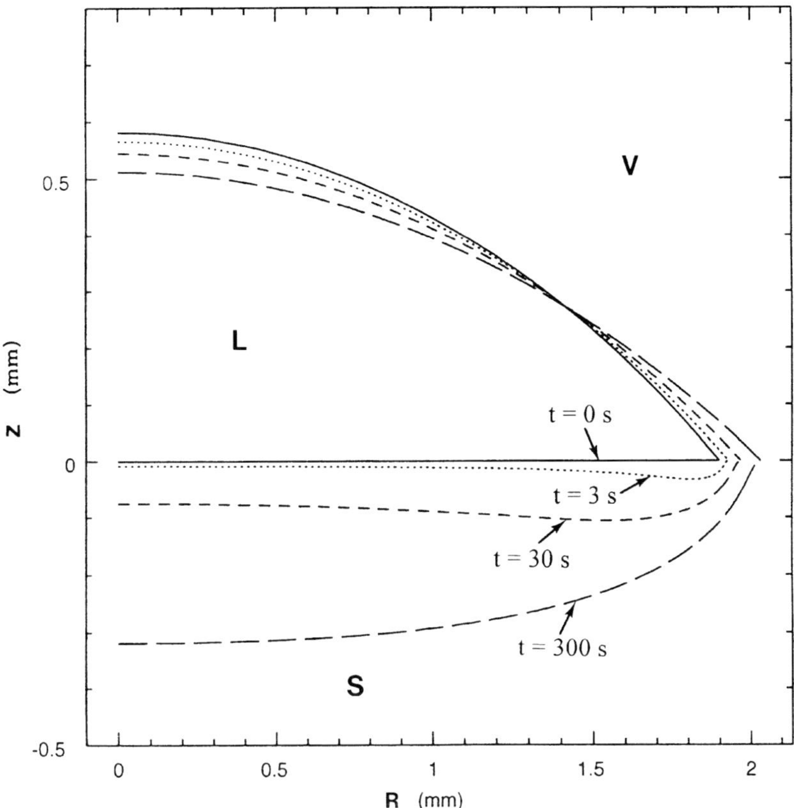

Figure 2.19. Calculated shape of the liquid region as a function of time for a Bi-30.5 at.% Sn drop on a Bi substrate at 245°C. Reprinted from (Warren et al. 1998) [7] with kind permission of the authors.

2.2.1.2 Dissolution sensitive σ_{LV} and σ_{SL} values. Much stronger increases in R are expected from dissolution of a tensio-active species able to decrease significantly σ_{SL} and/or σ_{LV}, even if the corresponding increase of the liquid volume is negligible. For example, for a droplet with $\theta_Y = 40°$, a decrease of σ_{LV} of 30% is enough to increase the contact radius by a factor 2.

Laurent (1988) studied the particular case where dissolution is high enough to modify σ_{SL} (or σ_{LV}) but small enough to allow neglect of any deviation of the S/L

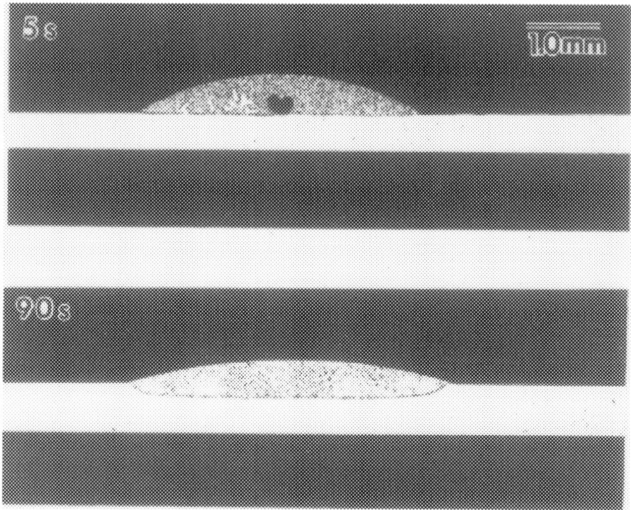

Figure 2.20. Cross sections of Bi-30.5 at.% Sn drops on pure Bi substrates quenched at different times (5s and 90s). Reprinted from (Warren et al. 1998) [7] with kind permission of the authors.

interface from being planar. A typical example is a metal/oxide system, like Cu/NiO, see Section 6.4.1 and Table 6.8, in which slight dissolution of the oxide produces strong changes in σ_{SL} and σ_{LV}. Laurent examined how the driving force, given by equation (2.1) for the non-reactive case, is modified by dissolution. Using a free energy variation treatment, Laurent found that when dissolution does not affect σ_{SV}^0 and σ_{LV}^0 (the superscript 0 denotes the values of these quantities before reaction), the driving force of wetting can be described formally:

$$f_d(t) = \sigma_{SV}^0 - (\sigma_{SL}^0 + \Delta\sigma(t) + \Delta G(t)) - \sigma_{LV}^0 \cos\theta(t) \qquad (2.13)$$

where $\Delta\sigma(t)$ takes into account the change in σ_{SL} brought about by the reaction $(\Delta\sigma(t) = \sigma_{SL}(t) - \sigma_{SL}^0)$ and $\Delta G(t)$ is the change of Gibbs energy per unit area released by the dissolution reaction at the interface.

Assuming that the mobility of the triple line (expressed by the velocity of the line in the corresponding non-reactive system for the same instantaneous contact angle) is high compared to the rate of interfacial transfer and to the diffusion rate in the bulk liquid, capillary equilibrium at the triple line is readily maintained, so that $f_d(t) = 0$ and the instantaneous contact angle is given by:

$$\cos \theta(t) = \cos \theta^0 - \frac{\Delta\sigma(t)}{\sigma_{LV}^0} - \frac{\Delta G(t)}{\sigma_{LV}^0} \tag{2.14}$$

where θ^0 is the equilibrium contact angle in the absence of reaction.

The reaction energy term $\Delta G(t)$ was first proposed by Aksay et al. (1974), both for dissolution reactions and for reactions with formation of a new phase at the interface. Aksay et al. considered that the energy produced by the reaction between the liquid and the solid at the periphery of the drop improves wettability. They argued that the effect of the $\Delta G(t)$ term is strongest during the early stages of contact because the interfacial reaction rate is at its maximum when the liquid contacts a fresh solid surface, but thereafter the reaction kinetics decrease. Thus, the effect of the reaction is to cause an initial decrease in the contact angle, which then increases and gradually approaches the equilibrium value (Figure 2.21). Major difficulties lie in the calculation of the $\Delta G(t)$ term because the coupling conditions of the time-dependent interfacial reaction with the kinetics of wetting are unknown. As a consequence, a calculation of the thickness of the reaction zone in "the immediate vicinity of the interface" is not yet possible.

The model of Aksay et al. predicts that the θ-t curve passes through a minimum. However, as a general rule, existing data for $\theta(t)$ in reactive couples show that θ decreases monotonically with time to a steady value. An exception is the Ag/SiC system in which transient decreases of θ were observed and explained by means of the model of Aksay et al. (Li 1994). However it will been seen in Section 2.2.2.1 (see Figure 2.28) that the transient decreases of θ in that system is not due to a $\Delta G(t)$ term (reactivity in Ag/SiC is indeed very weak) but can be explained easily on the basis of changes in interfacial energies of the system.

Experiments carried out in a model system (Cu-Ag binary alloy) (Sharps et al. 1981) show no contact angle minimum but the results have been interpreted using Aksay's model because wetting for reactive combinations (for instance the Ag-Cu eutectic alloy on pure solid Cu) was better than for non-reactive couples (Ag-Cu eutectic alloy on Cu presaturated in Ag). Similar results were obtained by Naidich et al. (1965, 1971) who found that pure molten Ni (and other Fe group metals) wets graphite better than molten Ni presaturated with C. The above experiments were recently reassessed (Eustathopoulos 1996) using additional information now available for the same system (Ni/C (Hara et al. 1995)) and for similar systems (Cu/Fe (Eremenko et al. 1995)). From this new analysis, it appears that the different wetting behaviour observed in these equilibrium and non-equilibrium systems can be explained by surface energy effects taking into account the actual configuration at the triple line, without any contribution of a ΔG term to the wetting driving force (Eustathopoulos 1996).

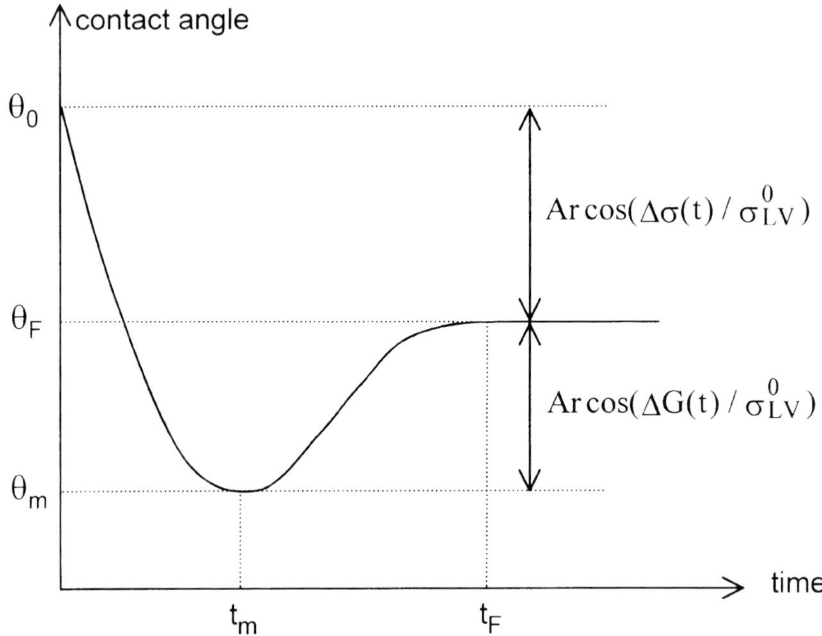

Figure 2.21. Contact angle versus time curve in a reactive system according to the model of Aksay et al. (1974).

Finally, the Aksay's model was used to explain the empirical correlation between reactivity and wettability observed in several metal/ceramic systems. It will be shown, for instance for metal/oxide couples (see Sections 6.4.1 and 6.5.2), that such a correlation can be explained by taking into account only the effect of reactions between liquid metals and oxides on interfacial energies.

According to the above discussion, the last term of the right hand side of equation (2.14) may be neglected, at least for systems with a weak or moderate reactivity. Therefore, any changes of contact angle for times much greater than 10^{-2} second reflect changes with time of at least one of the three characteristic interfacial energies of the system. An example is a Cu-Si alloy on a SiO_2 substrate studied by Rado (1997) using the dispensed drop technique (Figure 2.22). The first contact angle, $\theta_0 = 135°$, is obtained in less than 0.04 second (the frametime of the video technique used in this experiment). Thereafter, the modification of σ_{SL} due to the interaction between Si and the surface of SiO_2 (occurring as for the Si/Al_2O_3 couple, see Section 6.3) leads to a significant decrease of the contact angle in about

200 seconds. In this experiment, no significant dissolution of the substrate occurs, so that the S/L interface remains nearly planar.

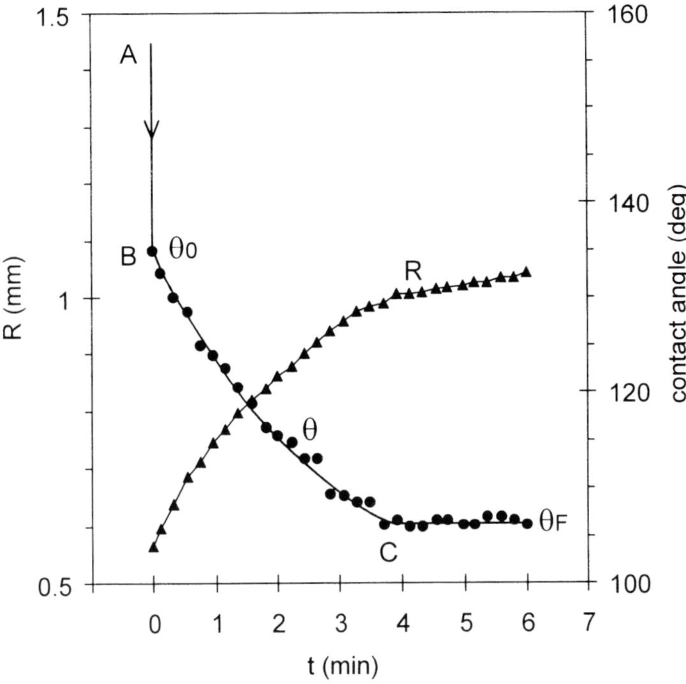

Figure 2.22. Variation of contact angle and drop base radius with time for a Cu-25 at.% Si alloy on SiO$_2$ at 1150°C. The duration of the A-B stage is shorter than 0.04 s. From (Rado 1997).

In many metal/ceramic systems with weak or moderate reactivity, interfacial reactions may be slow at relatively low temperatures, thus allowing the non-reactive stage (step A-B on Figure 2.22) and reactive stage (step B-C) to be separated. However, interfacial reactions at high temperature can be so fast for systems with a high reactivity, that the A-B and B-C stages cannot be separated (similarly, in some reactive glass/ceramic systems, these two stages can overlap because of a slow non-reactive stage due to the high viscosity of the glass). An example is Ni/SiC, a system in which Ni can dissolve up to 40 at.% Si at 1500C. In Figure 2.23, one can see that the spreading rate for this system, measured by the transferred drop technique, is as high as that for non-reactive systems, with an initial spreading rate close to 1 m/s (Grigorenko et al. 1998). The contact angle observed after 20 ms ($\theta \cong 20°$) is nearly equal to that measured for a Ni-40 at.% Si alloy on monocrystalline SiC (Rado et al. 1999). This good agreement can be explained for the Ni/SiC system by a rapid saturation of the interface in Si, and by the fact that the spreading time (20 ms) was too short to cause a heavy corrosion at the interface and creation of a significantly non-planar interface. Note that for the Ni/SiC system, high contact angles (in the range 50–70°) were observed by several investigators in experiments in which the first observation was possible only at $t > 1$s. At these times, interfacial reaction leading to graphite precipitation strongly perturbs the configuration of the triple line and therefore only apparent contact angles were observed.

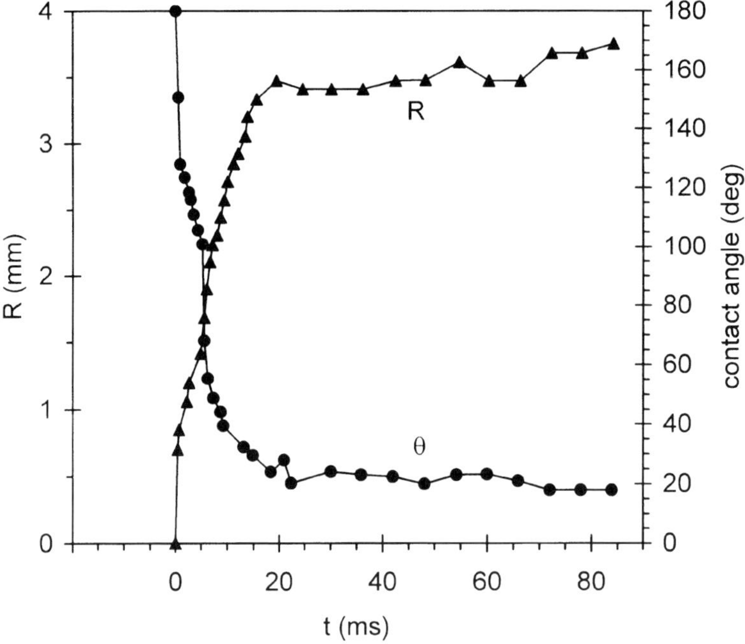

Figure 2.23. Variation of contact angle and drop base radius with time for Ni on SiC at 1500°C. Data from work reported in (Grigorenko et al. 1998).

2.2.2 Formation of 3D compounds

This Section focusses on interfacial reactions that cause the formation of *dense* 3D layers of solid reaction product. It will be assumed that the thickness of the layer is negligible compared to the drop size so that the layer does not disturb the measurement of the macroscopic contact angle. Moreover, although interfacial reactions occurring during spreading can be highly exothermic, the temperature is assumed to remain constant. This assumption seems reasonable in the sessile drop configuration, at least for spreading rates lower than 100 μm/s, but it is much more questionable for infiltration of a porous medium (Mortensen et al. 1998).

2.2.2.1 Driving force for reactive wetting. As with dissolutive wetting, the general expression for the driving force at any time t is given by expression (2.13) in which $\Delta\sigma(t)$ now represents the change in the S/V and S/L surface and interfacial energies brought about by the reaction. For the sake of simplicity, the L/V surface energy is taken to be unchanged. The $\Delta G(t)$ term is the change in Gibbs energy of the system due to the formation of a new compound.

During the last few decades, reactive wetting has been explained usually by reference to the $\Delta G(t)$ term. Accordingly, it has been concluded that the higher the reactivity in a system, the better the wettability. This conclusion has been tested recently by experiments in which the $\Delta G(t)$ term was varied but the surface energy term was kept constant (Espie et al. 1994, Landry et al. 1997). An example is the wetting behaviour of a CuPd-Ti alloy of fixed composition on two oxide substrates of different thermodynamic stability (Al_2O_3 and SiO_2). In these systems, Ti reacts with both oxides leading to the same interfacial product (Ti_2O_3) but reactivity, as measured by the thickness e of the Ti_2O_3 layer, differed by an order of magnitude. Despite this difference, the wetting of the two substrates was nearly the same (Table 2.1).

Table 2.1. Interfacial chemistry and final contact angle for CuPd-15 at.% Ti on Al_2O_3 and SiO_2 substrates at 1473K (Espie et al. 1994).

Substrate	Interfacial product	Thickness e (μm)	θ_F (deg)
Al_2O_3	Ti_2O_3	0.5	34 ± 2
SiO_2	Ti_2O_3	10	36 ± 2

From these results and other experiments (Kritsalis et al. 1994, Landry et al. 1996, Landry et al. 1997), it has been concluded that wetting in reactive systems is governed by the final interfacial chemistry at the triple line rather than by the intensity of interfacial reactions. The key question is to determine the interfacial configuration that defines the final contact angle θ_F of the system. Two cases will be considered depending on the difference in wetting between the reaction product and the substrate.

Reaction product more wettable than the substrate. Two configurations, one stable and the other metastable, can be produced at the triple line when the final contact angle θ_F is attained. The *stable* configuration depends on a layer of reaction product P extending on the free surface of the substrate S (Figure 2.24.a). Neglecting any effect of roughness of the reaction layer, θ_F is simply given by the Young equation (1.16) applied to the P/L/V system:

$$\cos \theta_F = \cos \theta_P \cong \frac{\sigma_{PV} - \sigma_{PL}}{\sigma_{LV}} \tag{2.15}$$

in which θ_P represents the equilibrium contact angle on the reaction product P.

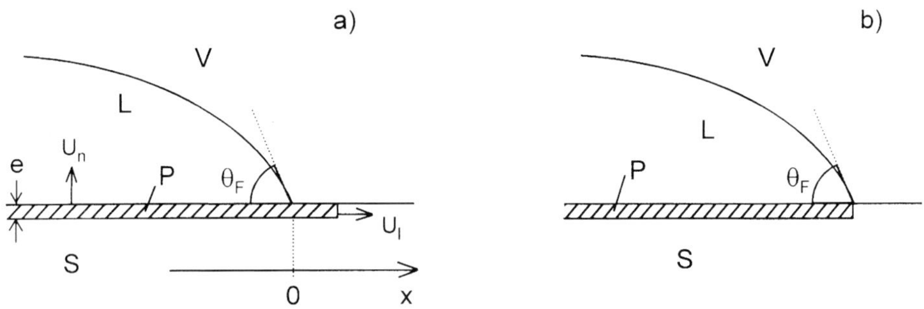

Figure 2.24. Possible configurations of a reacting system at the solid/liquid/vapour triple line when the steady-state contact angle is reached.

In the *metastable* configuration, the reaction product layer does not extend beyond the edge of the drop (Figure 2.24.b). As a consequence, the final contact angle θ_F, calculated by considering the effect of a small displacement of the triple line around the equilibrium position, is:

$$\cos \theta_F \cong \frac{\sigma_{SV} - (\sigma_{SP} + \sigma_{PL})}{\sigma_{LV}} \tag{2.16}$$

The configuration of Figure 2.24.b is possible if the extension of the reaction layer outside the P/L interface is blocked by thermodynamic barriers. The size of the barriers can be calculated by considering the change of the Gibbs energy G per unit length of triple line produced by a lateral extension δx of the reaction product layer outside the drop (Figure 2.24.a). Thus

$$\delta G = \Delta G_v e \delta x + (\sigma_{PV} + \sigma_{PS} - \sigma_{SV})\delta x = \Delta G_v e \delta x + \Delta\sigma\delta x \qquad (2.17)$$

where ΔG_v is the Gibbs energy of the reaction per unit volume of P. Setting $d(\delta G)/d(\delta x) = 0$, a critical layer thickness e^* is obtained such that:

$$e^* = -\frac{\Delta\sigma}{\Delta G_v} \qquad (2.18)$$

For $e > e^*$, no thermodynamic barrier to the extension of reaction product outside the drop exists, while for $e < e^*$, the configuration of Figure 2.24.b, corresponding to a metastable equilibrium, is possible. Taking typical values $\Delta\sigma = 1$ J.m^{-2} and $|\Delta G_v| = 10^9$ to 10^{10} J.m^{-3}, we find that e^* is of the order of 1 nm. Such values of e^* are too low to create a true barrier to the formation of product P outside the drop. Indeed, the slightest growth of P in a direction perpendicular to the interface i.e., towards the interior of the liquid, would be sufficient to destroy such barriers. Note that in the few cases where the thickness of the reaction layer in the vicinity of a triple line was measured, it was found to vary from several tens of nm to a few microns (Eustathopoulos 1998). In conclusion, except for very large and unrealistic $\Delta\sigma$ values, or extremely low values of $|\Delta G_v|$ (in this case even the condition of a continuous layer of P becomes questionable), thermodynamic barriers to the lateral growth of P on the substrate free surface are unlikely. Therefore, the final configuration expected in reactive wetting is that of Figure 2.24.a. Note that once this configuration is attained, the growth of P layer continues both inside the liquid (by diffusion through the P layer) and over the substrate free surface (for instance by evaporation/condensation or by surface diffusion). So far as this growth does not produce any significant change of liquid composition, θ_F (equation (2.15)) remains nearly constant with time.

An example which confirms the validity of equation (2.15) is Au-Ti alloy on Al$_2$O$_3$. Pure Au does not wet Al$_2$O$_3$ (at 1150°C, $\theta_0 = 135°$) but a Au-6 at.% Ti alloy wets this substrate ($\theta = 69°$) due to the formation of Ti$_2$O$_3$ at the interface (Zhuravlev and Turchanin 1997). Effectively, the same angle is displayed ($\theta = 67°$)

when Au-6 at.% Ti is used to wet a Ti_2O_3 substrate (Naidich 1981), in agreement with equation (2.15).

A slightly different procedure was used in another experiment performed with a Cu-1 at.% Cr alloy on a vitreous carbon substrate (Landry et al. 1997). Cr promotes wetting, forming at the interface a continuous layer a few micron thick of the wettable, metal-like, Cr_7C_3 carbide (Figure 2.25). After cooling, the solidified Cu-Cr drop was dissolved, a small quantity of a Cu-1 at.% Cr alloy was placed in the center of the Cr carbide layer and a second wetting experiment was performed at the same temperature. The plots in Figure 2.26 clearly show that the steady contact angle in the reactive Cu-Cr/C system is nearly equal to that in the non-reactive $Cu-Cr/Cr_7C_3$ system, but the wetting kinetics are very different. Those for the non-reactive couple are fast but those for the reactive couple are slow because the wettable Cr_7C_3 "substrate" is fabricated *in situ* and this takes a significant time.

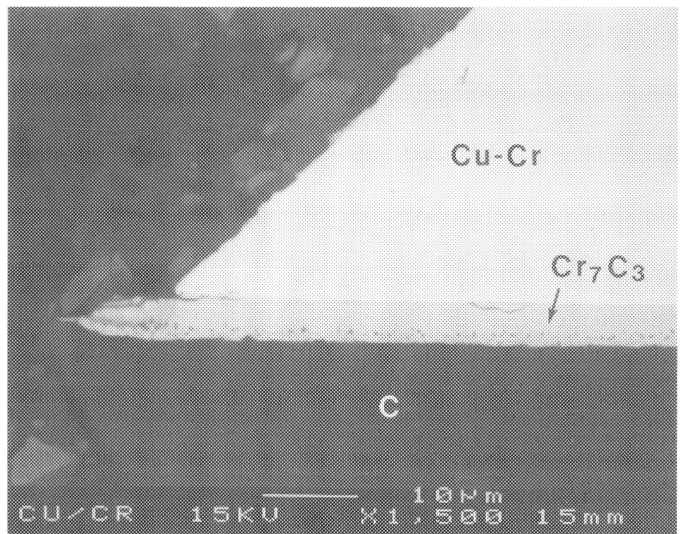

Figure 2.25. SEM micrograph of a cross-section perpendicular to the interface of a Cu-Cr alloy on vitreous carbon cooled from 1373K at steady state contact angle. From (Landry et al. 1997) [8].

Thus, the final contact angle in a reactive system leading to the formation at the interface of a dense 3D compound is given with good accuracy by the Young contact angle of the liquid on the reaction product, θ_P. Replacing θ_F by θ_P in equation (2.1), the reactive wetting driving force for $\theta < \theta_0$ is written:

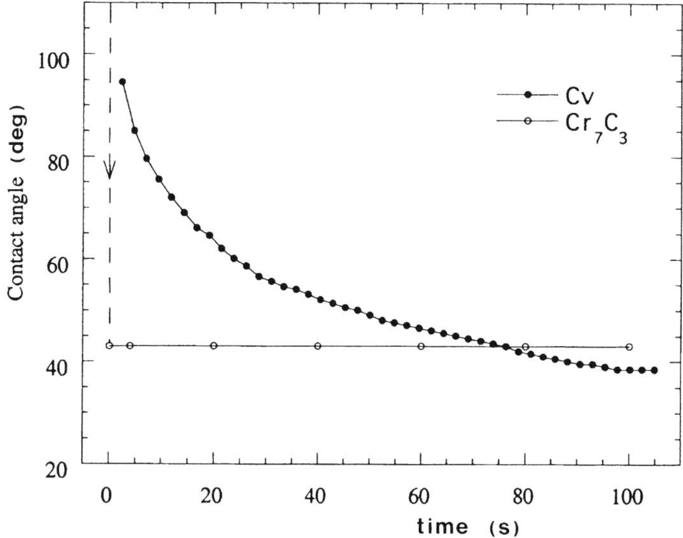

Figure 2.26. Contact angle versus time curves for a Cu-1 at.% Cr alloy on vitreous carbon and Cr_7C_3 at 1373K. From (Landry et al. 1997) [8].

$$f_d(t) = \sigma_{LV}^0[\cos \theta_P - \cos \theta(t)] \tag{2.19}$$

Several complications can produce differences between the value of the observed contact angle and the value θ_P given by equation (2.15). First, differences can result from roughness of the reaction product surface. In the case of the Cu-Cr/C system, the roughness of the Cr carbide layer formed at the interface is only a few tens of nanometers (Landry et al. 1997). As a result, the reproducibility of θ_F values (41 ± 4°) was found to be comparable with that for non-reactive systems (see Section 3.3). However, the roughness of Al_4C_3 formed by the Al/C system seems to be much higher (Landry and Eustathopoulos 1996b) and this can be one of the reasons for the comparatively high dispersion of experimental results obtained for this couple.

Another complication arises from the continuous change in the concentration of reactive solute resulting in changes of the L/V and L/S energies of the system. The concentration change depends on the value of the ratio HX/e, where H is the height of the drop, X the molar fraction of reactive element in the liquid and e the thickness of the reaction product P. The higher the value of this ratio, the smaller the change in X due to the growth of the compound P. Usually, when e becomes thicker than a few microns, the growth rate U_n normal to the S/L interface (see

Figure 2.24.a) is greatly decreased in accord with a parabolic law. Similarly, the lateral growth rate U_l decreases also with x because the source of reactive atoms for the spreading of P gets farther and farther away. For millimetre size droplets and not too low X values, it can be seen that in many cases changes in concentration in the liquid becomes negligible. For instance, wetting of a Cu-40 at.% Si droplet on vitreous carbon is due to the formation at the interface of a SiC layer of 0.1 to 0.5 μm in thickness (Landry et al. 1997). This requires a consumption of only about 10^{-3} of the Si molar fraction, a negligible change compared to the initial value of 0.4. For this reason, a nearly constant contact angle is observed at the end of spreading.

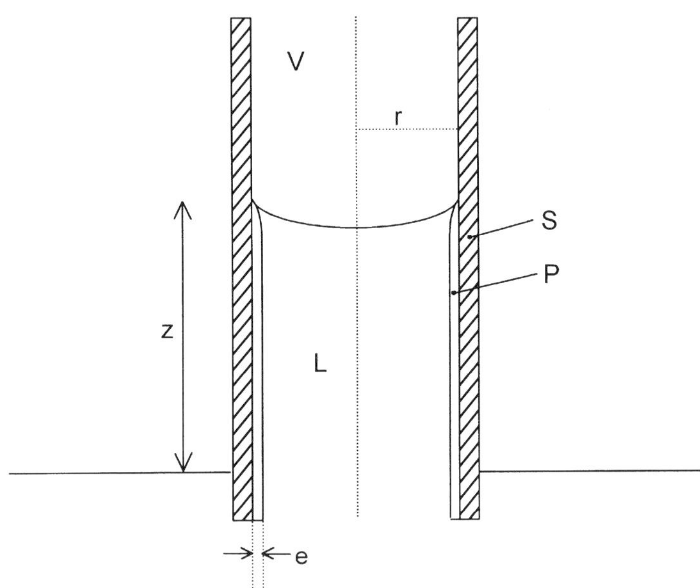

Figure 2.27. Capillary rise in a reactive solid/liquid system due to the formation of a wettable reaction product P.

However, concentration changes will be much higher in other configurations, for instance when the liquid wets a capillary tube of small radius r (Figure 2.27). In this case, a gradient of X will be established along the liquid column such that the final capillary height z would be lower than the nominal value $z^0 = 2\sigma_{LV} \cos \theta_F(X^0)/(r\rho g)$ given by equation (1.55), where $\theta_F(X^0)$ is the steady contact angle as measured in a classical sessile drop experiment with a millimetre size droplet and the nominal molar fraction of reactive solute X^0. When the capillary stands in a large volume of liquid, diffusion of reactive solute from the liquid to the infiltration front occurs, and at longer times z will tend towards the nominal value z^0. Otherwise, the capillary rise will be much lower than predicted by sessile drop experiments. Clearly, in reactive systems, one must pay attention when θ_F values measured for a sessile drop configuration have to be transferred to other configurations with different volume/interface ratios.

Reaction product less wettable than the substrate. The above discussion concerns systems in which the new compound formed at the interface is more wettable than the initial substrate. We will now consider the opposite case and take as an example the Ag/SiC couple (Figure 2.28). In this system, molten Ag reacts slightly with SiC forming Si dissolved in Ag and graphite (at 1100°C the molar fraction of Si in Ag in equilibrium with α-SiC is about 4×10^{-3}). Ag wets SiC before reaction ($\theta \ll 90°$) but not graphite ($\theta \gg 90°$) (see Table 7.2 and Section 8.1).

When a piece of Ag is placed on SiC and the temperature is raised and stabilized at a value higher than the metal melting point, the contact angle is close to 120°. Chemical analysis by Auger electron spectroscopy of the interface after cooling and elimination of the metallic phase indicates the formation of graphitic carbon, whose thickness is in the nanometre range, while the adjacent surface is found to be SiC. However, the configuration of Figure 2.29 is not stable. A slight perturbation can allow the liquid to contact fresh SiC substrate. This perturbation was produced by increasing the temperature, which led to a small decrease of θ of Ag on C but sufficient for the Ag to expand on to the SiC. This produced a substantial lateral displacement of the drop on the substrate (Figure 2.28.b), driven by the decrease of the surface energy of the system when the Ag/C interface is replaced by an Ag/SiC interface, followed by a fast spreading up to $\theta = 65°$ (Figure 2.28.c). Thereafter, a graphite layer was formed again at the interface and caused a dewetting process up to a contact angle close to 120° (Figure 2.28.d to f).

This example shows that, when the contact angle on the reaction product, θ_P, is higher than the contact angle on the substrate, θ_S, and in the absence of perturbation, the final contact angles θ_F are characteristic of the reaction product and not of the initial substrate ($\theta_F \approx \theta_P$). This is the same conclusion as that drawn in the previous Section for the case when θ_P is lower than θ_S (equation (2.15)).

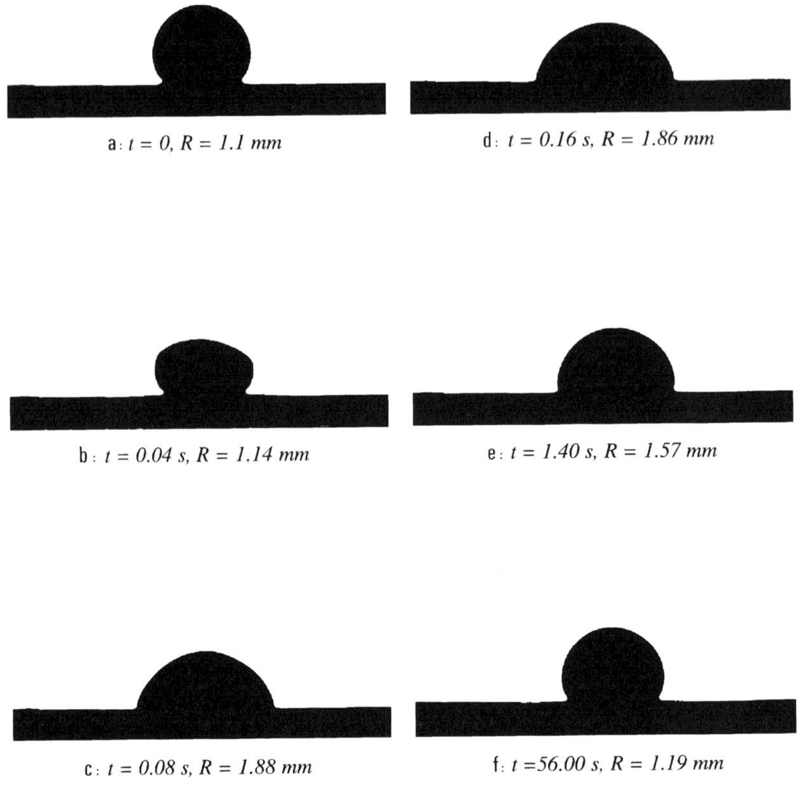

Figure 2.28. Successive images of a Ag drop on SiC (R : drop base radius).
(a): initial steady configuration corresponding to a time taken arbitrarily equal to zero. (b): the droplet moves rapidly (with a velocity higher than 20 mm/s) towards a fresh SiC surface until it is stopped by the edge of the substrate. (c): spreading of the drop on SiC. (d to f): dewetting of the drop up to a final contact angle characteristic of a graphite substrate. From (Rado et al., in press).

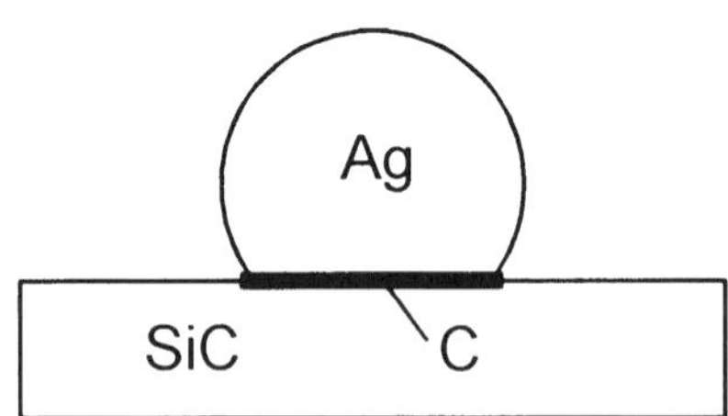

Figure 2.29. Schematic representation of a Ag sessile drop on a SiC substrate. Wetting is blocked by the non-wettable reaction product (graphite) formed at the interface.

2.2.2.2 Spreading kinetics. The general θ-t curve observed at constant temperature is given schematically on Figure 2.30. The A-B step corresponds to the non-reactive wetting. The B-C step is due to the formation of the new compound. In some cases, a drift (C-D′ step) is observed at long times resulting either from modification of the composition of the liquid (for instance by evaporation or interdiffusion with the solid) or from the modification of the solid surface (for instance by grain-boundary grooving, see Appendix H). For liquid metals on solid metals, θ_0 and θ_F are not very different (it can be seen in Chapter 5 that the contact angles of a liquid metal on another metal and on an intermetallic compound are relatively close). In contrast, in several metal/oxide or metal/C systems, θ_0 is much higher than θ_F. Moreover, in these systems, the two stages A-B and B-C usually have clearly different time scales and can be easily identified (for instance $t_0 \approx 10^{-2}$ s and $t_F \approx 10^2$–10^4 s). For this reason, the mechanisms of spreading kinetics have been mainly studied in metal/ceramic systems.

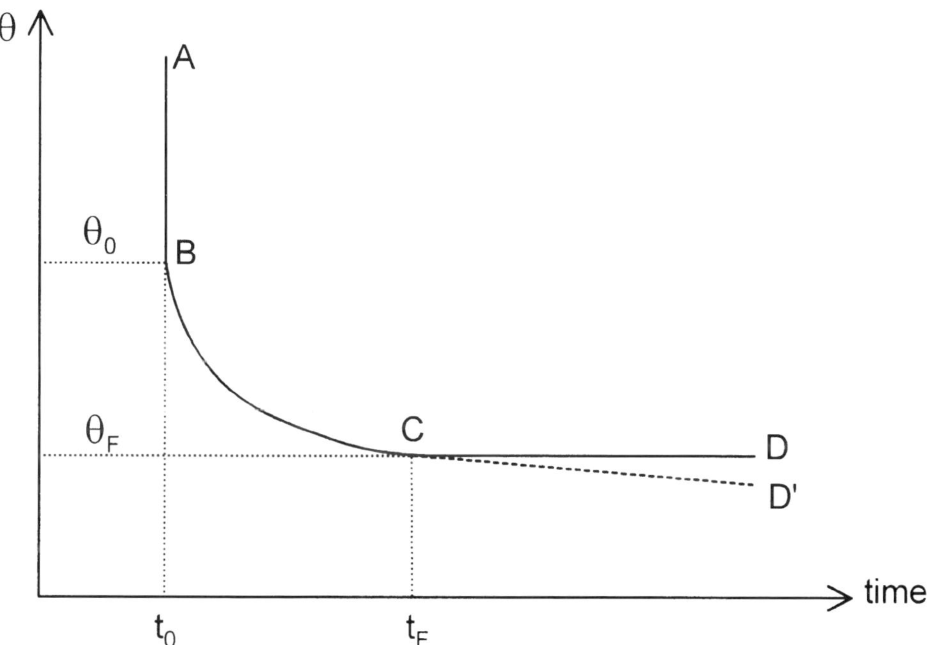

Figure 2.30. Schematic contact angle versus time curve during reactive wetting with formation of a new compound at the interface.

In the following, we consider a non-reactive liquid metal solvent A containing a reactive solute B that is in contact with a solid substrate. It is assumed that the reaction product P is better wetted by the A-B alloy than the initial substrate. Spreading kinetics will be discussed by means of the classical two-step scheme used in treating kinetic phenomena in materials science, consisting of a local process at the interface and transport phenomena in bulk materials. The only difference when considering reactive wetting, but an important one, is that the significant localised effects occur not at a 2D interface but at a 1D defect, the contact line between three phases S, L and V. At the line, *coupling between reaction and wetting processes* is established. Indeed, because the liquid wets the reaction product, the liquid is continuously in contact with (or close to) unreacted solid at the triple line. As a result, the reaction rate at this particular point is two to three orders of magnitude higher than that at the interface far from the triple line where the reaction occurs by slow diffusion through a solid layer (Landry and Eustathopoulos 1996b). On the other hand, reactivity is essential for wetting because it provides the wettable compound. Note that when $\theta = \theta_F$, spreading stops but the lateral extension of the reaction product continues. However, because the liquid source and the reaction front get farther and farther apart, the reaction slows down strongly (see Figure 2.31.b).

Within the framework of this general description, two limiting cases can be defined depending on the rate of the chemical reaction at the triple line compared to the rate of diffusion of reactive solute from the drop bulk to the triple line (or of a soluble reaction product from the triple line to the drop bulk). Note that this description does not need the knowledge of the detailed (and complicated) configuration close to the triple line, i.e., inside a small volume a few nm (or tens of nm) in diameter.

Reaction-limited spreading. In this first limiting case, chemical kinetics at the triple line are rate-limiting because diffusion within the droplet is comparatively rapid (or not needed when the drop is made of a pure reactive metal). In this case, (i) if the reaction does not change the global drop composition significantly so that the chemical environment of the triple line is constant with time and (ii) if a stationary configuration is established at the triple line during wetting, then the rate of reaction U_r (expressed in mole.m^{-2}.s^{-1}) and hence the triple line velocity are constant with time (Landry and Eustathopoulos 1996b, Dezellus et al. 1998). Then, writing a mass balance at the triple line and denoting by v_m^P the molar volume of P, the velocity of the triple line for a reaction of the type $M + S \rightarrow MS$ (where S is the solid substrate) is:

$$\frac{dR}{dt} = U_r \cdot v_m^P = K_2 v_m^P \, \exp\!\left(\frac{-\Delta G^*}{RT}\right)(a_B - a_B^{eq}) \qquad (2.20)$$

where a_B^{eq} is the value of the activity of reactive solute B in the liquid which is in equilibrium with the substrate and the reaction product P. ΔG^* is the activation Gibbs energy of the reaction and K_2 is a constant.

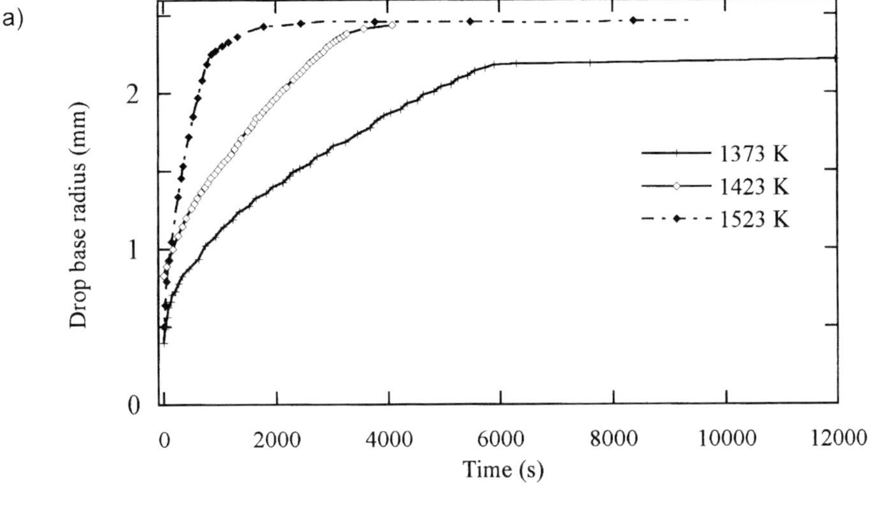

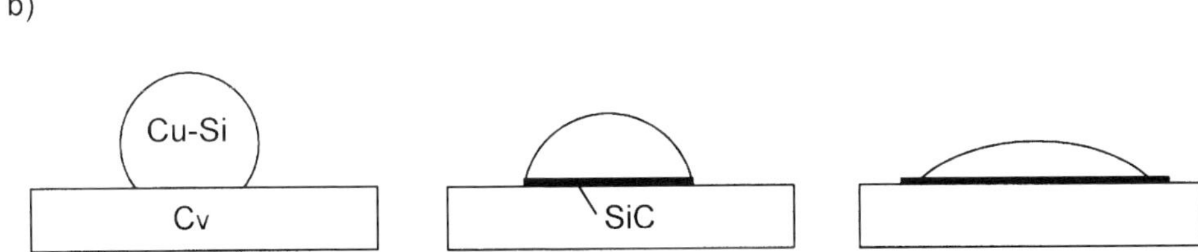

Figure 2.31. (a) Spreading kinetics of Cu-40 at.% Si alloy on vitreous carbon. (b) Schematic representation of interfaces at the beginning of spreading (left), during the linear stage (middle) and in the plateau (right). From (Dezellus et al. 1998).

An example is the behaviour of Cu-Si alloys on vitreous carbon (Dezellus et al. 1998). In this system, wetting is promoted by the formation of a continuous layer, a few tens of nm thick, of SiC, produced by Si concentrations higher than 15 at.%. Variations of drop base radius as a function of time at various temperatures are shown in Figure 2.31.a. The initial contact angle measured within 0.04 s is about 160° and thereafter, two main stages are observed for the R(t) curve. In the first stage, R increases parabolically with time. It is assumed that during this stage a steady-state growth is not still established due to incomplete covering of the S/L interface by the reaction product (Landry and Eustathopoulos 1996b). During the second stage, a steady configuration is established at the triple line and, as a result, the reaction rate and the triple line velocity are nearly constant with time. During this stage, the macroscopically observed contact angle is not related to capillary force equilibrium but is dictated by the drop volume and the radius of the reaction product layer (Figure 2.31.b middle). Indeed, the advance of the liquid is hindered by the presence of a non-wettable vitreous carbon surface in front of the triple line. Thus, the only way to move ahead is by lateral growth of the wettable SiC layer until the macroscopic angle equals the equilibrium contact angle of the liquid on SiC ($\cong 40°$). Then, the drop radius remains nearly constant while beyond the drop the reaction layer continues to extend but very slowly.

The Arrhenius activation energy ΔG^* in equation (2.20) obtained for Cu-Si alloys of constant composition is close to 250 kJ.mol^{-1} which is of the order of magnitude expected for a chemical reaction.

In this description, it is assumed that the formation of the reaction layer occurs only in the vicinity of the triple line. This does not take into account the possible effect of a reaction occurring ahead of the triple line to where the reactive species can be transferred by evaporation/condensation or by diffusion in the gas or by both processes. When the free surface of the droplet forms an acute angle through the vapour phase with the substrate, evaporation/condensation can provide a parallel transport path for Si from the drop to the triple line (see Figure 2.32.b). This should accelerate dR/dt to an extent that increases as θ exceeds 90° and may be partially responsible of the parabolic branch of R(t) observed at the beginning of spreading (Figure 2.31.a) (Dezellus et al. 1998).

"Linear wetting" can occur in different systems and for different types of reaction. Examples are Cu-Si alloys on oxidized SiC (Rado 1997), the reactive CuAg-Ti/Al$_2$O$_3$ system (Nicholas and Peteves 1995) or the Al/vitreous carbon system (Landry and Eustathopoulos 1996b). For the last system, small but significant deviations from linearity were observed when θ tends towards θ_F and attributed to roughness of the reaction product layer delaying the movement of the triple line by pinning.

For systems exhibiting linear wetting, it is expected that the spreading rate will be sensitive to the structure of the solid. Indeed, in the Al/C system, the spreading rate in the linear stage was found to vary by a factor 4 between vitreous carbon and pseudo-monocrystalline graphite (Drevet et al. 1996) (see also Figure 8.7).

Diffusion-limited spreading. When local reaction rates are comparatively high, the rate of lateral growth of the reaction product at the triple line is limited by the diffusive supply of reactant from the drop bulk to the triple line. Because the contact angle decreases continuously during wetting, the reduction in diffusion field will lead to a continuous decrease in the reaction rate and, as a result, of the rate of movement of the triple line itself (Figure 2.32.a). Therefore time-dependent spreading rates are expected in this case (Landry and Eustathopoulos 1996b).

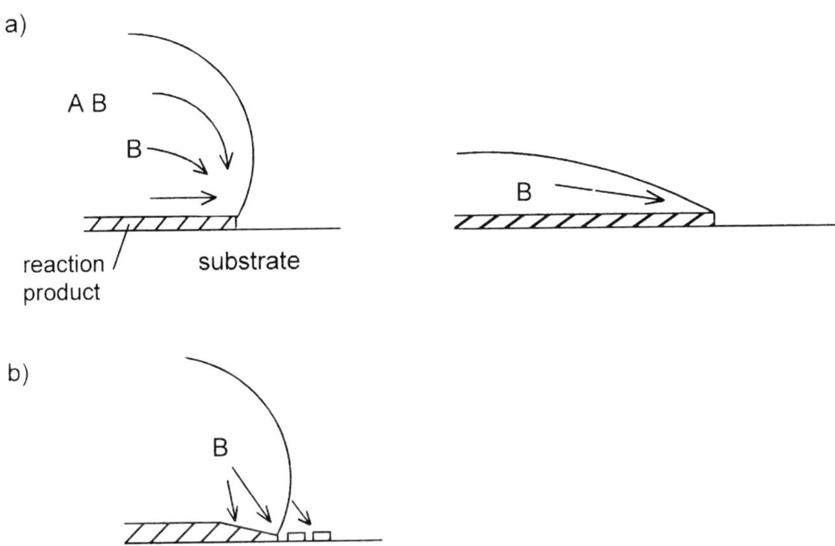

Figure 2.32. Diffusion limited spreading. (a) In the model configuration used in the analysis of Mortensen et al. (1997), the decrease of the triple line velocity with time is due to the reduction in diffusion field close to the triple line. (b) More realistic description of the reaction around the triple line.

In general, solutes are transported in the liquid by convection and diffusion and the governing equation is Fick's second law written in the referential of the triple line moving with a velocity U ($U = dR/dt$):

$$\frac{\partial C}{\partial t} + u.\nabla C = D\nabla^2 C + U.\nabla C \tag{2.21}$$

In this equation, the concentration C of reactive species is expressed as a mass fraction. u denotes the local velocity of the fluid and results both from the movement of the triple line and from convection caused by temperature and concentration gradients in the liquid.

Given the complexity of the real situation in the sessile drop configuration, a simplified analysis has been proposed by Mortensen et al. (1997). This neglects reactions at the interface far from the triple line and assumes that diffusion is the dominating mechanism for solute transport in a small volume near the triple line. Inside this volume, modelled as a straight wedge of angle θ, the velocity u is taken equal to U so that equation (2.21) reduces to

$$\frac{\partial C}{\partial t} = D\nabla^2 C \tag{2.22}$$

Solving this equation, the triple line velocity of a drop of base radius R, containing a reactive solute at a concentration C_0 (expressed in mole per unit volume), is found to depend on the instantaneous contact angle as follows:

$$\frac{dR}{dt} = \frac{2DK_3}{en_v}(C_0 - C_e)\theta \tag{2.23}$$

where D is the diffusion coefficient in the liquid phase, n_v is the number of moles of reactive solute per unit volume of the reaction product, e is the reaction product thickness at the triple line, C_e is the concentration of reactive solute in equilibrium with the reaction product and K_3 is a constant close to 0.04. For modest contact angles, $\theta < 60°$, the contact angle of a spherical cap can be approximated by $\theta = 4v/(\pi R^3)$. Introducing this into equation (2.23) yields:

$$R^4 - R_0^{\,4} = \text{const. } vt \tag{2.24}$$

where v is the drop volume and R_0 is the initial drop radius.

The most complete study so far used the Cu-Cr/vitreous carbon system and the transferred drop technique (Voitovich et al. 1998, 1999). In this system, the transition from large, non-wetting, contact angles to low wetting contact angles (Figure 2.26) is due to the formation of a continuous layer of a wettable Cr_7C_3 compound along the S/L interface. For all experiments, the final contact angle was nearly the same ($\theta_F \cong 40°$) and close to that of Cu-Cr alloy on a substrate of Cr_7C_3 (see Figure 2.26).

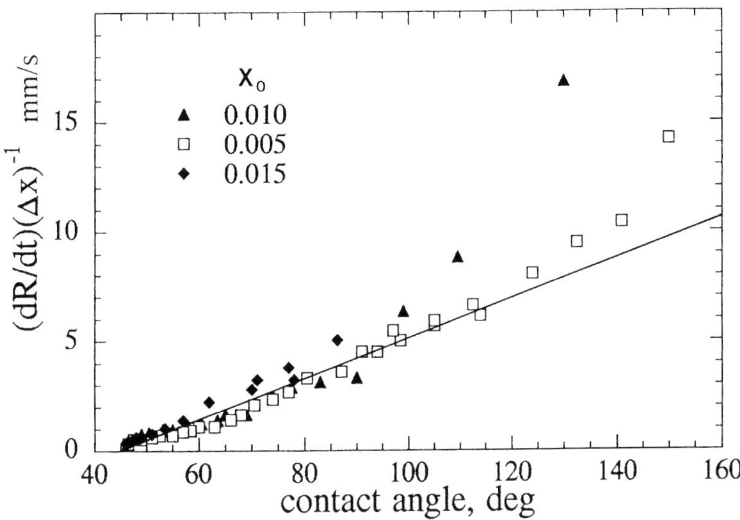

Figure 2.33. Composition normalised triple line velocity as a function of the instantaneous contact angle in the Cu-Cr/vitreous carbon system at 1373K with drop mass of about 100 mg. ΔX is equal to ($X_0 - X_e$) where X_0 and X_e are the initial and equilibrium molar fractions of Cr respectively. From data of (Voitovich et al. 1998) [9].

In agreement with the analysis of Mortensen et al. (1997) (equations (2.23) and (2.24)), the experimental data in the Cu-Cr/vitreous carbon system show that the spreading rate normalised by reference to the molar fraction of solute depends only on the contact angle (i.e. on the instantaneous geometry of the drop) in an approximately linear way (Figure 2.33). Discrepancies with the predictions of the analysis are observed, however, in that (i) at angles higher than 90°, there is an upward deviation in the spreading rate from the linear relation with θ observed at lower angles, and (ii) the intercept of the line of dR/dt versus θ with the θ axis is not at the origin ($\theta = 0$), but near $\theta = 45°$. The deviation from linearity observed at $\theta > 90°$ can be explained by a reaction occurring ahead of the triple line caused by evaporation of solute from the drop surface during the spreading process (Figure 2.32.b), as discussed above for the Cu-Si/vitreous carbon system. The second discrepancy at low contact angles is attributed to continued drop/substrate chemical reaction behind the triple line. Such a reaction can influence spreading kinetics first by causing a gradual lowering of the bulk drop concentration with time, and second by diverting solute flux lines away from the triple line, towards the solid/liquid interface (Figure 2.32.b) (Voitovich et al. 1999). Numerical calculations taking into account this phenomenon seem to confirm this interpretation (Hodaj et al., to be published).

Note that equations (2.23) and (2.24) neglect convection. In the sessile drop configuration, Marangoni convection is predominant with regard to bulk convection and can affect spreading kinetics (Eustathopoulos et al. 1998b). In practice, significant Marangoni convection can be generated in sessile droplets by relatively weak thermal gradients which are often present in nominally isothermal furnaces.

In diffusion-controlled wetting, substrate orientation is not expected to influence spreading parameters. An example is the behaviour of Cu-Ti alloy on α-Al_2O_3 (Kritsalis et al. 1991, Drevet et al. 1996). The addition of about 10 at.% Ti in Cu decreases the contact angle from the 130° measured for pure Cu to about 30°. However, the $\theta(t)$ curve exhibits two stages. The first one is completed in less than 1 second and can be attributed to adsorption of Ti at the interface (see Figure 6.37). The second stage is much slower (200 seconds) and corresponds to the formation of a wettable TiO layer, 1 μm thick, at the interface. Non-linear spreading has been observed for the reactive stage, indicating that it is controlled by diffusion of Ti from the drop bulk to the triple line. Experiments with Cu-Ti droplets with a fixed composition and mass on four differently orientated substrates of Al_2O_3 led to nearly equal spreading times. This behaviour is very different from that of Al/C system with linear spreading kinetics (see Figure 8.7).

Returning to the first stage of rapid spreading observed in the Cu-Ti/Al_2O_3 system in which adsorption of Ti decreases significantly the contact angle. The observed fall is from an initial value of 130° to close to or lower than 90° depending on the Ti concentration in the melt. It is reasonable to think that for certain systems, especially for metal/oxide systems, the formation of a 3D layer at the interface is preceded by formation of an adsorption layer (i.e. of a 2D layer rich in reactive solute with a composition different from that of the adjacent bulk phases) leading to a first fall of the contact angle (Eustathopoulos and Drevet 1994, Tomsia et al. 1998).

Unfortunately, there are no specific experimental studies of spreading kinetics of alloy/oxide systems in which only adsorption occurs without formation of a 3D compound. However, when analysing R(t) curves in reactive systems, it may be useful to calculate the spreading rate in the case where the decrease of the contact angle is due to the lateral extension of an adsorption layer. This needs diffusion of interface-active solute from the drop bulk to the liquid adjacent to the adsorption layer at the triple line, followed by a transfer from the liquid to the adsorption layer. Assuming that the overall process is limited by the diffusion stage, the spreading rate can be estimated from equation (2.23) taking $e = 0.5$ nm, a typical value for an adsorption layer, $D = 10^{-9}$ m^2.s^{-1}, $n_v = 10^{-7}$ mole.m^{-3} (assuming that the adsorption layer is constituted by pure interface-active solute), $C_0 = 10^{-8}$ mole.m^{-3} and $C_e = 0$ (because transfer at the interface is assumed to be much faster than diffusion, the equilibrium concentration in the liquid adjacent to the adsorption layer is reasonably taken equal to 0). Then, for an instantaneous contact angle of $\pi/2$ rad, the calculated spreading rate is 30 mm/s which corresponds to a spreading time of the order of 10^{-1} second. This explains the fast initial spreading observed in the Cu-Ti/Al$_2$O$_3$ system (Figure 6.37). Note that such spreading times are two to three orders of magnitude shorter than those observed in diffusion-limited reactive wetting during the formation of 3D compound. This is due to the value of e in equation (2.23) being typically 10^2–10^3 nm as compared to only 1 nm for 2D adsorption.

A different interpretation of reactive wetting for alloy/oxide systems was proposed recently by Tomsia et al. (1998). According to these authors, reactive wetting is due to adsorption of an interface-active solute at the liquid/solid interface rather than to formation of a new 3D compound. As adsorption is a fast process, the comparatively slow spreading kinetics observed experimentally have been considered to be controlled by the propagation of "wetting ridges" pinning the triple line (see Section 2.1.2 and Figure 2.14).

2.2.3 Concluding remarks

Dissolution of the solid in the liquid leads to an increase of the contact radius R and to a decrease of the *apparent*, or visible, contact angle. However, the increase of R is rather limited in the absence of tensio-active effects of dissolved elements. For metal/metal systems, because dissolution is generally controlled by diffusion in the liquid alloy, a nearly stable configuration for millimetre size droplets can be reached only after times of the order of 10^2 seconds, i.e., several orders of magnitude higher than for wetting of non-reactive metals. Much stronger changes in the contact radius are expected when dissolved species modify either the S/L interface or L/V surface energies, as for some metal/oxide couples.

Wetting is promoted by the formation of a 3D compound that is more wettable than the initial substrate. In this case, two main regimes have been identified. In the first, spreading kinetics are controlled by the *chemical reaction* at the triple line and this results in a constant spreading rate. This regime has been observed in metal/ceramic couples and for millimetre size droplets it has characteristic times of the order of 10^3–10^4 seconds. In the second regime occurring in systems exhibiting comparatively fast reactions (spreading times of 10^1–10^2 seconds), spreading is limited by *diffusion* of a reactive species from the drop bulk to the triple line, resulting in a spreading rate which scales with the instantaneous contact angle. More work is needed to specify the effect on reactive spreading kinetics of reactions ahead and behind the triple line as well as of "wetting ridges".

REFERENCES FOR CHAPTER 2

Aksay, I., Hoye, C. and Pask, J. (1974) *J. Phys. Chem.*, **78**, 1178

Blake, T. D. (1993) in *Wettability*, ed. by J. C. Berg and M. Dekker, New York, p. 251

Carré, A. and Shanahan, M. E. R. (1995) *Langmuir*, **11**, 3572

de Gennes, P. G. (1985) *Reviews of Modern Physics*, **57**, 827

Denesuk, M., Cronin, J. P., Zelinski, B. J. J., Kreidl, N. J. and Uhlmann, D. R. (1993) *Phys. Chem. Glasses*, **34**, 203

Dezellus, O., Hodaj, F. and Eustathopoulos, N. (1998) in *Proc. 2nd Int. Conf. on High Temperature Capillarity*, Cracow (Poland), 29 June-2 July 1997, ed. N. Eustathopoulos and N. Sobczak, published by Foundry Research Institute (Cracow), p. 18

Drevet, B., Landry, K., Vikner, P. and Eustathopoulos, N. (1996) *Scripta Mater.*, **35**, 1265

Dussan V., E. B. and Davis, S. (1974) *J. Fluid Mech.*, **65**, 71

Ebrill, N. (1999) Ph.D. Thesis, University of Newcastle, Australia

Eremenko, V., Kostrova, L. and Lesnic, N. (1995) in *Proc. Int. Conf. High Temperature Capillarity*, Smolenice Castle, May 1994, ed. N. Eustathopoulos, Reproint Bratislava, p. 113

Espie, L., Drevet, B. and Eustathopoulos, N. (1994) *Metall. Trans. A*, **25**, 599

Eustathopoulos, N. (1983) *Int. Metals Rev.*, **28**, 189

Eustathopoulos, N. and Drevet, B. (1994) *J. Phys. III (Fr.)*, **4**, 1965

Eustathopoulos, N. (1996) in *Proc. of JIMIS-8*, Interface Science and Materials Interconnection, Toyama, Japan 1-4 July, The Japan Institute of Metals, p. 61

Eustathopoulos, N. (1998) *Acta Mater.*, **46**, 2319

Eustathopoulos, N., Garandet, J. P. and Drevet, B. (1998b) *Phil. Trans. Roy. Soc. Lond. A*, **356**, 871

Grigorenko, N., Poluyanskaya, V., Eustathopoulos, N. and Naidich, Y. V. (1998) in *Proc. 2nd Int. Conf. on High Temperature Capillarity*, Cracow (Poland), 29 June-2 July 1997, ed. N. Eustathopoulos and N. Sobczak, published by Foundry Research Institute (Cracow), p. 27

Hara, S., Nogi, K. and Ogino, K. (1995) in *Proc. Int. Conf. High Temperature Capillarity*, Smolenice Castle, May 1994, ed. N. Eustathopoulos, Reprint Bratislava, p. 43

Hocking, L. M. and Rivers, A. D. (1982) *J. Fluid Mech.*, **121**, 425

Hodaj, F., Barbier, J. N., Dezellus, O., Mortensen, A. and Eustathopoulos, N., to be published

Hoffman, R. (1975) *J. Colloid Interface Sci.*, **50**, 228

Ishimov, V. I., Hlinov, V. V. and Esin, O. A. (1971) in *Kinetics of Wetting of Solid Oxides by Liquid Metals / Physical Chemistry of Surface Phenomena in Melts*, Kiev, Naukova Dumka, p. 213 (in Russian)

Joanny, J. F. (1985) Thesis, Université Paris VI, France

Kistler, S. F. (1993) in *Wettability*, ed. by J. C. Berg and M. Dekker, New York, p. 311

Kritsalis, P., Coudurier, L. and Eustathopoulos, N. (1991) *J. Mater. Sci.*, **26**, 3400

Kritsalis, P., Drevet, B., Valignat, N. and Eustathopoulos, N. (1994) *Scripta Metall. Mater.*, **30**, 1127

Landry, K., Rado, C. and Eustathopoulos, N. (1996) *Metall. Mater. Trans. A*, **27**, 318

Landry, K. and Eustathopoulos, N. (1996b) *Acta Mater.*, **44**, 3923

Landry, K., Rado, C., Voitovich, R. and Eustathopoulos, N. (1997) *Acta Mater.*, **45**, 3079

Laurent, V. (1988) Ph.D. Thesis, INP Grenoble, France

Li, J. G. (1994) *Materials Letters*, **18**, 291

Marmur, A. (1983) *Advances in Colloid and Interface Science*, **19**, 75

Mortensen, A., Drevet, B. and Eustathopoulos, N. (1997) *Scripta Mater.*, **36**, 645

Mortensen, A., Hodaj, F. and Eustathopoulos, N. (1998) *Scripta Mater.*, **38**, 1411

Mullins, W. W. (1958) *Acta Met.*, **6**, 414

Mullins, W. W. (1960) *Trans. Met. Soc. AIME*, **218**, 354

Naidich, Y. V. and Kolesnichenko, G. (1965) in *Surface Phenomena in Metallurgical Processes*, ed. A. Belyaev, Consultants Bureau, New York, p. 218

Naidich, Y. V., Perevertailo, V. and Nevodnik, G. (1971) *Poroshk. Metall.*, **97**, 58 (in Russian)

Naidich, Y. V., Perevertailo, V. M. and Nevodnik, G. M. (1972) *Poroshkovaya Metallurgiya*, **7**(117), 51

Naidich, Y. V. (1981) in *Progress in Surface and Membrane Science*, vol. 14, ed. by D. A. Cadenhead and J. F. Danielli, Academic Press, New York, p. 353

Naidich, Y. V., Sabuga, W. and Perevertailo, V. (1992) *Adgeziya Raspl. Pajka Mater.*, **27**, 23 (in Russian)

Nakae, H. and Goto, A. (1998) in *Proc. 2nd Int. Conf. on High Temperature Capillarity*, Cracow (Poland), 29 June-2 July 1997, ed. N. Eustathopoulos and N. Sobczak, published by Foundry Research Institute (Cracow), p. 12

Nicholas, M. and Peteves, S. (1995) in *Proc. Int. Conf. High Temperature Capillarity*, Smolenice Castle, May 1994, ed. N. Eustathopoulos, Reproprint Bratislava, p. 143

Ownby, P. D., Weirauch, D. A. and Lazaroff, J. E. (1995) in *Proc. Int. Conf. High Temperature Capillarity*, Smolenice Castle, May 1994, ed. N. Eustathopoulos, Reproprint Bratislava, p. 330

Popel, S. I., Zakharova, T. V. and Pavlov, V. V. (1974) in *Adhesion of Melts*, Kiev, Naukova Dumka, p. 53 (in Russian)

Popel, S. I. (1994) in *Surface Phenomena in Melts*, published by Metallurgiya, Moscow, p. 273-321 (in Russian)

Rado, C. (1997) Ph.D. Thesis, INP Grenoble, France

Rado, C., Kalogeropoulou, S. and Eustathopoulos, N. (1999) *Acta Mater.*, **47**, 461

Rado, C., Kalogeropoulou, S. and Eustathopoulos, N., *Mater. Sci. Eng.*, in press

Sabuga, W. (1990) "Capillary Transport Processes on the Surface of Metal Melts and Spreading", Ph.D. Thesis, Institute for Problems of Materials Science, Kiev (in Russian)

Saiz, E., Tomsia, A. P. and Cannon, R. M. (1998) *Acta Mater.*, **46**, 2349

Samsonov, V. M. and Muravyev, S. D. (1998) in *Proc. 2nd Int. Conf. on High Temperature Capillarity*, Cracow (Poland), 29 June-2 July 1997, ed. N. Eustathopoulos and N. Sobczak, published by Foundry Research Institute (Cracow), p. 45

Sharps, P., Tomsia, A. and Pask, J. (1981) *Acta Metall.*, **29**, 855

Sorokin, Y. V., Hlinov, V. V. and Esin, O. A. (1968) in *Spreading of Molten Slags under the Influence of Surface Forces / Surface Phenomena in Melts (Poverhnostnie Yavleniya v Rasplavov)*, Kiev, Naukova Dumka, p. 359 (in Russian)

Summ, B. D., Yushchenko, V. S. and Shchukin, E. D. (1987) *Colloids and Surfaces*, **27**, 43

Tanner, L. H. (1979) *Appl. Phys.*, **12**, 1473

Tomsia, A. P., Saiz, E., Foppiano, S. and Cannon, R. M. (1998) in *Proc. 2nd Int. Conf. on High Temperature Capillarity*, Cracow (Poland), 29 June-2 July 1997, ed. N. Eustathopoulos and N. Sobczak, published by Foundry Research Institute (Cracow), p. 59

Voitovich, R., Mortensen, A. and Eustathopoulos, N. (1998) in *Proc. 2nd Int. Conf. on High Temperature Capillarity*, Cracow (Poland), 29 June-2 July 1997, ed. N. Eustathopoulos and N. Sobczak, published by Foundry Research Institute (Cracow), p. 81

Voitovich, R., Mortensen, A., Hodaj, F. and Eustathopoulos, N. (1999) *Acta Mater.*, **47**, 1117

Warren, J. A., Boettinger, W. J. and Roosen, A. R. (1998) *Acta Mater.*, **46**, 3247

Yost, F. G. and O'Toole, E. J. (1998) *Acta Mater.*, **46**, 5143

Zhuravlev, V. S. and Turchanin, M. A. (1997) *Powder Metallurgy and Metal Ceramics*, **36**, 141

Chapter 3
Methods of measuring wettability parameters

The main part of this Chapter is dedicated to the different variants of the sessile drop method used in more than 90% of wetting studies at high temperatures. A detailed description is also given of the *wetting balance* technique, a simplified version of which is being widely used to control soldering processes in electronic industries. Another aspect of the wetting balance is that it allows study wetting on non-planar solids.

An attempt is made to evaluate the accuracy of data obtained by these techniques, to determine the main causes of scattering and the optimal conditions for contact angle and work of adhesion measurements.

A Section is devoted to "wetting" of a substrate by *solid* particles. This is particularly important in processing of nanomaterials for which capillary equilibrium is expected in short times because of the reduced size of the particles.

3.1. SESSILE DROP EXPERIMENTS

Sessile drop experiments are simple in principle and are the most commonly used means of quantitatively measuring wetting behaviour for both chemically inert and reactive systems. Attention will be paid first to the conduct and use of sessile drop experiments using inert systems, but for both inert and reactive systems the experiment consists essentially of permitting a drop of liquid, usually ranging in volume from 0.005 to 5 ml, to spread over a horizontal solid substrate until an equilibrium configuration is achieved. Thus the area of the liquid surface is increased and that of the substrate is decreased by conversion to liquid/solid interface. This change is sketched schematically in Figure 3.1 and continues until further advance of the liquid front is energetically unfavourable as discussed in Section 1.2 (see Figure 1.4). The rate at which the liquid front advances over the substrate surface to assume the equilibrium configuration is discussed in Chapter 2.

The contact angle is by far the most often quoted characteristic to be derived from sessile drop experiments, but reference will be made later in this Section to other liquid parameters and particularly to the liquid surface energies, σ_{LV}. While values for both θ and σ_{LV} can be determined from measurements made during a single sessile drop experiment it is often convenient to use differently sized drops. Small drops will assume the profiles of nearly spherical caps and this regularity assists the estimation of θ values from their dimensions, while the gravitational

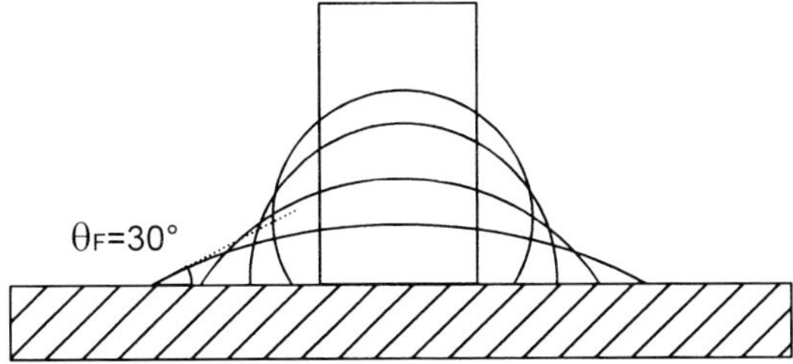

Figure 3.1. Schematic illustration of the profiles assumed as a solid cylinder melts to form a sessile drop, with an initial contact angle of 120°, which then spreads over the substrate, displaying transient contact angles to achieve a final equilibrium contact angle θ_F of 30°.

flattening of large drops assists calculation of σ_{LV} values. When selecting a suitable small or large size for a particular liquid, guidance can be provided by reference to the "capillary length" which defines the boundary between size regimes in which gravity is and is not of major importance. This critical size, equal to $(2\sigma_{LV}/(\rho g))^{1/2}$ in which ρ is the liquid density and g is the acceleration due to gravity (see equation (1.23)), is a few millimetres for most metals and glasses. Thus a drop of volume 0.005 ml with a spherical diameter of 2 mm will be suitable for θ measurements but one with a volume of several ml would be more convenient for the derivation of σ_{LV} values.

3.1.1 Materials and equipment requirements

The five main requirements for conduct of a sessile drop experiment relevant to high temperature capillary phenomena are: characterisation of the materials, a flat horizontal substrate, a test chamber to provide a controlled and generally inert gaseous environment, a facility that heats the sample to a predetermined temperature and a means of measuring the geometry and size of the sessile drop. Satisfying these requirements demands careful and precise experimental procedures.

Thorough characterisation of the materials used is essential if the experimental results are to be reproducible and usable by other workers. The prime requirement is a clear specification of their compositions, with particular attention being paid to surface and interface active components that will change the energies and hence the contact angle of the system as defined by the Young equation.

It is also important to have an understanding of the microstructure of the substrate, and of its surface texture. Thus preferential wetting of thermal grain-boundary grooves (Figure 3.2) can distort the triple line (see Appendix H) and the

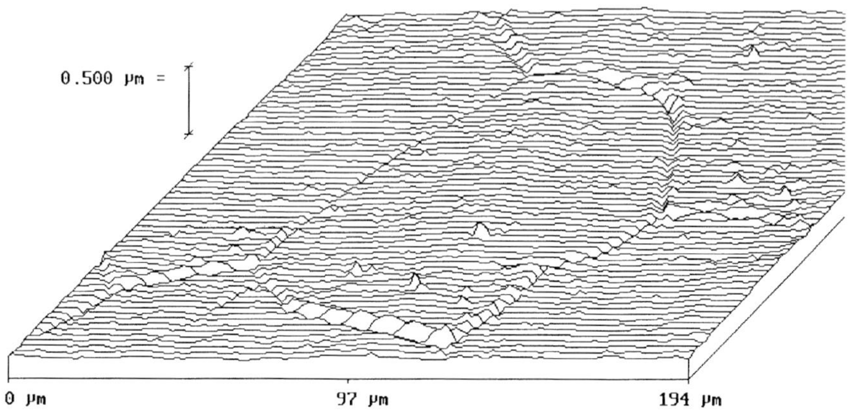

Figure 3.2. Grain-boundary grooves formed on polycrystalline Ni surface (grain diameter of the order of 100 μm) in two hours at 900°C. For wetting liquids, these grooves can act as capillaries. The profile was obtained by high-resolution optical profilometry.

presence of a second phase at the surface of the substrate will complicate the energy balance achieved by the equilibrated sessile drop and may give rise to pinning of the triple line as explained in Section 1.3.2. Using rough substrates increases the contact angle values for non-wetting systems, those for wetting systems also sometimes increase but generally decrease as discussed in detail in Section 1.3.1. What is meant by a "smooth" substrate depends on the wetting characteristics of the system, but in the past this has usually meant a roughness amplitude, R_a, of no more than 1 μm and an average asperity slope of no more than 5°. More demanding standards of smoothness, an R_a of no more than 0.1 μm and a slope of less than 2°, are used frequently at present in laboratory work. In the case of non-wetting systems, even very small defects in the substrate surfaces can pin the triple line. Thus surfaces with R_a values of less than 0.01 μm have to be used if an accuracy of ± 2–$3°$ is needed for θ values, (Hitchcock et al. 1981, De Jonghe et al. 1990). Further, it is almost certain that even higher standards will be needed in future for the reproducible characterisation of wetting behaviour and striving for these standards will enhance the attractiveness of using single crystals as substrates because they permit the easier preparation of smooth surfaces and also because they are of higher and more reproducible purity than most polycrystalline materials. In all cases, not only is statistical characterisation of the roughness necessary but also optical microscopy of the substrate surface for evidence of scratches which may not significantly affect R_a values but could cause profile distortions where they intersect the drop peripheries.

Unless the substrate is horizontal, the drop will not be circularly symmetrical and the contact angles assumed at diametrically opposed locations will not be equal as illustrated in Figure 3.7.f. Liquid will flow to the lower location and this excess will cause the drop to be distorted to two values of contact angle, θ_a and θ_r, respectively greater and smaller than that required by the equilibrium condition defined by the Young equation (1.16) or, in the case of a very smooth surface, to run down the slope. However, using a tilting substrate can be a means of determining the hysteresis domain of contact angle of a sessile drop, as explained in Section 3.1.2. Similar effects on a smaller scale can be caused if the substrate is convex or concave rather than flat. Distortions in the symmetry of the drops can also have major effects in degrading the reliability of derived values of liquid surface energies, and can have some effect on the measurement of density values.

No real substrate will be perfectly flat or horizontal, but experience suggests some practical standards of acceptance that are required for experiments to yield reliable sessile drop data. The substrate surfaces must be at least sufficiently horizontal for no movement to occur when a solid cylinder melts to form a sessile drop. A tilt of less than 1–2° is necessary if the drop profiles are to be acceptably symmetrical. An adequate horizontal standard can be assured at the start of the experiment by using a spirit level placed on top of the substrate, but there is no simple method of being certain that the standard is maintained during the course of the experiment when the test chamber furniture as well as the sample are heated. In practice it has to be accepted that the substrate is sufficiently horizontal if the sessile drop melts without moving over the substrate and produces an apparently symmetrical profile. This can be verified *in situ* by means of a rotating stage on which the substrate is placed (see Figure 3.6).

Achieving a controlled gaseous environment within the test chamber requires it to be leak-tight and to be purged with an inert gas or evacuated and, sometimes, to be back filled with an inert gas. The purity of the inert gas and the quality of the vacuum required for true, materials determined, characterisation of wettability depends on the chemical reactivity of the sessile drop and its substrate but it is common to employ inert gases containing only a few ppm of oxidising gases such as O_2, H_2O and CO_2. This level of impurity, however, corresponds to a partial pressure of 10^{-5}–10^{-6} atm of oxidising agents in 1 atm of gas and is enough to cause oxidation of most common pure metals at their melting temperatures (Table 3.1). It is good practice, therefore, to purify the inert gases by passing them slowly over heated powder or shavings of a reactive metal such as Ti or Zr. If concern is felt about the use of such metals, an alternative is to mix a small amount of a reducing gas, normally H_2, in with the inert gas. Such mixtures also enable environments to be used with previously selected O_2 partial pressures by controlling the H_2/H_2O or CO/CO_2 ratios, and the value of P_{O2} for such environments can be verified by using electrochemical sensors, (Taimatsu and Sangiorgi 1992). Additionally, for inert gas or evacuated environments, verification of P_{O2} values can be obtained by measuring the σ_{LV} of a well characterised metal such as Cu and interpolating these values with established $\sigma_{LV} = f(P_{O2})$ relationships (Gallois and Lupis 1981) (see Figure 4.3).

Table 3.1. Partial pressure of oxygen in equilibrium with the oxide of the metal at 1000°C and at the melting point of the metal T_F.

Metal	P_{O2} (atm) at 1000°C	P_{O2} (atm) at T_F
Al	10^{-35}	3×10^{-52}
Cu	10^{-7}	2×10^{-6}
Fe	10^{-15}	2×10^{-9}
Mg	10^{-38}	10^{-57}
Ni	10^{-10}	8×10^{-6}
Pb	8×10^{-9}	2×10^{-28}
Si	10^{-28}	10^{-19}
Sn	6×10^{-14}	10^{-50}
Ti	2×10^{-30}	8×10^{-17}

Particularly in the past, many measurements of wettability have been made using evacuated chambers, although this can sometimes cause problems due to evaporation from the liquid or its substrate. Typically, vacua lying in the range 5×10^{-7}–5×10^{-5} mbar have been used in which the partial pressures of the oxidising gases in oil diffusion pumped systems are one to three orders of magnitude smaller because of the reducing atmosphere generated by cracked oil molecules. However this cracking can also result in C contamination of the sessile drop and the substrate surface, so in general it is better to strive for a clean system with a low partial pressure of oxidising gases by using liquid N_2 traps to minimise back diffusion of pump oils or better still to employ oil free equipment such as turbo-molecular pumps.

At an equal P_{O2} value, a high vacuum environment is preferable to inert gas because it promotes evaporation and dissociation of oxide films on the liquid or substrate surfaces by forming volatile sub-oxides, as elaborated in Section 6.4.2. However, the use of high vacuum environments also promotes evaporation of the liquid and this can lead to anomalously low measured contact angles since the drop peripheries tend to be fixed, and hence the measured contact angle changes from that of an advancing liquid front to that of one which is receding, from θ_a to θ_r as illustrated in Figure 3.3. This effect occurs quite frequently but the change in the contact angle values usually lies within the range of measurement precision unless the loss exceeds several percent and the observation of major effects usually requires a loss of about 10%. Such losses can occur, however, when conducting sessile drop experiments using volatile metals such as Mg or Zn or materials containing volatile components, such as the alkali metals present in some glasses, rendering the measurements obtained of doubtful validity.

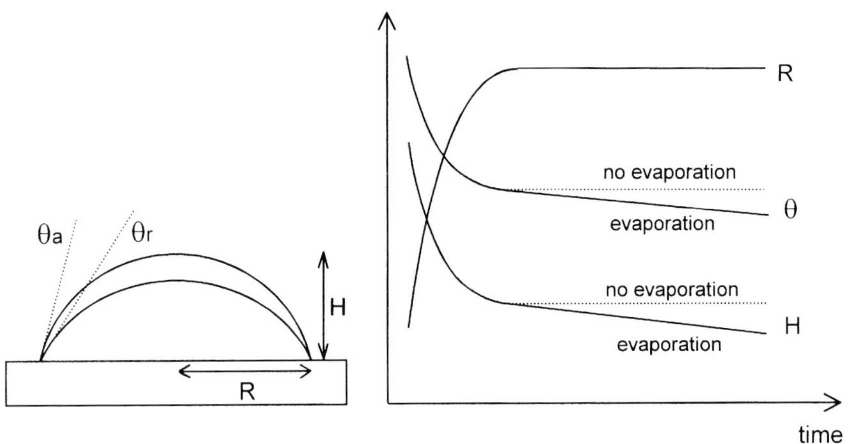

Figure 3.3. Schematic illustration of the effects of evaporation on the profile and parameters of a sessile drop.

Some form of mechanical pumping will be required to purge or evacuate and back fill the test chamber and care must be taken to minimise transmission of vibration to the sample. Vibration that causes movement of the sessile drop over its substrate is unacceptable and generally any vibration that can be felt by placing a hand on the test chamber should be avoided.

The methods of heating the samples and of controlling their temperatures are not specific to sessile drop tests and many methods have been employed in practice but most experiments are conducted using electrical resistance heating elements which, ideally, should be outside the test chamber to eliminate the possibility of contamination caused by evaporation or desorption from their hot surfaces. This possibility is particularly real when graphite heating elements are used which are often porous. Additionally, care must be taken to achieve melting of the liquid in times that are short compared to those allowed for subsequent spreading. In particular, when using alloys with wide melting ranges, the heating rates should be fast through that temperature range to minimize premature wetting of a substrate by migration of a liquid phase from a partially molten drop.

As already said, the commonest measurement made on sessile drops is the value of their contact angles. The experiments, therefore usually require the use of some form of goniometer to measure the angle directly when viewing the drop profile, or a means of recording the drop size and geometry with still, cine-, video or X-ray cameras (Figure 3.4). It is essential, therefore, that the chamber be fitted with observation ports, some of which should be in the same plane as the sample, and that the axes of the cameras should be on the plane of the substrate surface. Failure to satisfy this last requirement will cause distortions in the drop image. For many experiments, it is also essential to use background illumination of the drop profile, and to ensure that the beam is collimated and in the same plane as the substrate. At temperatures of about 700°C or above, the sessile drops will be sufficiently luminous to be photographed optically, but it is still desirable to use background illumination if sharp images are to be obtained (Figure 3.5). This technique can to some extent mitigate the clouding of the cool observation ports by condensation of volatile species from the liquid drops, and it can also be helpful to back-fill the test chamber with an inert gas rather than leave it evacuated.

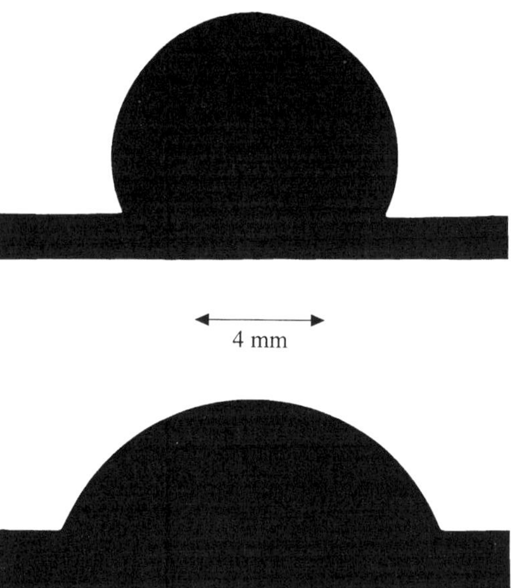

Figure 3.4. Photographs taken with background illumination of sessile drops of pure Al on polycrystalline alumina in high vacuum at 850°C (top) and 1200°C (bottom). From (Coudurier et al. 1984).

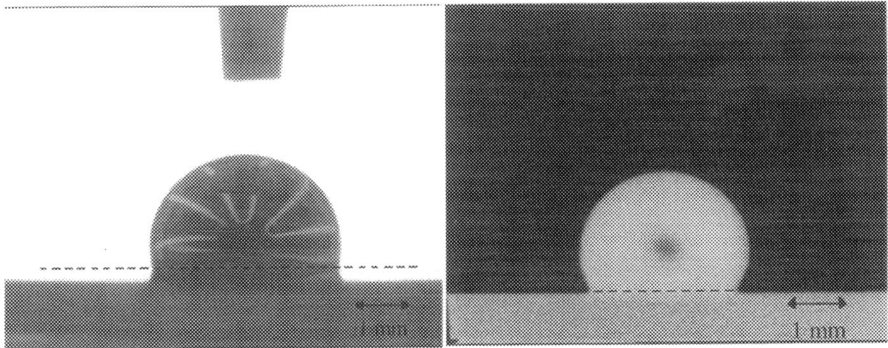

Figure 3.5. Video images of sessile drops of two Ni alloys on Al_2O_3 at 1500°C with (left) and without (right) background illumination (Labrousse 1998). On the left image, one can distinguish the reflection of the resistance heater on the droplet as well as the capillary used to dispense the droplet. By tilting slightly the substrate, it is possible to determine accurately the position of the triple line (dashed line) by the intersection of the drop profile and the drop shadow (or reflection) on the substrate surface.

Thus conduct of even a simple sessile drop experiment can be difficult and require a complex array of sophisticated equipment, such as that shown in Figure 3.6.

3.1.2 Types of experiment

A sessile drop experiment can take one of several forms, as shown schematically in Figure 3.7. The classic form shown in Figure 3.7.a calls for a small piece of solid sessile drop material, typically 0.01 ml in volume, to be placed on a substrate and then heated above its melting temperature. The sessile drop can be composed of a pure material or it can be an alloy or a chemical compound. It is not always convenient to prepare alloys or compounds prior to conducting the sessile drop experiment, and a variant of the classic technique, shown in Figure 3.7.b, is to form the alloy or compound *in situ* from a mechanical mixture of the components, (Mortimer and Nicholas 1973). In some work done in the former Soviet Union, metal alloys were formed *in situ* by plunging a piece of solid solute into a molten sessile drop of solvent (Naidich and Zhuravlev 1971).

Another variant calls for a metal or alloy to be melted in an unwetted and chemically inert closed ceramic tube positioned and to be dispensed on to the substrate surface through a small hole in the tube end by applying a back pressure of inert gas, Figure 3.7.c, (Nicholas et al. 1984), or using a piston (Naidich et al. 1983, Rado et al. 1998). One advantage of this technique is that oxide films on

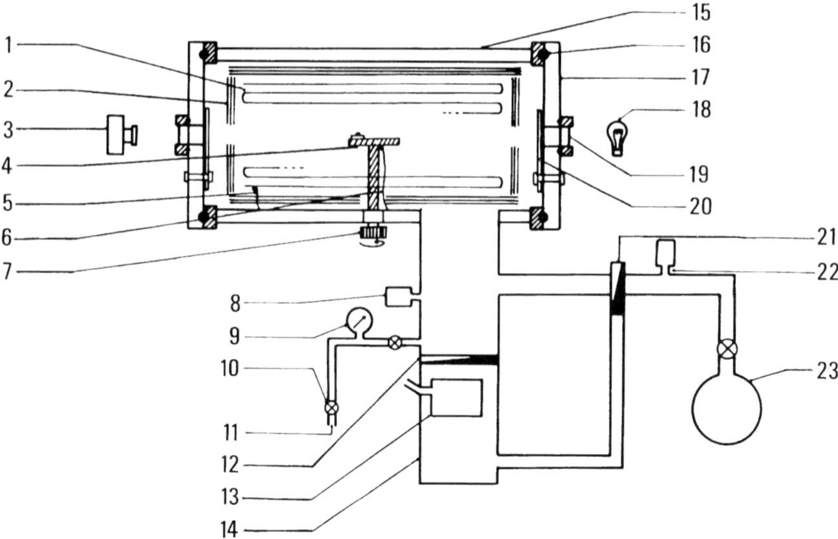

Figure 3.6. Laboratory equipment used to conduct sessile drop experiments at high temperatures (Rivollet et al. 1987) [10]. 1:molybdenum resistance heater, 2:molybdenum radiation shields, 3:camera, 4:alumina rotating stage, 5:PID regulation thermocouple, 6:thermocouple, 7:rotating vacuum seal, 8:ion-gauge, 9:manometer, 10:micrometer-valve, 11:gas inlet, 12:pneumatic gate valve, 13:N_2-cooled trap, 14:diffusion pump, 15:water-cooled stainless steel enclosure, 16:copper gasket, 17:removable disk, 18:incandescent lamp, 19:view-port, 20:view-port mask, 21:3-way automatic gate valve, 22:Pirani gauge, 23:zeolite pump.

liquid metals are disrupted during dispensing. However, care must be taken to ensure that the distance between the tube and the substrate is sufficient to avoid bridging by the dispensed drop but not so great as to result in the drop bouncing off the substrate or splattering. Again, a sessile drop can be melted on an inert substrate which is then raised so that the top surface of the drop contacts a fresh and possibly reactive solid surface and subsequently lowered, Figures 3.7.d and 3.7.e, (Naidich et al. 1995, Hara et al. 1995, Voitovich et al. 1998). Liquid can be transferred to the top substrate provided this is better wetted than the donor and this configuration is particularly useful for the study of wetting kinetics because it avoids interactions between the liquid and its receiving substrate that can occur during heating to the experimental temperature, Figure 3.7.d. The configuration shown in Figure 3.7.e can be used to study hysteresis effects because both advancing and receding angles are formed as the bottom substrate is raised and lowered. Contact angles assumed by both advancing and receding liquid fronts can be measured also by tilting a substrate until the sessile drop begins to run down hill, Figure 3.7.f, a technique usually used only with low-melting temperature liquids.

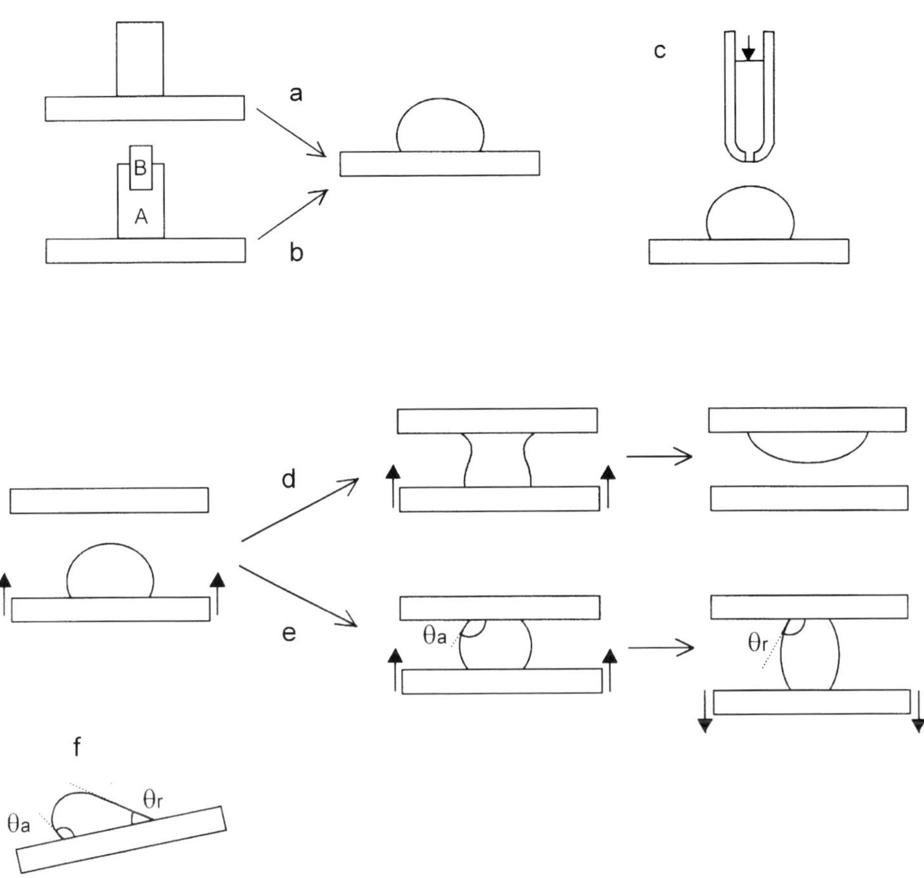

Figure 3.7. Methods of conducting sessile drop experiments : (a) classic technique, (b) *in situ* formation of an alloy, (c) dispensed drop, (d) transferred drop, (e) double substrate, (f) tilted plate.

3.1.3 Information that can be derived

3.1.3.1 Contact angles. According to the Young equation (1.16), the equilibrium contact angle, θ_Y, is a unique characteristic for each particular materials combination, determined by the surface and interfacial energies of the system components, and hence is independent of the liquid and solid configuration involved. This argument leads to the conclusion that a relatively simple sessile drop experiment can be used to provide information about the capillary behaviour of inert systems in situations in which direct measurements are not practical or easy, such as during the penetration of capillary gaps or the formation of braze fillets. For chemically reactive systems, care must be taken when contact angle values measured in the sessile drop configuration have to be transferred to other configurations, as discussed in Section 2.2.2.1 (see also Figure 2.27).

Sessile drop experiments permit the changes in wettability with temperature to be observed by ramping the temperature, or preferably by holding the sample at a series of increasing temperatures so that several measurements can be made during each constant temperature hold to test whether stable contact angle values have been achieved. Generally, increasing the temperature produces a small decline in contact angle values for inert systems, such as 10 degrees or so for a temperature rise of several hundred degrees Celsius. In principle, it should also be possible to check the effect of temperature on contact angle values by concluding the experiment with a series of holds at decreasing temperatures but this is seldom done in practice because receding liquid fronts often exhibit hysteresis effects.

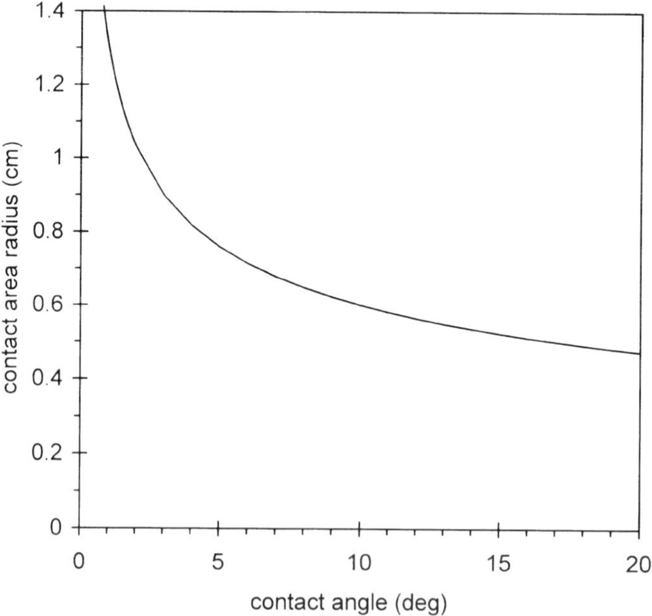

Figure 3.8. Relationship between the contact radius, R, and the contact angle, θ, of a sessile drop of volume 0.03 ml that has the form of a spherical cap.

If care is taken, angular measurements can be made from drop profiles with an accuracy of 1–2° for θ values ranging from about 20 to 160°. Measurements of very small or high contact angles are difficult because precise alignment of the goniometer or camera with the plane of the substrate surface becomes critically important despite increasing uncertainty about the exact location of the contact line when viewing the drop profile. It is a common practice to view sessile drops from above during or after the experiment if they have very low contact angles and to use the radius of the contact area, identified as R in Figure 3.9, as a measure of wettability. This radius is sensitive to small changes in wettability as illustrated in Figure 3.8 and, if so wished, can be used to derive the contact angle by assuming the drop to be a spherical cap of volume v and substituting in the expression

$$v = \frac{\pi R^3}{3} \frac{(2 - 3\cos\theta + \cos^3\theta)}{\sin^3\theta} \tag{3.1}$$

If the measurement of R is made after the experiment is completed, the volume used should still be that of the liquid because it is unusual for there to be significant contraction in the contact area when the drop solidifies.

To have unreserved confidence in such θ data it is advisable to measure several geometric and dimensional characteristics of the sessile drop and to test for their internal consistency. Thus the θ values assumed by the left hand and right hand sides of the drop profile should be measured and compared for consistency. Additionally, it is prudent to measure the height of the drop apex, H, and the radius R of the substrate contact area. If the drop has the profile of a spherical cap then a value can be calculated for the contact angle, θ_{calc}, by substitution in

$$\tan\left(\frac{\theta_{calc}}{2}\right) = \frac{H}{R} \tag{3.2}$$

and comparison with θ again provides a test of the assumed symmetry of the drop. On some occasions, θ_{calc} may be preferred to that of θ since a small and inconsequential perturbation at the drop periphery could cause a large change in θ without much affecting θ_{calc}.

It is also important to verify the consistency of changes in the geometric and dimensional characteristics occurring during the course of an experiment (Table 3.2). Thus a decrease in the contact angle and drop height that is not mirrored by an increase in the radius of the contact area suggests that liquid has been lost, by infiltration of the substrate or evaporation for example, while an increase in the contact angle that is not mirrored by a decrease in the radius of the contact area demonstrates pinning of the advancing liquid front, triple line, by some localised substrate heterogeneity, (Rivollet et al. 1987).

Table 3.2. The different possible cases of changes with time in the shape of the drop at constant temperature (Rivollet et al. 1987).

$\theta(t)$	R(t)	H(t)	Comments
↓ or =	↑	↓	at steady state $\theta = \theta_a$
↑ or =	↓	↑	at steady state $\theta = \theta_r$
↓	=	↓	θ tends towards θ_r (ex.: decrease of drop volume)
↑	=	↑	pinning of an advancing triple line (ex.: increase of drop volume)

↑: increases, ↓: decreases, =: constant, θ_a: advancing contact angle, θ_r: receding contact angle

Workers concerned with liquid/solid systems with exceptionally good wettability, such as some braze/metal workpiece systems, find that the drops do not preserve a perfectly circular area of contact because of the sensitivity of the liquid to changes in capillary attraction caused by minor variations in the surface texture of the workpiece. Because of such irregularity, some workers have used the contact area of a small volume of braze as a measure of wettability. Thus Feduska used area measurements to differentiate between the wetting of different stainless steels by a wide range of metals and alloys (Feduska 1959).

Not all aspects of wetting behaviour are sensitive to the precise values of small contact angles. For example, the rise up a vertical capillary is a function of the *cosine* of the contact angle (equation (1.55)), and therefore a decrease in θ from $10°$ to $0°$ causes an increase in penetration of only 1.5%.

3.1.3.2 Liquid surface energies. The surface energy of a liquid can be derived from very accurate measurements of the size and shape of a sessile drop (Passerone and Ricci 1998). This is not in fact the spherical cap which has been assumed or implied up to now but is somewhat flattened due to gravity. This is of particular importance for large drops with diameters substantially larger than the capillary length, $(2\sigma_{LV} / (\rho g))^{1/2}$. This flattening causes the principal radii R_1 and R_2 identified in Figure 3.9 to vary from point to point on the drop surface, reflecting the balance between the hydrostatic pressure and the capillary pressure due to the liquid surface energy and it is this deviation from sphericity that is used to determine surface energy values.

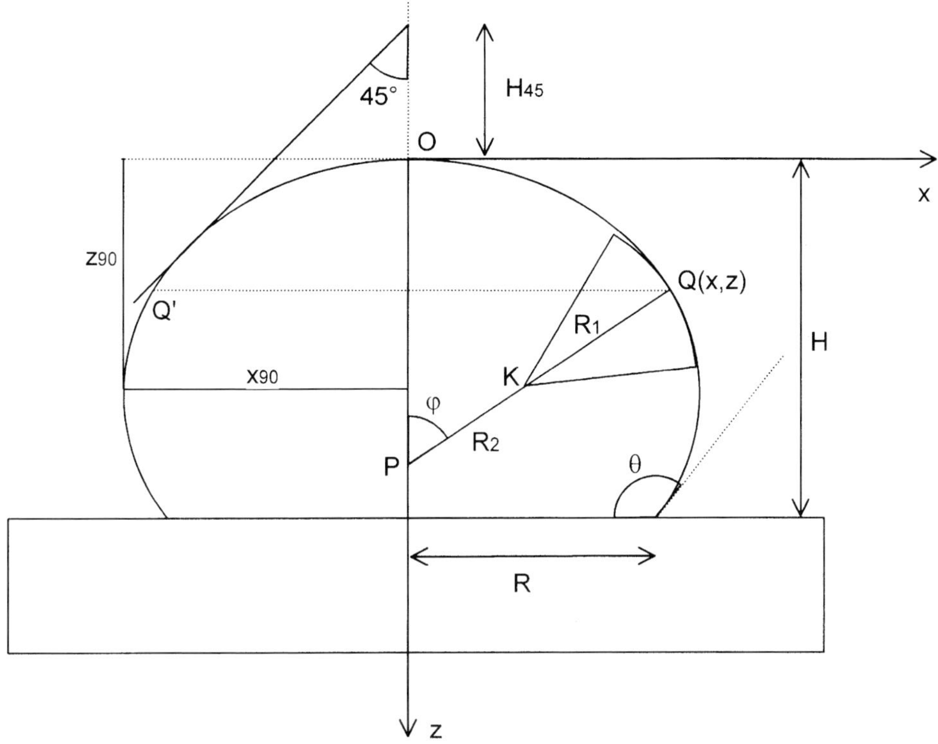

Figure 3.9. Profile of a sessile drop showing the measurements used to derive liquid surface energy values. The principal radii of curvature at point Q are $R_1 = QK$ and $R_2 = QP$.

For the sessile drop sketched in Figure 3.9, the principal radii of the surface at the opposed positions Q and Q′ are R_1 in the plane of the paper ($R_1 = QK$) and R_2 in a plane perpendicular to the paper ($R_2 = QP = x/\sin\varphi$) so that

$$P_L^Q - P_V^Q = \sigma_{LV}\left(\frac{1}{R_1} + \frac{\sin\varphi}{x}\right) \tag{3.3}$$

Similarly, at the drop apex (point O)

$$P_L^O - P_V^O = \frac{2\sigma_{LV}}{b} \qquad (3.4)$$

because at that point $b = R_1 = R_2$. Neglecting the difference in vapour pressure between points Q and O enables us to write

$$P_L^Q - P_L^O = \sigma_{LV}\left(\frac{1}{R_1} + \frac{\sin\varphi}{x}\right) - \frac{2\sigma_{LV}}{b} \qquad (3.5)$$

But the difference between P_L^Q and P_L^O must be equal to the hydrostatic pressure of a column of height z so that

$$P_L^Q - P_L^O = \sigma_{LV}\left(\frac{1}{R_1} + \frac{\sin\varphi}{x}\right) - \frac{2\sigma_{LV}}{b} = \rho g z \qquad (3.6)$$

In applying this equation it should be noted that the density of the liquid, ρ, is used instead of the more rigorously correct difference between it and the density of the vapour.

The radius of curvature R_1 and $\sin\varphi$ can be expressed in terms of the first and second derivatives of z with respect to x. Substituting this into equation (3.6) we obtain a differential equation for which numerical solutions are necessary. The best approach is by curve fitting using computerised iterative calculations such as those suggested by Maze and Burnet, (1969), until there is a convergence of σ_{LV} values. This approach can yield surface energy values with high estimated accuracies and a reproducibility of a few percent, but great care has to be taken to use contaminant free materials and environments and to ensure that the substrate is horizontal and not subject to vibration. Recording the sessile drop image is relatively simple, but its subsequent evaluation can be tedious, typically requiring measurement of x with a precision of about ±1 μm at perhaps 50 or more different locations with z values incrementally increasing by 10–20 μm. These measurements have to be made on enlargements of the recorded images of the drop profiles and high-quality equipment must be used to retain sharpness. To test that the experimental procedures for recording and measuring images are of the desired standard, it is good practice to use them first to measure the surface energy of Hg at room temperature, established as 0.487 ± 0.005 J.m^{-2}, and then of Au at its melting temperature, equal to 1.138 ± 0.015 J.m^{-2} (see Table 4.1).

Today, the development of computerised numerical calculation and computer-aided imaging techniques (Rotenberg et al. 1983, Liggieri and Passerone 1989,

Naidich and Grigorenko 1992, Passerone and Ricci 1998) is making the sessile drop method more and more reliable and accurate. However, it should be noted that substantial effort has been required historically to derive a surface energy by curve fitting, and many authors have suggested simplifications which require far fewer measurements. Thus, Bashforth and Adams, (1883), rewrote equation (3.6) in the form:

$$\frac{b}{R_1} + \frac{b \sin \varphi}{x} = 2 + \beta \frac{z}{b} \tag{3.7}$$

with

$$\beta = \frac{b^2 \rho g}{\sigma_{LV}} \tag{3.8}$$

Bashforth and Adams generated tables of β and x/z for $\varphi = 90°$ as well as tables of the ratios x/b and z/b for differing values of b and φ. By measurement of x_{90} and z_{90} at $\varphi = 90°$, b and β can be determined from these tables and the liquid surface energy can then be derived from equation (3.8) with an accuracy of about $\pm 2\%$ for $\beta > 2$ if the droplet coordinates are measured with an accuracy better than 0.1% (Sangiorgi et al. 1982). A method is presented in Appendix D allowing calculation of the mass of a droplet for an optimised σ_{LV} measurement.

For drops with an x_{90} three or four times greater than the capillary constant, Worthington, (1885), derived the approximate expression

$$\sigma_{LV} = 0.5 \rho g z_{90}^2 \left(\frac{1.641 x_{90}}{1.641 x_{90} + z_{90}} \right) \tag{3.9}$$

which for big droplets i.e., pancakes ($x_{90} > 10 l_c$ where l_c is the capillary length), reduces to $\sigma_{LV} = 0.5 \rho g z_{90}^2$. Finally, many workers have made use of the empirical relationship:

$$\sigma_{LV} = \rho g x_{90}^2 \left(\frac{0.052}{f_1} - 0.1227 + 0.0481 f_1 \right) \tag{3.10}$$

suggested by Dorsey, (1928), which is the most accurate of these simplified expressions, making use of a factor f_1 equal to $(H_{45}/x_{90}) - 0.4142$ (see Figure 3.9 for the definition of H_{45}).

While convenient, it should be noted that most of these approximations require the same precision of measurement as do curve fitting techniques and obviously lead to less accurate results. They have been particularly used to derive values for high melting temperature metals and alloys, yielding values that have an author to author reproducibility of typically 10%, as demonstrated by examination of collected data for Cu and Fe, (Iida and Guthrie 1988). It is, however, often difficult when making such comparisons to decide whether the differences reflect varying precisions of measurement, differences in experimental procedure or qualities of materials.

Most of the approximation methods require the sessile drop to be non-wetting if they are to yield values with errors of less than about 10%. Curve fitting techniques can be applied to wetting drops, but these become increasing imprecise as the contact angles decrease below $90°$. For such wetting systems, the "large drop" approach of Naidich et al. (1963) shown in Figure 3.10 can be of value in that it produces a large and readily measurable apparent contact angle, $\theta_{app} = \theta + \alpha$, and hence facilitates the derivation of accurate σ_{LV} values.

Another approach to increasing the contact angle and hence improving the accuracy of σ_{LV} measurements is to change the substrate to one that is less wettable, and indeed the use of several different substrate materials is desirable if there is to be confidence that there has been no chemical interaction that may have affected the liquid surface.

Sessile drop experiments are also used to measure the effects of temperature on liquid surface energies. Because the temperature coefficient $d\sigma_{LV}/dT$ of liquid metals and oxides is usually a very small, negative, value (-0.05 to -0.5 $mJ.m^{-2}.K^{-1}$), a temperature rise of several hundred degrees is necessary to produce decreases in the surface energy that can be reliably detected by measurements of drop profiles. Even in this case, the error on the temperature coefficient lies between 30% and 100% (see Section 4.1.1).

3.1.3.3 Liquid density and other characteristics. Profile measurements for a symmetrical sessile drop can be used to calculate its volume and hence its density if the original mass is known and there has been no significant loss due to evaporation. As for the measurement of σ_{LV}, the volume can be estimated by measurements of the drop profile coordinates and numerical integration by using image analysis techniques (Naidich and Grigorenko 1992). Alternatively, the

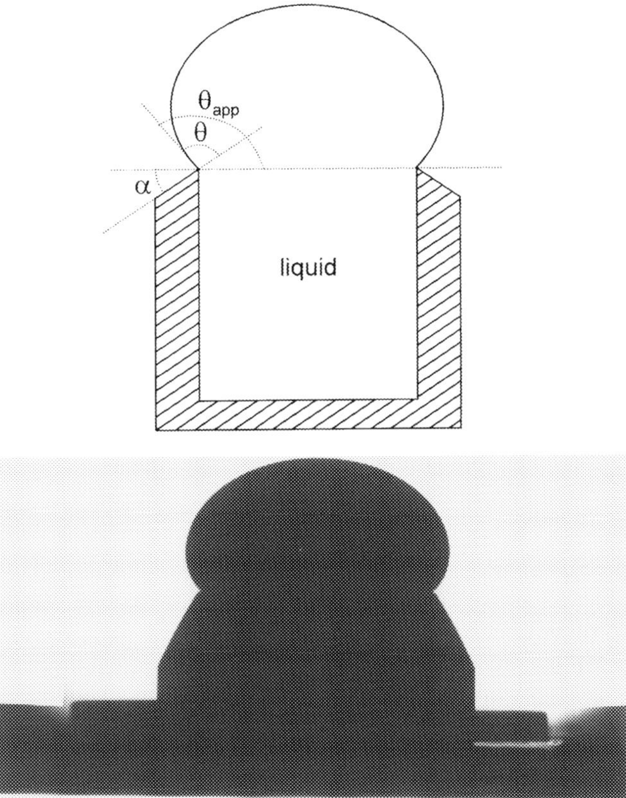

Figure 3.10. Top: arrangement of a "large drop" experiment, after (Naidich et al. 1963). Bottom: photograph of a large drop of gold alloy formed on a sapphire cup at 900°C in Ar atmosphere (cup diameter: 12 mm). Printed from (Ricci and Passerone, private communication) with kind permission of the authors.

volume can be roughly estimated from a few measurements and use of the Bashforth and Adams tables, (1883), by substitution in the expression

$$v = \pi R^2 \left(\frac{2b}{\beta} - \frac{2b^2 \sin \theta}{x_{90}\beta} + H \right) \tag{3.11}$$

where H is the drop height. Both methods of calculating volumes call for great experimental care to form symmetrical drops and for considerable precision in the dimensional measurements. Nevertheless, densities can be derived from sessile drop experiments with modest estimated errors, less than 1% (Naidich and Grigorenko 1992).

Sessile drop experiments have been used extensively to derive quantities characterising spreading and penetration phenomena, such as the work of adhesion and work of immersion, given by equations (1.45) and (1.54), using a single sessile drop experiment to measure both the contact angle and the liquid surface energy.

3.1.4 Information derivable from reactive systems

3.1.4.1 Contact angles. The conduct of sessile drop experiments using chemically reactive materials, and environments, is essentially the same as that for inert systems but the configurations shown in Figures 3.7.c and 3.7.d are the most suitable because the first contact of the liquid drop is with a fresh, unreacted, substrate at the experimental temperature. Once more, well characterised materials should be used with flat horizontal substrates, as should means for making very precise measurements of the geometry and size of the drops. However, there are some differences of emphasis when considering the significance of the various requirements. Thus, control of heating rates and temperature hold times is more important with reactive than inert systems because these parameters can affect the progress of chemical interactions. Similarly, for the interpretation of results in reactive systems, it is particularly important to record not only the changes with time of contact angle values but also of drop base radii.

Strictly speaking, equilibrium wetting parameters cannot be measured for reacting systems because there is a continuous change in the chemistry of solid and liquid phases caused by dissolution reaction or by formation of new phases at the interface. Nevertheless, as discussed in Section 2.2.2, a local equilibrium can be established at the liquid front for many systems of practical interest and this results in the observation and measurement of an unchanging contact angle, θ_F, that is close to but slightly different from the equilibrium value. For instance, in systems in which a continuous layer of new phase is formed at the interface, θ_F is close to the Young contact angle of the liquid on the reaction product. In systems in which the substrate A is slightly dissolved in the liquid B, θ_F tends towards the Young contact angle of a BA alloy on solid A. Differences between θ_F and θ_Y can be caused by a variety of effects, such as the roughness of the reaction product layers formed at the liquid/solid interface or a dissolution of the substrate to create a pit (see Figure 2.18). In all cases, interpretation of θ_F values requires interface characterisation close to and far from the drop periphery.

3.1.4.2 Liquid surface energies. The parameters needed to derive liquid surface energy values in reactive systems are almost always measured from photographic or video images, the changing size and shapes of drops rendering it impractical to employ the slow process of real time measurements using a goniometer telescope. The photographs provide a record of the conditions prevailing at a particular instant and hence can be used to monitor the effect of the change in composition induced by the reactions on the surface properties of the liquid drop.

Liquid surface energy values can be decreased by segregation of surface active components that enter the liquid from the substrate or from the gaseous environment. In the case of metal sessile drops, these components will be low melting temperature elements such as Pb or Sn or non-metals such as O, 1% of which halves the liquid surface energy of Fe (see Figure 4.4). In contrast, the surface energy values of molten metal alloys can be increased if the reaction depletes them of a low surface energy component that had segregated to the free surface. This could occur if the substrate has a stronger affinity with the surface active solute than does the liquid, as when Cu contaminated with O is brought into contact with Ti or W (Nicholas and Poole 1967).

These alterations in liquid surface energy values are genuine reflections of the reactions, but spurious changes can be caused if the reaction with the substrate or the gaseous environment results in the formation of a second, solid, phase in the drop or on its surface. The creation of an oxide skin on the surface of a metal drop can cause a dramatic increase in the apparent surface energy value, sometimes as much as ten to one hundred times, but this is without any physical or chemical meaning because the surface is no longer that of a liquid drop. Similar distortions, and even in the authors experience decreases to apparently meaningless negative σ_{LV} values, can be caused if a second solid phase is dispersed within the liquid due to supersaturation of some component released by reaction with the substrate or physical detachment of grains of the solid substrate.

3.1.5 Solidified sessile drop

It is not always convenient or practical to measure contact angles while the sessile drops are liquid : this requires special equipmemt and is imprecise for very small contact angles. Many workers, therefore, have chosen to derive contact angle data from measurements of the size and geometry of solidified small sessile drops (Figure 3.11). The most common measurements made are of the base drop radii of well-wetting drops of a known volume, from which contact angles are derived by substitution into equation (3.1). The validity of this method depends not only on the reasonableness of assuming that the drop profile can be approximated by a spherical cap, but also that the volume of the sessile drop was not altered during

Figure 3.11. Solidified Cu drop (mass 107 mg) on an oxidised W substrate. $\theta = 120 \pm 4°$. From (Lorrain 1996).

the experiment by evaporation or reaction with the substrate. Alternatively, if the solidified sessile drop is only moderately wetting, an approximate contact angle can be derived also by substitution into equation (3.2).

In principle, estimates of the contact angles of non-wetting sessile drops can be made similarly, but shape and size distortions caused by solidification make it increasingly difficult to derive precise estimates or even, for example, to rank the behaviour of families of alloys in a semi-quantitative manner.

Thus, measurements of solidified sessile drops can be and are used to derive estimated values of contact angles, but these estimates are less precise than those derived from direct observation of liquid drop profiles except for very well wetting systems. The estimates for solidified drops tend to be smaller, typically by up to 5°, and more variable than those derived from observation of liquid profiles. Larger differences can be observed when, during cooling, dewetting of the liquid drop occurs before the resolidification temperature is reached. In this case, traces of the location of the triple line before dewetting are easily seen on the substrate surface.

3.1.6 Contact angles of solid particles on a substrate
The shape of *solidified*, millimetre size, droplets reflect wetting of the liquid on the solid substrate. The equilibrium shape of a solid particle M on a solid S in a vapour

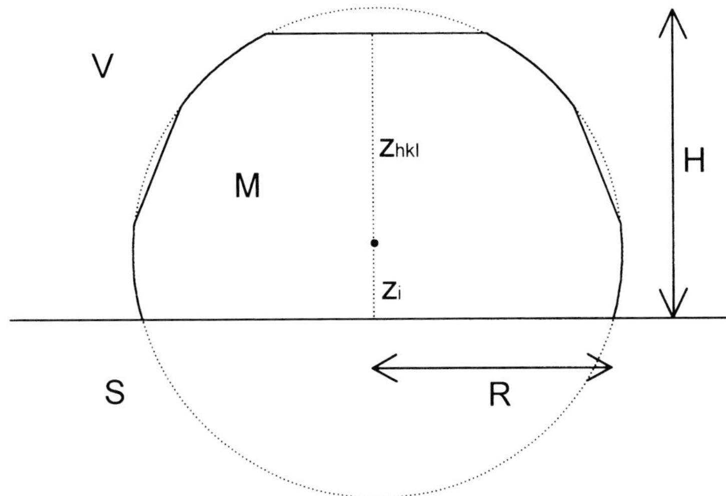

Figure 3.12. A liquid drop is spherical in shape (dashed line) while the equilibrium shape of a solid monocrystalline particle of cubic structure metal is a truncated sphere (full line).

V provides information on the wetting of the *solid* M on S. Micron size particles are needed to attain the equilibrium shape by surface self-diffusion in reasonable times (several hours or tens of hours depending on the value T/T_F^M where T_F^M is the melting temperature of M). For small isotropic solid particles, the equilibrium shape dictated only by surface forces is a spherical cap with a contact angle given by the Young-Dupré equation (1.45). Solids are usually anisotropic so that the surfaces of small particles at high temperatures display facets or both facets and curved regions (Figure 3.12). The equilibrium shape of such a particle lying on a substrate S, obtained by minimizing the total surface energy of the M/S system at constant temperature and M volume and assuming that the S/V surface and S/M interface are co-planar, is given by the generalised Wulff equation (Winterbottom 1967):

$$\frac{\sigma_{MV}^{hkl}}{z_{hkl}} = \frac{\sigma_{SM} - \sigma_{SV}}{z_i} \tag{3.12}$$

where σ_{MV}^{hkl} is the solid M/vapour surface energy for the (hkl) plane and z_{hkl} and z_i are respectively the distances from the (hkl) surface plane and the M/S interface to the centre of the crystallite. By introducing into equation (3.12) the expression (1.8) of work of adhesion W_a, we obtain

$$\frac{W_a}{\sigma_{MV}^{hkl}} = 1 - \frac{z_i}{z_{hkl}} \tag{3.13}$$

It is possible to write this expression in terms of an equivalent contact angle, θ_{equiv}, by comparing with the classical Young-Dupré equation (1.45). Then one gets

$$\cos\theta_{equiv} = -\frac{z_i}{z_{hkl}} \tag{3.14}$$

The equilibrium shape of a metallic particle is a sphere with small facets. Thus, the value of θ_{equiv} can also be estimated from the expression $\tan(\theta_{equiv}/2) = H/R$.

In practice, a metallic film, 100–300 nm thick, is deposited on the substrate S. The coated substrate is then annealed in high vacuum or a neutral atmosphere at high temperature, usually between 0.7 and 0.95 T_F^M, for several tens of hours to break up the film into individual particles (Figure 3.13.a) and for these particles to attain their equilibrium morphology (Figure 3.13.b) (Soper et al. 1996). Another practice is to heat the coated substrate at a temperature slightly higher than T_F^M to obtain rapid breaking of the film into small droplets and then to decrease the temperature below T_F^M and anneal the solidified micro-droplets. Measurement of the linear dimensions of the particles by scanning electron microscopy leads to θ_{equiv} values reproducible within a few degrees. This method has been applied with success to different non-reactive metal/oxide (Pilliar and Nutting 1967, Murr 1973, Soper et al. 1996) and metal/graphite (Heyraud and Metois 1980, Gangopadhyay and Wynblatt 1994) systems. The effect of alloying on θ_{equiv} has also been studied by Gangopadhyay and Wynblatt (1994).

A variant of the equilibrium solid "drop" configuration is presented in Figure 3.14. This shows a ceramic particle embedded in a solid metallic matrix which had previously been subjected to deformation so that the particle decoheres from the matrix. Following a high-temperature treatment, the void in contact with the particle assumes an equilibrium form from which the contact angle θ can be measured by transmission electron microscopy. This approach is interesting, principally because the surfaces should be less contaminated than in the sessile drop configuration. Examples of this technique for metal/oxide systems are given in (Hondros 1978).

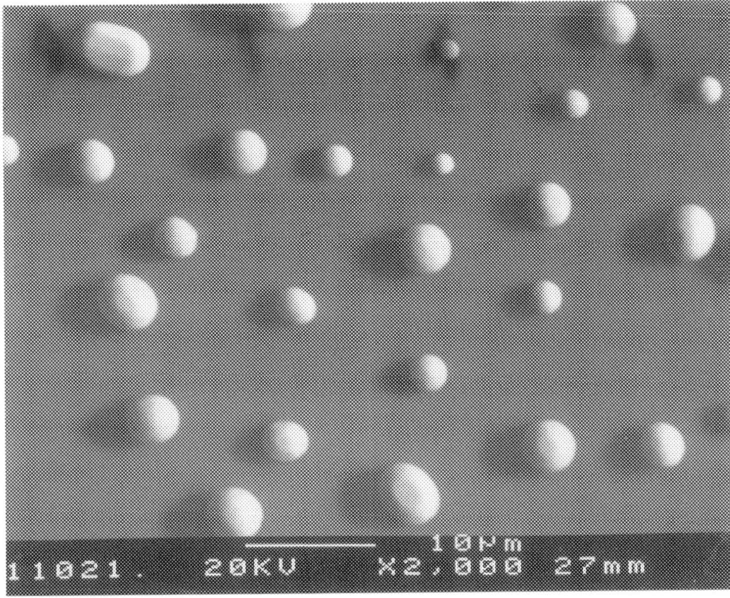

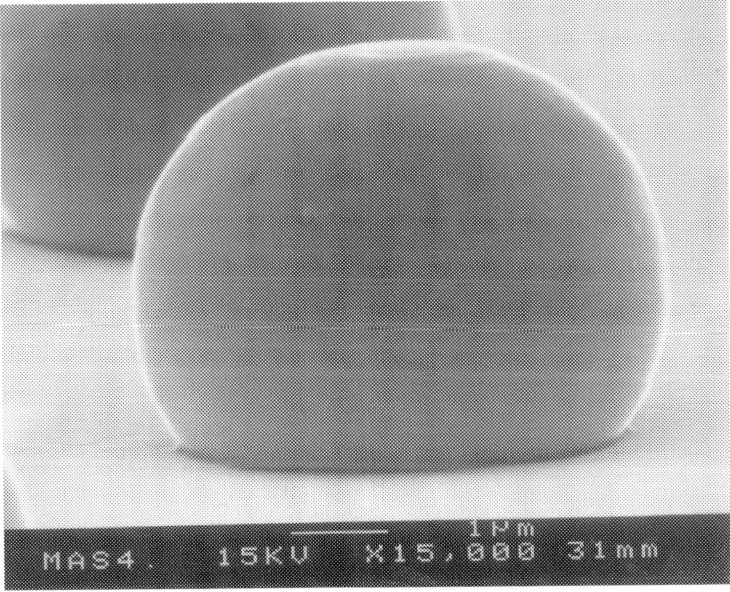

Figure 3.13. Top: when annealed at high temperature, a thin film of Cu on monocrystalline alumina breaks and forms micron-size particles. Bottom: SEM micrograph of a Cu monocrystalline particle on (0001) face of alumina formed after annealing at 1040°C. The facet on the top of the particle is a (111) plane. From (Soper et al. 1996) [11].

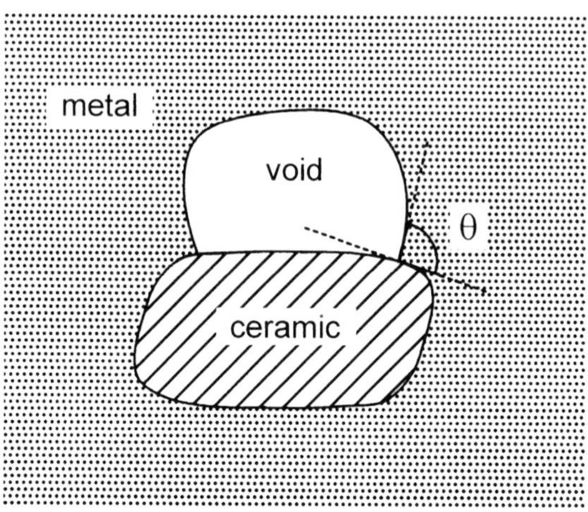

Figure 3.14. Equilibrium configuration of a void at a ceramic particle contained in a metal matrix. According to Hondros (1978).

3.2. THE WETTING BALANCE TECHNIQUE

The wetting balance technique is a variant of the maximum pull (or detachment) method used to measure liquid-vapour surface tensions (Keene 1993). It is nowadays widely employed in the electronics industry to quantify wetting of solders, but has also been used for wetting studies in metal/ceramic systems (Naidich and Chuvashov 1983b, Nakae et al. 1989, Rivollet et al. 1990). As compared to the sessile drop method which needs planar substrates, solids of various geometry can be studied by this technique.

The method described in detail in (Chappuis 1982) consists of continuously monitoring the force acting on a solid, cylinder or blade, moving into and out of a liquid bath during an immersion-emersion cycle. The apparatus developed by Rivollet et al. (1986, 1990) for high temperature studies (Figure 3.15) consists mainly of a furnace and a pumping group, a translation system allowing the displacement of the crucible containing the liquid bath at a constant rate and an electronic microbalance connected to the solid. For wetting liquids, additional information on wetting parameters can be obtained by measuring optically the height of the meniscus formed by the liquid on the solid surface (Figure 3.16.a). The changes in weight, equal to the sum of capillary and buoyancy forces, versus the depth of immersion are analysed by means of the Laplace equation giving the shape of the meniscus formed around a vertical solid. This method allows a simultaneous measurement of the surface tension of the liquid and the contact

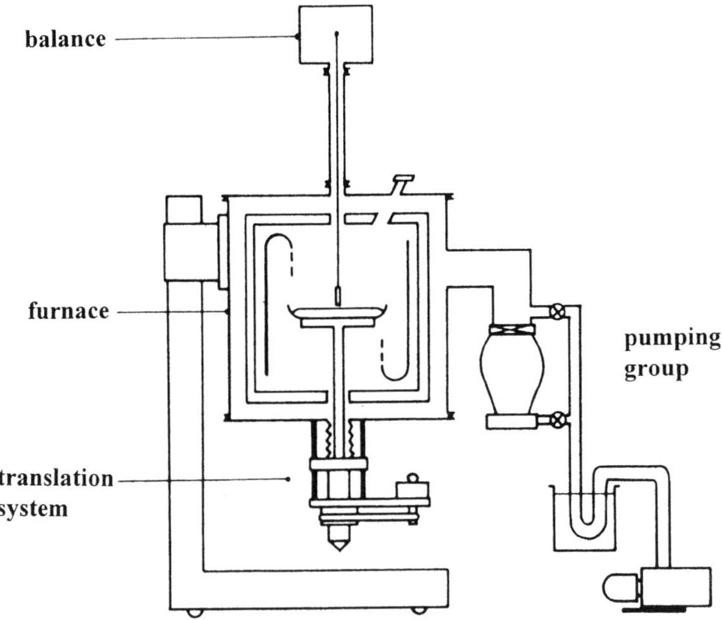

Figure 3.15. Sketch of the wetting balance (Rivollet et al. 1990) [2].

angles. Both advancing and receding contact angles can be deduced from the weight changes during immersion and emersion respectively (Figure 3.16).

In the following, and for the sake of simplicity, the vertical translation rate of the solid in the liquid is supposed to be infinitely slow. Then, if the origin of the weight measurements is taken as the weight of the solid in the gas, the force f exerted on a solid partially immersed in a liquid, whose base is located at a level z_b with respect to the flat horizontal surface of the liquid, is equal to the sum of the weight of the meniscus w_m and the buoyancy force:

$$f = w_m + \Omega \rho g z_b \tag{3.15}$$

where Ω is the solid base area, ρ the liquid density and g the acceleration due to gravity. Orr et al. (1977) demonstrated that for a prism of any right section of perimeter p, the weight of the meniscus is:

$$w_m = p \gamma_{LV} \cos \phi \tag{3.16}$$

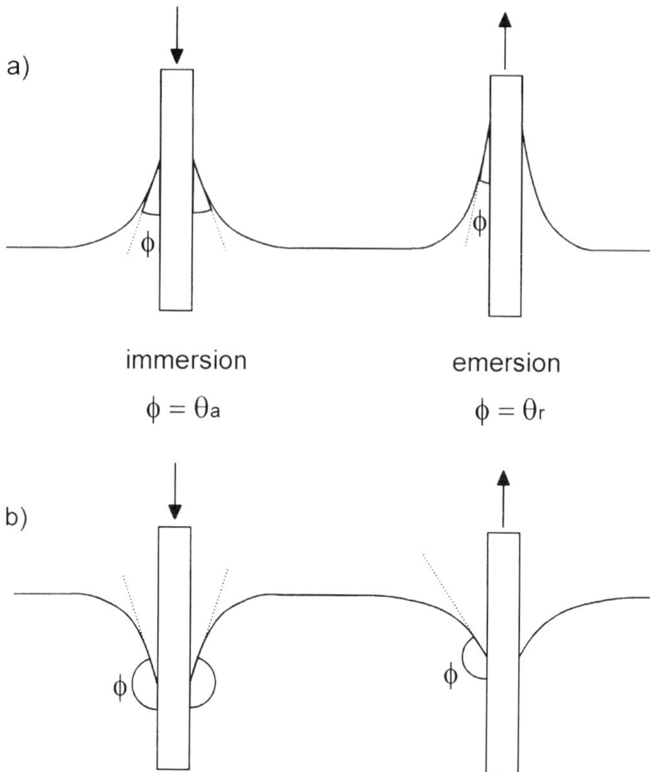

Figure 3.16. Meniscus formed with a partially immersed vertical plate and a (a) wetting and (b) non-wetting liquid.

ϕ being the *joining angle* of the liquid meniscus on the solid surface, defined by the tangent to the liquid surface at the triple line and the vertical (see Figure 3.16), and γ_{LV} the surface tension of the liquid (as w_m is a force, surface tension instead of surface energy is used in equation (3.16)). As will be seen below, ϕ corresponds to well-defined contact angles only at some particular points of the $f(z_b)$ immersion-emersion curve.

The curve of the variation in the force exerted on the solid during its immersion-emersion in a *non-wetting liquid* ($\theta_Y > 90°$) as a function of the depth of immersion is shown schematically on Figure 3.17 for successive positions of the solid in the liquid. As the solid is lowered, it contacts the liquid (A) and thereafter depresses its surface to form a downward curving meniscus (B). Then, the joining angle ϕ becomes larger and larger until it equals the advancing contact angle θ_a and

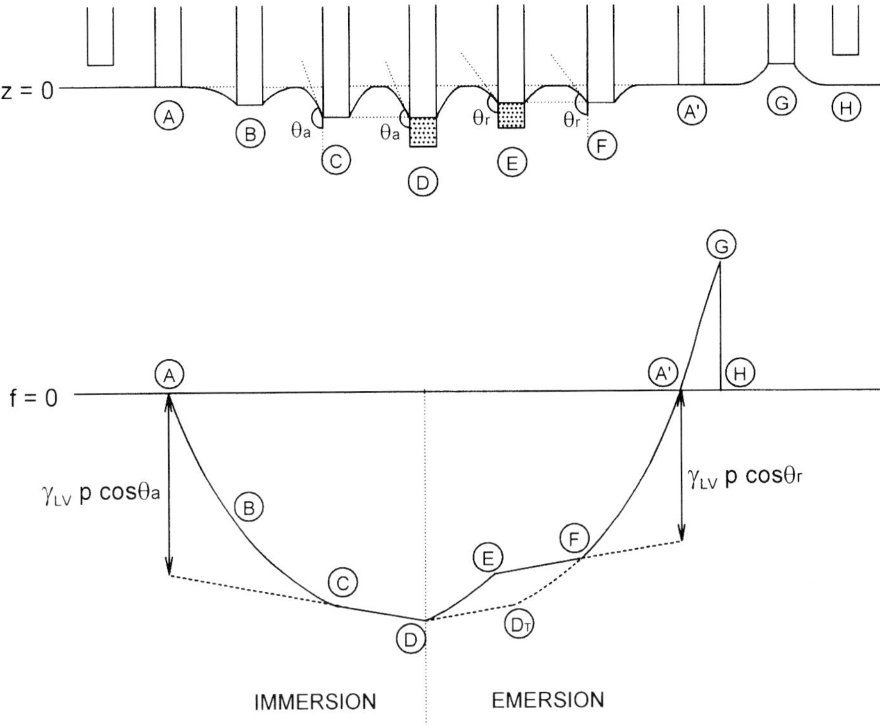

Figure 3.17. Common shape of an experimental curve of variations in the force exerted on a cylindrical solid during its immersion-emersion in a *non-wetting* liquid plotted as a function of the depth of immersion of the solid. The translation rate of the solid is supposed to be infinitely slow. From (Rivollet et al. 1990) [2].

the solid enters the liquid (C). After C, a deeper immersion of the solid does not give rise to any further change in the shape of the meniscus and the change in force is only due to the variation in the buoyancy force, which is linear with z_b according to equation (3.15) (C-D). The emersion of the solid begins in D. The joining angle changes continuously from θ_a to the receding contact angle θ_r (D-E). The meniscus has an equilibrium shape from E to F and the change in force is only due to buoyancy. From F to G, the joining angle changes from θ_r to $(\theta_r - 90°)$ where the meniscus falls off the solid. If the solid motion is reversed between points F and A', the curve does not retrace the same path as during the emersion but goes through points F-D_T-D, equivalent to B-C-D of the first immersion. This is a direct consequence of wetting hysteresis.

A slightly different series of events occurs when the liquid is *wetting* ($\theta_Y < 90°$), as indicated in Figure 3.18. First, the advancing contact angle θ_a is

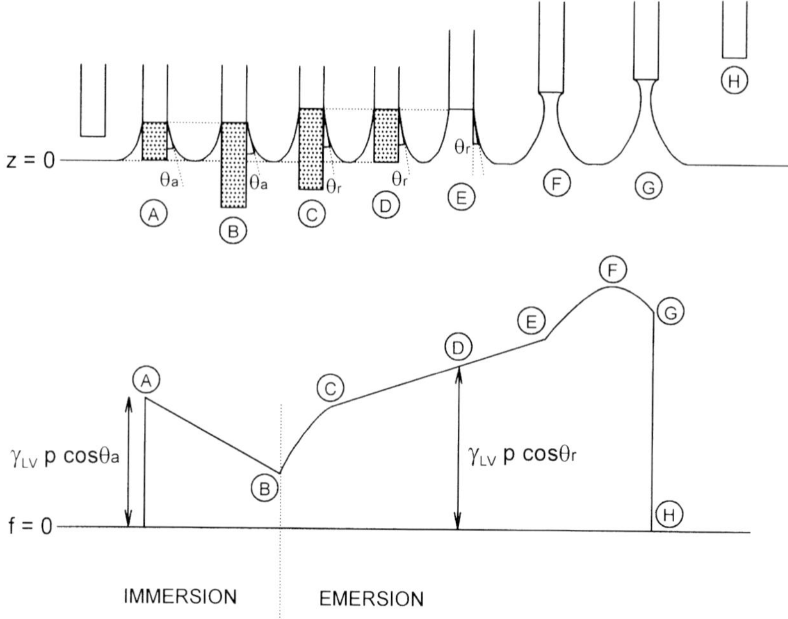

Figure 3.18. Common shape of an experimental curve of variations in the force exerted on a cylindrical solid during its immersion-emersion in a *wetting* liquid plotted as a function of the depth of immersion of the solid. The translation rate of the solid is supposed to be infinitely slow. From (Rivollet et al. 1990) [2].

spontaneously reached as soon as the solid touches the liquid. This contact angle is similar to the advancing contact angle measured in a typical sessile drop experiment i.e., an irreversible contact angle. Hence, for smooth ($R_a < 10$ nm) and chemically homogeneous solid surfaces, this angle is a good estimate (within a few degrees) of the thermodynamic contact angle of the system. Moreover, this step can be used not only to determine θ_a values but also to study the spreading kinetics. The variation of the force is linear during the immersion (A-B), as it is only due to the buoyancy contribution. Secondly, from point E, the meniscus is connected to the base of the solid rather than its sides and the joining angle ϕ becomes negative relative to the sides ($\theta_r - 90° < \phi < \theta_r$). When ϕ becomes less than ($\theta_r - 90°$) i.e., less than the receding contact angle on the base of the solid, a rupture of the meniscus occurs (G). Note that the curve goes through a maximum at point F when the increase of the buoyancy force is exactly balanced by the decrease of the weight of the meniscus. Although no solid is immersed in the liquid at point F, it can be verified that the hydrostatic pressure is not null in the liquid at z_b at point F because this pressure is zero for $z_b = 0$.

Using equations (3.15) and (3.16), the product $\gamma_{LV}\cos\phi$ can be determined for each point of the curve A-C of Figure 3.17, as f and z_b are measured. Furthermore, the products $\gamma_{LV}\cos\theta_a$ and $\gamma_{LV}\cos\theta_r$ can be determined at some specific points of the curves. In the wetting case (Figure 3.18), they can be measured directly at A ($f = p\gamma_{LV}\cos\theta_a$) and D ($f = p\gamma_{LV}\cos\theta_r$) as there is no buoyancy force ($z_b = 0$). In the non-wetting case (Figure 3.17), these values can be deduced by extrapolation of the linear parts C-D and E-F to the vertical of points A and A' respectively where $z_b = 0$.

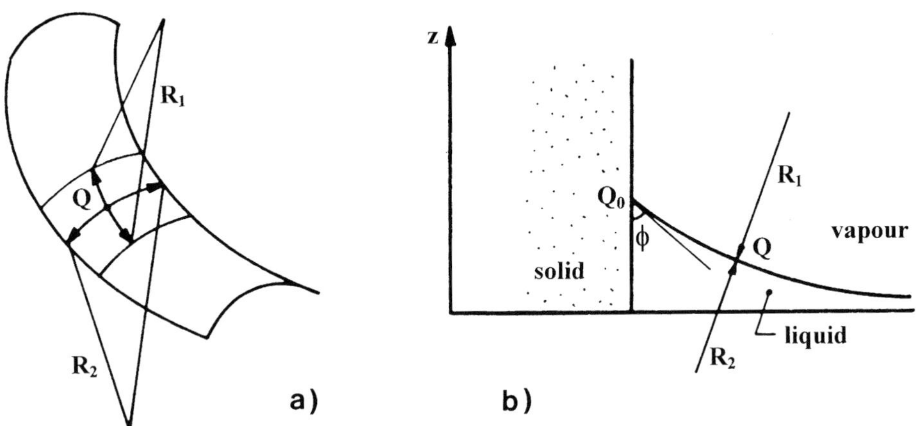

Figure 3.19. Geometrical quantities used to describe the shape of a liquid surface. (a) principal curvature radii; (b) vertical position z of the point Q of the surface and joining angle ϕ of the liquid profile with the solid. From (Rivollet et al. 1990) [2].

As shown previously, the product $\gamma_{LV}\cos\phi$ can be measured from experimental values of the force f (equation (3.15)). To determine θ_a and θ_r, one must at first calculate γ_{LV}. To this end, the Laplace equation is introduced, which applies at each point Q of the liquid meniscus surface with a vertical coordinate z relative to the flat horizontal surface of the bulk liquid (Figure 3.19.b):

$$\Delta P = \gamma_{LV}\left(\frac{1}{R_1} + \frac{1}{R_2}\right) = \rho gz \tag{3.17}$$

Here, ΔP is the pressure difference between the sides of the liquid surface and in the case of a meniscus, ΔP is equal to the hydrostatic pressure. R_1 and R_2 are the principal radii of the surface (Figure 3.19.a). Then, γ_{LV} depends on the three unknown quantities z, R_1 and R_2. However, the Laplace equation can be applied at point Q_0 of the triple line (see Figure 3.19.b). Thus, one of the three unknown quantities can be suppressed if the height (or the depth) of the meniscus z^*, a function of γ_{LV} and ϕ, is measured. This can be done either by a direct optical measurement (Mozzo 1970) or by examining the $f(z_b)$ curve (Rivollet 1986).

The value of z^* being known, R_1 and R_2 have now to be determined. This is possible only in a few simple cases featuring a geometry of the solid with a high degree of symmetry.

For a vertical blade of infinite width, the Laplace equation becomes:

$$\frac{\gamma_{LV}}{R_1} = \rho g z \tag{3.18}$$

as $R_2 \rightarrow \infty$. According to equation (1.23), the height of the meniscus is given by:

$$z^* = \left(\frac{2\gamma_{LV}}{\rho g}\right)^{1/2} (1 - \sin \phi)^{1/2} \tag{3.19}$$

Here, we recall that the force is written:

$$f = p\gamma_{LV} \cos \phi + \Omega \rho g z_b \tag{3.20}$$

Thus, if f and z_b are given by the experimental $f(z_b)$ curve and z^* measured by the means of an optical method, the two unknown quantities γ_{LV} and ϕ can be obtained by solving equations (3.19) and (3.20). Note that this procedure can be applied for different z_b values to increase the accuracy.

For non-wetting liquids, the measurement of the height of the meniscus z^* is difficult to perform by an optical technique. In this case, one can use the method developed by Rivollet (1986) based on the non-linear parts of the $f(z_b)$ curve corresponding to the situation where the meniscus is connected to the solid base (part A-C of Figure 3.17). The $f(z_b)$ curve is obtained by solving equations (3.19) and (3.20) with $z_b = z^*$. The value of γ_{LV} is then determined by fitting the calculated curve to the experimental one.

However, although the mathematical problem is solved, interpretation of experiments with non-wetted blade-shaped solids can be problematical. This is because the corners of the blade induce variations in the triple line level and an erratic experimental curve $f(z_b)$ in comparison to that calculated using equation

(3.18). For that reason, the use of solids of cylindrical shape is preferable in the non-wetting case (Appendix E).

Once γ_{LV} is known, θ_a and θ_r can be determined from the forces measured at points A and D (see Figure 3.18). For non-wetting liquids, θ_a and θ_r are quasi-reversible contact angles and are close to the maximum advancing and minimum receding contact angles for the solid substrate. Consequently, these values can differ greatly from the Young contact angle even for rather smooth solid surfaces (Table 3.3). Recalling that the stable equilibrium contact angle on a rough surface is the Wenzel contact angle θ_W (equation (1.35)), this angle can be calculated from θ_a and θ_r values using (Rivollet 1986):

$$\theta_W = \mathrm{Arc\,cos}\left(\frac{\cos\theta_a + \cos\theta_r}{2}\right) \tag{3.21}$$

Then, the Young contact angle is calculated combining equation (3.21) and the equation of Wenzel, yielding:

$$\theta_Y = \mathrm{Arc\,cos}\left(\frac{\cos\theta_a + \cos\theta_r}{2s_r}\right) \tag{3.22}$$

where s_r is the Wenzel parameter equal to the ratio of the actual to the planar area. For well-polished surfaces ($R_a < 100$ nm), s_r is very close to unity and equation (3.22) can be applied with a good accuracy taking $s_r = 1$. For substrates with a high roughness and for which the value of s_r is not known accurately, neither are values of θ_Y calculated by equation (3.22) accurate (De Jonghe et al. 1990).

Table 3.3. Capillary properties of molten $Sn/\alpha\text{-}Al_2O_3$ system measured by the wetting balance technique at 1370K using Al_2O_3 cylinders (De Jonghe et al. 1990). θ_Y is calculated by equation (3.22).

R_a (nm)	s_r	σ_{LV} (mJ/m^2)	θ_a (deg)	θ_r (deg)	θ_Y (deg)
5	1.0000	485	131	126	128
8.5	1.0001	465	137	116	126

Until now, the translation rate of the solid has been considered to be infinitely slow. In practice, the contact angle measured during an immersion-emersion cycle performed with a constant translation rate U, is a function of U (see Figures 2.1 and 2.12). For instance, during the *immersion* of a solid in a non-wetting liquid performed with U $\neq$ 0, a contact angle θ_i greater than θ_a was observed (Rivollet et al. 1990). If the motion of the solid is stopped between points C and D of Figure 3.17, then θ_i reaches the static contact angle θ_a. Similarly, the contact angle θ_e measured during the *emersion* of the solid is less than θ_r. It has been shown that for non-wetting systems, the difference between θ_i and θ_a, and θ_e and θ_r, is less than 5° for U $\leq$ 1 mm/min and smooth solids (Rivollet 1986, Rivollet et al. 1990). In practice, it is preferable to work with finite but low translation rates to determine wetting parameters because dynamic contact angles are more reproducible than static contact angles measured with U $=$ 0.

Examples of measurements of wetting parameters carried out by the wetting balance technique are given in Table 3.4 for three non-reactive liquid/solid couples. Liquid metals were either non-wetting or wetting and the solids were cylinders. For non-wetting liquids (Sn and Al on Al_2O_3), θ_Y was calculated from θ_a and θ_r by means of equation (3.22). For molten Pb on solid Fe, θ_Y was estimated by the irreversible advancing contact angle measured at point A (Figure 3.18). In all cases, contact angle values are in good agreement with those obtained by the sessile drop technique on solid plates. σ_{LV} values (equal to γ_{LV} values) measured by the wetting balance technique differ somewhat from values measured by other methods (see Table 4.1) but the differences remain reasonable.

Table 3.4. Comparison of θ_Y and σ_{LV} values measured by the wetting balance technique and other methods.

			Wetting balance		Sessile drop	σ_{LV} (mJ/m^2)
Liquid	Solid	T (K)	σ_{LV} (mJ/m^2)	θ_Y (deg)	θ_Y (deg)	(from Table 4.1)
Sn	α-Al_2O_3	1373	487 [a]	128 [a]	126 $\pm$ 4 [c] [d]	474
Al	α-Al_2O_3	1073	890 [a]	92 [a]	90 [e] [f]	846
Pb	Fe	1073	440 [b]	44 [b]	48 [g]	410

[a] (De Jonghe et al. 1990), [b] (Rivollet 1986), [c] (Rivollet et al. 1987), [d] (Naidich 1981), [e] (Landry and Eustathopoulos 1996), [f] (Naidich et al. 1983), [g] (Gomez-Moreno et al. 1982)

A standard electronic-zero microbalance can measure variations of masses of hundreds of milligrammes with an accuracy of 0.01%. Consequently, solid plates or cylinders of a few millimeters in perimeter are large enough to yield an excellent accuracy. However, large quantities of liquid (several tens of grammes) are needed to ensure a nearly flat liquid-vapour surface. Hence, this method is not appropriate for study of expensive liquid materials. Nevertheless, the wetting balance technique offers several experimental advantages. It allows the study of solids with non-planar surfaces, and particularly those with cylindrical shapes, such as the fibres or monofilaments used in composites. Additionally, fundamental studies on contact angle hysteresis can be performed and the effect on the contact angle of the relative velocity of the solid and the liquid can be examined. This method can also be used to follow spreading kinetics, albeit only for wetting liquids. Wetting in reactive liquid/solid systems can be studied, providing that the variation of the solid perimeter owing to interfacial reaction remains negligible in the course of an immersion-emersion cycle.

3.3. ACCURACY OF CONTACT ANGLE DATA

Almost all the data on contact angles relevant to high temperature materials have been obtained since 1950. The subsequent decades have seen great progress in the chemical and morphological characterization of surfaces and gaseous atmospheres as well as the greater availability of pure materials, and these changes have allowed contact angle measurements to be made for increasingly well defined solid/liquid/vapour systems. However, if all contact angle data in the literature measured by the sessile drop method for a given liquid/solid system at the same temperature are considered, the experimental values of θ can vary by several tens of degrees. In the Sn/Al_2O_3 system for instance, values of θ lie between 108° and 174°, and this leads to a variation of the work of adhesion of two orders of magnitude! This example is not exceptional as shown by similar dispersion observed for Ni, Fe or Cu on Al_2O_3 (Table 3.5). This scatter is due principally to the uncontrolled effects of substrate roughness and of oxygen partial pressure in the furnace rather than errors in the measurement of the contact angle, which are only a few degrees (typically ±3°).

We saw in Section 3.1.1 that an average roughness much less than 100 nm is needed if the value of θ_Y is to be measured accurately to within a few degrees. However, many of the published data have been obtained for "polished" surfaces, i.e., without any quantification of surface roughness. In practice, many of polished surfaces have roughness in the micron scale and this can lead to values of the advancing contact angle θ_a being higher than θ_Y by tens of degrees.

Generally, smooth surfaces are more easily obtained with monocrystals. The same average roughness R_a can be obtained by mechanical polishing of polycrystals but this operation often leads to an "accidental" roughness on this type

Table 3.5. Minimum and maximum values of contact angle measured for some pure metals on alumina. Results come from data compilations given in references.

Metal	Temperature (K)	$\theta_{min} - \theta_{max}$ (deg)	Reference
Ni	1823	100 – 141	(Eustathopoulos and Passerone 1978, Beruto et al. 1981)
Fe	1823	109 – 141	(Eustathopoulos and Passerone 1978, Beruto et al. 1981)
Cu	1423	114 – 170	(Li 1988, Tomsia et al. 1998)
Sn	1373	108 – 174	(Rivollet et al. 1987)

of substrate. For example, holes caused by the decohesion of grains can have a negligible influence on the average roughness but can pin the triple line and prevent spreading of the drop on the substrate. Moreover, use of polycrystals can lead to roughening at high temperature by grain boundary grooving, a process which occurs easily with metallic substrates (Figure 3.2). Solids obtained by sintering, as are most ceramics, have some porosities in the micrometre scale which also can pin the triple line, especially of non-wetting liquids (Figure 3.20). An additional advantage of monocrystals is that they minimize and even prevent the segregation of impurities to ceramic surfaces in short-time experiments which can occur through a rapid diffusion in the grain boundaries of polycrystalline materials. For all these reasons, contact angle values measured on monocrystals have to be preferred when the other experimental conditions (purity of materials, atmosphere, etc) are similar.

The second important cause of scattering of contact angle data is pollution from the atmosphere, mainly oxygen. For instance, a liquid metal on a ceramic surface in the low temperature range and/or in an atmosphere with a high P_{O2} (a "bad" vacuum or impure neutral gas) can yield very high apparent contact angles due to an oxide film covering the liquid metal. In these conditions, no real interface between the metal and the substrate exists. Typical examples are Al on Al_2O_3 or C substrates (Landry et al. 1996b). Similar effects can be produced by uncontrolled oxidation of the solid surface, for instance of metallic solids or non-oxide ceramics like SiC or Si_3N_4 (see Sections 7.1.1.1 and 7.1.2.2).

At high temperatures or in an atmosphere with a low P_{O2}, oxide films are often eliminated by dissolution in the metal, by thermodynamic decomposition or by formation of volatile sub-oxides. Even in these conditions, oxygen can still affect wettability, for instance by adsorption at metal/oxide interfaces. This effect leads to contact angles lower than the nominal contact angle of the inert system. Some examples of these oxygen effects are given in Section 6.4. Therefore, when contact

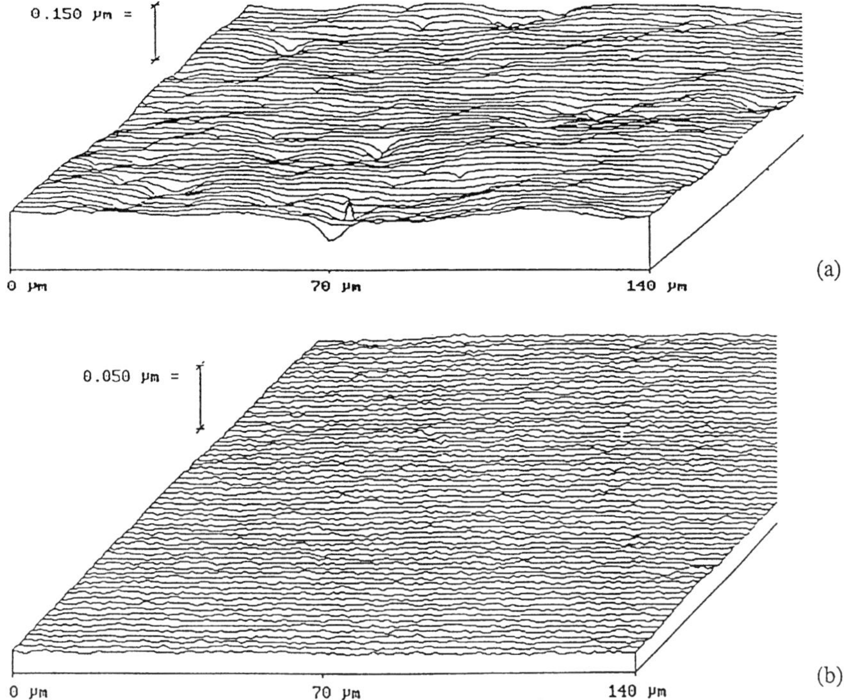

Figure 3.20. (a) Surface profile of sintered TiC after polishing. Pores opening on the surface can be seen and are responsible for the relatively high roughness of this substrate (average roughness $R_a = 10$ nm, maximum roughness $R_t = 200$ nm). For comparison, the surface profile of monocrystalline SiC after polishing is shown in (b) ($R_a = 2$ nm, $R_t = 10$ nm). From (Landry 1995).

angle values of the literature are used in a particular application, one has to verify that the working atmosphere is similar to that used for the determination of the contact angle.

To estimate the best accuracy that can be expected for contact angle values, results obtained for the non-reactive Au/monocrystalline Al_2O_3 system either by the same team for different runs or by different laboratories are assembled, as shown in Table 3.6. This system is particularly simple as the surface properties of neither gold nor alumina seem to depend on values of oxygen partial pressure (Gallois 1980, Chatain et al. 1993). Moreover, the use of monocrystalline substrates allows reduction of surface roughness to the nanometre scale. The data in Table 3.6 show that, even for this very simple system, the accuracy of contact angle values is hardly better than $\pm 5°$, leading to an uncertainty in work of adhesion values of ±17 %. As substrates with different orientations were used, part of this scatter may be due to crystallographic effects.

Table 3.6. Contact angle values in the Au/monocrystalline Al_2O_3 system at 1373K.

Atmosphere	Substrate orientation	θ (deg)	Reference
high vacuum, He, He-O_2	random	131 ± 2	(Chatain et al. 1993)
high vacuum	$(11\bar{2}3)$	141	(Brennan and Pask 1968)
high vacuum, Ar, Ar-H_2	$(11\bar{2}3)$	140 ± 3	(Gallois 1980)
high vacuum	$(0001), (11\bar{2}3)$	138	(Naidich 1981)
He	random	139 ± 2	(Rivollet et al. 1987)
air	not specified	136 ± 6 [*]	(Grigorenko et al. 1995)

[*] at 1473K

If gold is replaced by another metal with a greater affinity for oxygen and if rougher substrates are used, the scatter of contact angle values in non-wetting systems can be higher, as observed in Sn/Al_2O_3 and Cu/Al_2O_3 couples (Table 3.5). For wetting systems, the effect of roughness on the apparent contact angle is weaker as shown by the data given in Table 3.7 for a Si-rich Co-Si alloy on different types of SiC (Gasse 1996). Varying the average roughness by more than two orders of magnitude caused the contact angle to vary by less than 10 degrees, corresponding to a scattering of work of adhesion values of only $\pm4\%$. This lower roughness sensitivity for a low contact angle system explains the good agreement observed for values of θ for pure molten Si on the basal plane of α-SiC (Table 3.8). This agreement is obtained despite the use of different polarity faces of SiC and different furnace atmospheres for a system in which both the liquid (Si) and the solid (SiC) have a high affinity for oxygen. In fact, owing to the high melting temperature of Si, both the Si and SiC surfaces are easily deoxidised by formation of volatile SiO, even in commercialy pure argon containing several ppm of oxygen and water. At the molten Si/vapour surface as well as at the solid/liquid/vapour triple line, P_{O2} is fixed locally at very low values by the Si itself rather than by the impinging oxygen molecules desorbed by the furnace walls (Drevet et al. 1993).

Table 3.7. Contact angle values in the non-reactive Co-66.6 at.% Si/SiC system at 1623K in a high vacuum (Gasse 1996).

SiC	R_a (nm)	θ (deg)
sintered	50	30
sintered	900	20
monocrystal	2	29
monocrystal	60	22

Table 3.8. Contact angle values in the Si/monocrystalline α-SiC system.

Atmosphere	Temperature (K)	θ (deg)	Reference
vacuum	1753	36	(Naidich and Nevodnik 1969)
vacuum	1713	30 – 45	(Yupko and Gnesin 1973)
Ar	1703	38	(Li and Hausner 1991)
Ar	1770	38	(Nikolopoulos et al. 1992)
He	1690	34 ± 2	(Drevet et al. 1993)
vacuum	1697	39 (Si face of SiC)	(Rado 1997)
		47 (C face of SiC)	

Contact angles measured by different teams for the reactive Cu-1 at.% Cr/carbon system are given in Table 3.9. In this system, Cr reacts with carbon to form a layer of wettable chromium carbide. In sessile drop experiments at about 1400 K, a quasi-stationary contact angle θ_F is reached after about 100 seconds (see Figure 2.26). It is noteworthy that the reproducibility of θ_F values measured by the same team for eight different runs ($41 \pm 4°$ (Landry et al. 1997)) is comparable to that found for θ values of non-reactive systems. Moreover, the agreement between results obtained at 1400 K by different teams is satisfactory. The lower contact angles obtained by Nogi et al. (1990) can be explained by the much higher temperature used by these authors.

At present, when working with high-purity materials, smooth solid surfaces and low P_{O2} atmospheres, the thermodynamic contact angle in a particular system can be determined at best within about five degrees. Roughness must be very low, particularly in non-wetting systems in order to obtain such an accuracy. The control of P_{O2} is critical for oxidisable liquids and solids, specially at relatively low temperatures, and dynamic vacuum is often preferable to a static neutral gas atmosphere.

Table 3.9. Stationary contact angle values in the reactive Cu-1 at.% Cr/carbon system.

Substrate	Atmosphere	Temperature (K)	θ_F (deg)	Reference
vitreous carbon	high vacuum	1423	40	(Mortimer and Nicholas 1973)
graphite	Ar	1403	45	(Devincent and Michal 1993)
vitreous carbon	high vacuum	1373	41 ± 4	(Landry et al. 1997)
graphite	Ar-5% H_2	1773	20–30	(Nogi et al. 1990)

3.4. CONCLUDING REMARKS

The variants of the sessile drop method remain the simplest and most employed techniques for wetting studies despite the great amount of information that can be derived from the wetting balance technique. When only contact angle data are needed, small droplets weighing typically 10^{-2}–10^{-1} g are preferable because they allow deformation of the drop by gravity to be neglected. For the measurement of the liquid/vapour surface energy, larger droplets usually weighing 1g or more are necessary to achieve a good accuracy. In all cases, an automatic system for data acquisition and image analysis is desirable.

When the optimal conditions are satisfied, i.e., a well-controlled atmosphere, a pure, homogeneous and smooth solid surface and an accurate measurement system, the Young contact angle and the liquid surface energy can be derived with an accuracy of ± 3 deg and $\pm 2\%$ respectively allowing the work of adhesion ($W_a = \sigma_{LV}(1 + \cos\theta)$) to be determined with an accuracy of about 10%.

REFERENCES FOR CHAPTER 3

Bashforth, F. and Adams, J. C. (1883) *An Attempt to Test Theory of Capillary Action*, Cambridge University Press

Beruto, D., Barco, L. and Passerone, A. (1981) *Oxides and Oxide Films*, ed. A. K. Vijh, vol. 1, Dekker, New York

Brennan, J. J. and Pask, J. A (1968) *J. Am. Ceram. Soc.*, **51**, 569

Chappuis, J. (1982) in *Multiphase Science and Technology*, vol. 1, ed. G. F. Hewitt, J. M. Delhaye and N. Zuber, Hemisphere Publishing Corporation, p. 387

Chatain, D., Chabert, F., Ghetta, V. and Fouletier, J. (1993) *J. Am. Ceram. Soc.*, **76**, 1568

Coudurier, L., Adorian, J., Pique, D. and Eustathopoulos, N. (1984) *Rev. Int. Hautes Tempér. Réfract.*, Fr., **21**, 81

De Jonghe, V., Chatain, D., Rivollet, I. and Eustathopoulos, N. (1990) *J. Chim. Phys. F.*, **87**, 1623

Devincent, S. M. and Michal, G. M. (1993) *Met. Trans.*, **24A**, 53

Dorsey, N. E. (1928) *J. Washington Acad. Sci.*, **18**, 505

Drevet, B., Kalogeropoulou, S. and Eustathopoulos, N. (1993) *Acta Metall. Mater.*, **41**, 3119

Eustathopoulos, N. and Passerone, A. (1978) in *Proc. 7emes Journées Int. Physicochimie et Sidérurgie*, Versailles, France, p. 61

Feduska, W. (1959) *Weld. J.*, **38**, 122s

Gallois, B. (1980) Ph.D. Thesis, Carnegie Mellon Univ., Pittsburg, USA

Gallois, B. and Lupis, C. H. P. (1981) *Metall. Trans. B*, **12**, 549

Gangopadhyay, U. and Wynblatt, P. (1994) *Metall. Mater. Trans. A*, **25**, 607

Gasse, A. (1996) Ph.D. Thesis, INP Grenoble, France

Gomez-Moreno, O., Coudurier, L. and Eustathopoulos, N. (1982) *Acta Metall.*, **30**, 831

Grigorenko, N. F., Stegny, A. I., Kasich-Pilipenko, I. E., Naidich, Y. V. and Pasichny, V. V. (1995) in *Proc. Int. Conf. High Temperature Capillarity*, Smolenice Castle, May 1994, ed. N. Eustathopoulos, Reprint Bratislava, p. 123

Hara, S., Nogi, K. and Ogino, K. (1995) in *Proc. Int. Conf. High Temperature Capillarity*, Smolenice Castle, May 1994, ed. N. Eustathopoulos, Reprint Bratislava, p. 43

Hcyraud, J. C. and Metois, J. J. (1980) *Acta Met.*, **28**, 1789

Hitchcock, S. J., Carroll, N. T. and Nicholas, M. G. (1981) *J. Mater. Sci.*, **16**, 714

Hondros, E. D. (1978) in *Precipitation Processes in Solids*, Warrendale, PA, The Metallurgical Society of AIME, p. 1

Iida, T. and Guthrie, R. I. L. (1988) *The Physical Properties of Liquid Metals*, Clarendon Press, Oxford

Keene, B. J. (1993) *International Materials Reviews*, **38**, 157

Labrousse, L. (1998) Internal Report, LTPCM, INP Grenoble, France

Landry, K. (1995) Ph.D. Thesis, INP Grenoble, France

Landry, K. and Eustathopoulos, N. (1996) *Acta Mater.*, **44**, 39

Landry, K., Kalogeropoulou, S., Eustathopoulos, N., Naidich, Y. and Krasovsky, V. (1996b) *Scripta Mater.*, **34**, 841

Landry, K., Rado, C., Voitovich, R. and Eustathopoulos, N. (1997) *Acta Mater.*, **45**, 3079

Li, J. G. (1988) Ph.D. Thesis, INP Grenoble, France

Li, J. G. and Hausner, H. (1991) *J. Mater. Sci. Lett.*, **10**, 1275

Liggieri, L. and Passerone, A. (1989) *High Temp. Techn.*, **7**, 80

Lorrain, V. (1996) Ph.D. Thesis, INP Grenoble, France

Maze, C. and Burnet, G. (1969) *Surf. Sci.*, **13**, 451

Mortimer, D. A. and Nicholas, M. G. (1973) *J. Mater. Sci.*, **8**, 640

Mozzo, G. (1970) *Bull. Soc. Chim.*, Numéro Spécial, Colloque Mulhouse, p. 3219

Murr, L. E. (1973) *Mater. Sci. Eng.*, **12**, 277

Naidich, Y. V., Eremenko, V. N., Fesenko, V. V., Vasiliu, M. I. and Kirichenko, L. F. (1963) in *The Role of Surface Phenomena in Metallurgy*, ed. V. N. Eremenko, Consultants Bureau, New York, p. 41

Naidich, Y. V. and Nevodnik, G. M. (1969) *Inorganic Materials*, **5**, 1759

Naidich, Y. V. and Zhuralev, V. S. (1971) *Industrial Laboratory*, **37**, 452

Naidich, Y. V. (1981) in *Progress in Surface and Membrane Science*, vol. 14, ed. by D. A. Cadenhead and J. F. Danielli, Academic Press, New York, p. 353

Naidich, Y. V., Chubashov, Y. N., Ishchuk, N. F. and Krasovskii, V. P. (1983) *Poroshkovaya Metallurgiya*, **246**, 67 (English translation : Plenum Publishing Corporation p. 481)

Naidich, J. V. and Chuvashov, J. V. (1983b) *J. Mater. Sci.*, **18**, 2071

Naidich, Y. V. and Grigorenko, N. F. (1992) *J. Mater. Sci.*, **27**, 3092

Naidich, Y. V., Voitovich, R. P. and Zabuga, V. V. (1995) *J. Colloid Interface Sci.*, **174**, 104

Nakae, H., Yamaura, H., Sinohara, T., Yamamoto, K. and Oosawa, Y. (1989) *Mat. Trans. JIM*, **30**, 423

Nicholas, M. G. and Poole, D. M. (1967) *J. Mater. Sci.*, **2**, 269

Nicholas, M. G., Crispin, R. M. and Ford, D. A. (1984) *Brit. Ceram. Proc.*, **34**, 163

Nikolopoulos, P., Agathopoulos, S., Angelopoulos, G. N., Naoumidis, A. and Grubmeier, H. (1992) *J. Mater. Sci.*, **27**, 139

Nogi, K., Osugi, Y. and Ogino, K. (1990) *ISIJ International*, **30**, 64

Orr, F. M., Scriven, L. E. and Chu, T. Y. (1977) *J. Colloid Interface Science*, **60**, 402

Passerone, A. and Ricci, E. (1998) in *Drops and Bubbles in Interfacial Research*, ed. R. Miller, Elsevier, Amsterdam, p. 475

Pilliar, R. M. and Nutting, J. (1967) *Phil. Mag.*, **16**, 181

Rado, C. (1997) Ph.D. Thesis, INP Grenoble, France

Rado, C., Senillou, C. and Eustathopoulos, N. (1998) in *Proc. 2nd Int. Conf. on High Temperature Capillarity*, Cracow (Poland), 29 June-2 July 1997, ed. N. Eustathopoulos and N. Sobczak, published by Foundry Research Institute (Cracow), p. 53

Ricci, E. and Passerone, A., ICFAM-CNR, Genova, Italy, private communication

Rivollet, I. (1986) Ph.D. Thesis, INP Grenoble, France

Rivollet, I., Chatain, D. and Eustathopoulos, N. (1987) *Acta Metall.*, **35**, 835

Rivollet, I., Chatain, D. and Eustathopoulos, N. (1990) *J. Mater. Sci.*, **25**, 3179

Rotenberg, Y., Boruvka, L. and Neumann, A. W. (1983) *J. Colloid Interf. Sci.*, **93**, 169

Sangiorgi, R., Caracciolo, G. and Passerone, A. (1982) *J. Mater. Sci.*, **17**, 2895

Soper, A., Gilles, B. and Eustathopoulos, N. (1996) *Materials Science Forum*, Transtec Publications, Switzerland, **207-209**, 433

Taimatsu, H. and Sangiorgi, R. (1992) *Surface Science*, **261**, 375

Tomsia, A. P., Saiz, E., Foppiano, S. and Cannon, R. M. (1998) in *Proc. 2nd Int. Conf. on High Temperature Capillarity*, Cracow (Poland), 29 June-2 July 1997, ed. N. Eustathopoulos and N. Sobczak, published by Foundry Research Institute (Cracow), p. 59

Voitovich, R., Mortensen, A. and Eustathopoulos, N. (1998) in *Proc. 2nd Int. Conf. on High Temperature Capillarity*, Cracow (Poland), 29 June-2 July 1997, ed. N. Eustathopoulos and N. Sobczak, published by Foundry Research Institute (Cracow), p. 81

Winterbottom, W. L. (1967) *Acta Metall.*, **15**, 303

Worthington, A. M. (1885) *Phil. Mag.*, 5th series, **20**, 51

Yupko, V. L. and Gnesin, G. G. (1973) *Poroshkovaya Metallurgiya*, **10**, 97

Chapter 4
Surface energies

As shown by the Young-Dupré equation (1.45), the equilibrium contact angle value results from the competition between adhesion forces between the liquid and the solid, as measured by the work of adhesion W_a, and cohesion forces in the liquid, expressed by its surface energy, σ_{LV}. Thus knowledge of this last quantity and of its variation with temperature, composition of the liquid and environmental parameters (mainly P_{O2}), is necessary to explain wetting behaviour. For this reason, the bulk of data given below are for σ_{LV} values. However, a limited amount of data will be presented also for two other quantities included implicitly in the Young-Dupré equation, i.e., σ_{SV} and σ_{SL}.

4.1. DATA FOR METALS AND ALLOYS

Reliable values of σ_{LV} of pure metals for most commonly used metals have been obtained during the last forty years, using mainly:

1) the "maximum bubble pressure" technique in which the pressure necessary to push a gas bubble through a capillary into the liquid under investigation is measured,

2) the sessile drop technique described in Section 3.1 and

3) the "drop weight" technique in which a drop is formed at the end of a rod of the metal heated to its melting point. The σ_{LV} value is deduced from the weight of the drop after its detachment.

Only the two first methods allow measurement of the temperature coefficient of the surface energy. The maximum bubble pressure technique is well-adapted for metals with low and intermediate melting points and specially for oxidizable metals, while the sessile drop technique has been applied with success to measure σ_{LV} values up to 1500°C. The drop weight method is particularly useful for very high melting-point metals because it avoids liquid contact with container materials. This is also true for the recently developed "levitation drop" technique that analyses the oscillation spectrum of a magnetically levitated droplet.

4.1.1 Pure liquid metals
Selected values of σ_{LV} and the temperature coefficient σ'_{LV} ($= d\sigma_{LV}/dT$) are given in Table 4.1, taken from the reference compilation (Eustathopoulos et al. 1999) in

which the selection criteria are described in detail. The metals have been divided into two classes A and B. For a metal of *class A*, the accuracy of σ_{LV} at the melting temperature T_F, estimated from the experimental measurements of the literature reviewed by Keene (1993), is better than $\pm 5\%$ and the accuracy of σ'_{LV} is better than $\pm 50\%$. For a metal of *class B*, the accuracies are worse than $\pm 5\%$ and $\pm 50\%$.

In the case of Fe and Si, although σ_{LV} values at the melting point measured by several authors are not very different, a high difference in σ'_{LV} values is observed. For this reason two values of σ'_{LV} are given in Table 4.1 characteristic of "low" and "high" values quoted in the literature.

Table 4.1. Experimental values of σ_{LV} vs temperature, T, taken from the compilation of Eustathopoulos et al. (1999) and values of σ'_{LV} calculated according to (Eustathopoulos et al. 1998) for Fe and Si, for class B metals except Te and for class A metals for which experimental values have not been measured $(\sigma_{LV}(T) = \sigma_{LV}(T_F) + \sigma'_{LV}[T - T_F])$ where T_F is the melting temperature).

Element	T_F (K)	Class	Experimental data $\sigma_{LV}(T)$ (mJ.m^{-2})	Calculated data σ'_{LV} (mJ.m^{-2}.K^{-1})
Ag	1234	A	910 - 0.17 (T - T_F)	
Al	933	A	867 - 0.15 (T - T_F)	
Au	1336	A	1138 - 0.19 (T - T_F)	
Ba	1002	B	267 - 0.07 (T - T_F)	-0.06
Be	1560	B	1320	-0.25
Bi	544	A	374 - 0.08 (T - T_F)	
Ca	1112	A	366 - 0.11 (T - T_F)	
Cd	594	B	642 - 0.20 (T - T_F)	-0.16
Ce	1071	B	794 - 0.07 (T - T_F)	-0.09
Co	1768	A	1884 - 0.37 (T - T_F)	
Cr	2130	B	1628 - 0.20 (T - T_F)	-0.33
Cs	302	A	69 - 0.06 (T - T_F)	
Cu	1357	A	1355 - 0.19 (T - T_F)	
Dy	1682	B	648 - 0.13 (T - T_F)	-0.15
Er	1795	B	637 - 0.12 (T - T_F)	-0.16
Eu	1090	B	264 - 0.05 (T - T_F)	-0.16
Fe	1809	A	1855 - 0.23 (T - T_F)	-0.29
			1844 - 0.5 (T - T_F)	
Ga	303	A	708 - 0.07 (T - T_F)	
Gd	1585	B	664 - 0.06 (T - T_F)	-0.13
Ge	1210	A	587 - 0.105 (T - T_F)	
Hf	2500	A	1490	-0.19
Hg	234	A	500 - 0.215 (T - T_F)	

Ho	1743	B	$650 - 0.12 (T - T_F)$	-0.15
In	430	A	$555 - 0.12 (T - T_F)$	
Ir	2716	A	2140	-0.18
K	336	A	$112 - 0.08 (T - T_F)$	
La	1193	B	$737 - 0.11 (T - T_F)$	-0.09
Li	454	B	$404 - 0.16 (T - T_F)$	-0.14
Lu	1936	B	$940 - 0.07 (T - T_F)$	-0.21
Mg	922	B	$583 - 0.15 (T - T_F)$	-0.16
Mn	1517	B	$1219 - 0.35 (T - T_F)$	-0.23
Mo	2890	B	1915	-0.23
Na	371	A	$192 - 0.095 (T - T_F)$	
Nb	2740	A	1840	-0.19
Nd	1289	B	$689 - 0.09 (T - T_F)$	-0.11
Ni	1726	A	$1838 - 0.42 (T - T_F)$	
Os	3300	B	2500	-0.29
Pb	601	A	$462 - 0.11 (T - T_F)$	
Pd	1825	B	$1475 - 0.28 (T - T_F)$	-0.24
Pr	1204	B	$743 - 0.09 (T - T_F)$	-0.14
Pt	2042	A	1707	-0.29
Pu	913	B	550	-0.13
Rb	313	B	$85 - 0.07 (T - T_F)$	-0.05
Re	3453	A	2520	-0.21
Rh	2233	B	2000	-0.24
Ru	2523	B	2180	-0.28
Sb	904	B	$366 - 0.04 (T - T_F)$	-0.10
Sc	1812	B	$939 - 0.12 (T - T_F)$	-0.21
Se	494	B	$103 - 0.09 (T - T_F)$	
Si	1685	B	$827 - 0.48 (T - T_F)$	-0.19
			$749 - 0.15 (T - T_F)$	
Sm	1345	B	$430 - 0.07 (T - T_F)$	-0.10
Sn	505	A	$570 - 0.11 (T - T_F)$	
Sr	1041	A	$289 - 0.075 (T - T_F)$	
Ta	3287	A	2010	-0.24
Tb	1630	B	$669 - 0.06 (T - T_F)$	-0.16
Te	723	B	178	
Th	2028	B	978	-0.14
Ti	1943	B	1525	-0.28
Tl	577	A	$461 - 0.09 (T - T_F)$	
U	1405	B	$1550 - 0.14 (T - T_F)$	-0.16
V	2175	A	1855	-0.26
W	3653	A	2310	-0.20
Y	1799	B	$872 - 0.09 (T - T_F)$	-0.11
Yb	1097	B	$320 - 0.10 (T - T_F)$	-0.17
Zn	693	B	$815 - 0.25 (T - T_F)$	-0.22
Zr	2125	A	1435	-0.19

Equation (1.15), when applied to monoatomic liquids, predicts that the molar surface energy $\sigma_{LV}.\Omega_m$ (with $\Omega_m = N_a.\omega$) is proportional to the heat of evaporation of the liquid, L_e, and to a structural parameter m_1. If m_1 is the same for all metals, σ_{LV} is expected to scale with the quantity L_e / Ω_m. It was shown in (Eustathopoulos et al. 1998) that equation (1.15), valid in principle only at 0K, holds also at T_F. As Ω_m is proportional to $v_m^{2/3}$ (v_m denoting the molar volume), one obtains:

$$\sigma_{LV}(T_F) = C_1 \frac{L_e}{v_m^{2/3}} \tag{4.1}$$

where C_1 is a dimensionless constant equal to 0.25×10^{-8}.

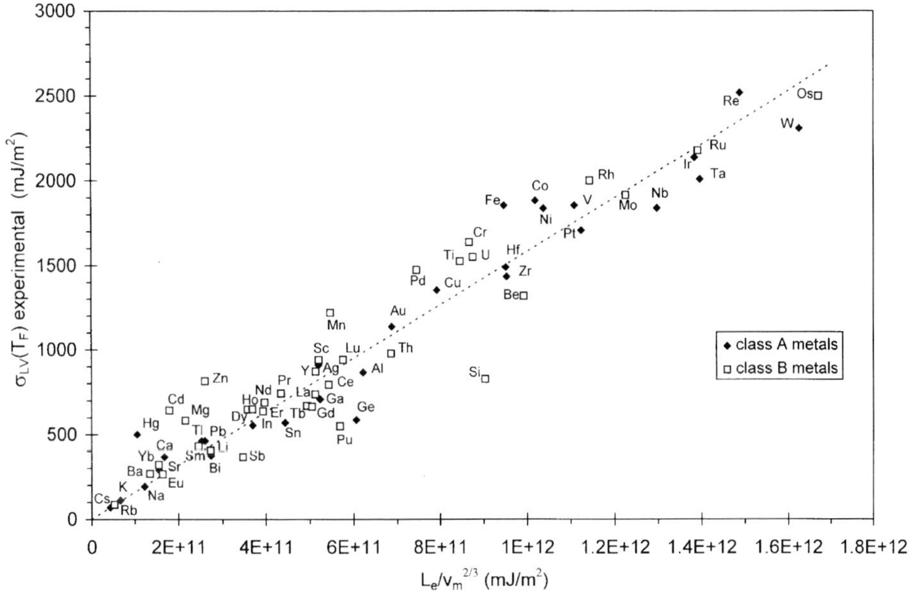

Figure 4.1. Experimental values of surface energy of pure liquid metals at their melting temperature plotted as a function of $L_e / v_m^{2/3}$. From (Eustathopoulos et al. 1998) [12].

Figure 4.1 shows that the correlation predicted by equation (4.1) is satisfied rather well. The slope of the straight line is 0.16×10^{-8} in good agreement with the value of C_1, considering the crudeness of the structural and energetical model of

Skapski (1948, 1956) used as a basis in (Eustathopoulos et al. 1998) to derive equation (4.1). Note that because the ratio L_e/T_F for pure metals is approximately constant, correlations between $\sigma_{LV}(T_F)$ and $T_F/v_m^{2/3}$ are also to be expected and have been found to hold (Keene 1993).

The temperature coefficient of surface energy, σ'_{LV}, is given by (Eustathopoulos et al. 1998):

$$\sigma'_{LV}(T) = -\frac{C_2 S_S}{v_m^{2/3}} + \frac{2}{3}\frac{\sigma_{LV}(T_F)}{\rho}\frac{d\rho}{dT} \qquad (4.2)$$

where C_2 is a constant and ρ is the liquid density. S_S, a quantity found to be nearly constant for all metals, is the molar surface excess entropy and results from an atomic disorder at the surface higher than in the bulk liquid ($S_S > 0$). Thus, the first term of equation (4.2) is the entropy contribution of σ'_{LV} while the second term reflects the fact that the number of atoms per unit area contributing to the surface energy decreases with temperature. The contribution of the second term varies from 30% of the total for low σ_{LV} metals to 55% for high σ_{LV} metals. A comparison of experimental σ'_{LV} data for class A metals and calculated values from equation (4.2) with $C_2 S_S = 5.7 \times 10^{-5}$ mJ.mole^{-1}.K^{-1} is shown on Figure 4.2. In view of the good agreement, equation (4.2) was used to calculate σ'_{LV} values for class B metals and for metals for which σ'_{LV} has not been measured up to now (Table 4.1).

The surface energies of pure metals can be decreased by segregation of oxygen. This impurity can come from within the bulk of the liquid, from the gaseous environment or from the solid substrate used in wetting experiments. As will be discussed in more detail in Section 6.4.1, oxygen dissolved in bulk metal develops strong interactions accompanied by charge transfers from the metal to oxygen. Because these interactions create strong perturbations of the metallic bond adjacent to an oxygen atom, there is a marked tendency for oxygen to segregate at the free surface where the perturbation is partially relaxed.

Figure 4.3 shows the change of σ_{LV} of Cu as a function of the partial pressure of oxygen, P_{O2}, set by using gas mixtures of Ar/H$_2$ or CO/CO$_2$. The σ_{LV} values remain constant for a range of P_{O2} from 10^{-20} to 10^{-12} atm at a value of 1320 mJ.m^{-2}, the surface energy of pure Cu. Above $P_{O2} \cong 10^{-12}$ atm, σ_{LV} decreases smoothly and at $P_{O2} \approx 10^{-9}$ atm the slope reaches a near constant value corresponding to a surface saturated with oxygen. It is worthwhile noting that the saturation level is attained for a bulk molar fraction of oxygen as low as 1.5×10^{-4} i.e., about 40 ppm. At higher P_{O2} values, σ_{LV} continues to decrease, even though

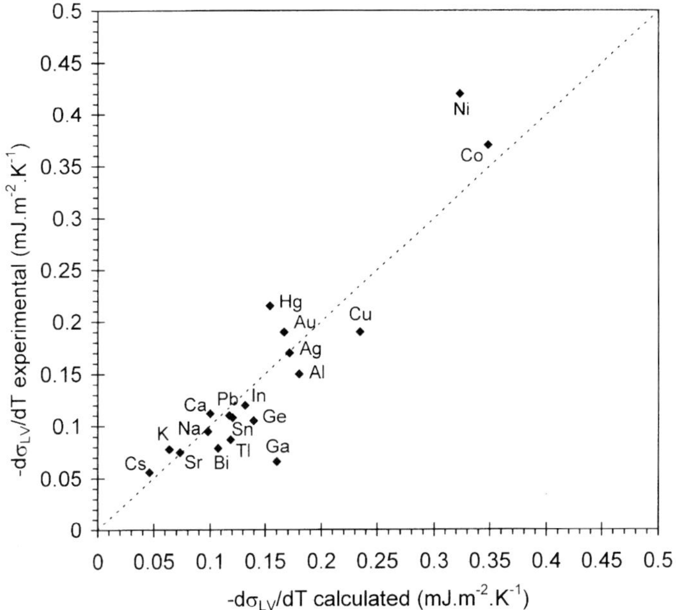

Figure 4.2. Comparison between experimental and calculated values of the temperature coefficient of surface energy using equation (4.2) for class A liquid metals. From (Eustathopoulos et al. 1998) [12].

the surface composition remains nearly constant, because of the increase of the chemical potential of oxygen. No measurements were performed at P_{O2} higher than 10^{-5} atm to avoid precipitation of Cu oxide.

An analysis of this type of σ_{LV}-P_{O2} curve using the Gibbs adsorption equation is given in Section 6.4.1 for Ag which exhibits a similar behaviour and further discussion of the surface energy variations of metal-oxygen systems is given in (Joud and Passerone 1995). Experimental data for the effect of oxygen on σ_{LV} exist for Ag (Sangiorgi et al. 1982), Al (Goumiri and Joud 1982), Bi (Hasouna et al. 1991), Co (Ogino et al. 1982), Cr (Chung et al. 1992), Fe (Nakashima et al. 1992), Hg (Nicholas et al. 1961), Ni (Sharan and Cramb 1997), Pb (Passerone et al. 1983), Si (Niu et al. 1998) and Sn (Taimatsu and Sangiorgi 1992).

Figure 4.4 shows that O and other elements developing strong interactions with Fe such as S, Se and Te decrease its surface energy markedly (Kozakevitch 1968, Ogino et al. 1983). The effect of N is much smaller while that of dissolved C and P is very weak or even negligible (Kozakevitch 1968).

Kozakevitch and Urbain (1961) discussed the nature of the surface layer saturated with adsorbed oxygen (or another non-metallic element) by comparing

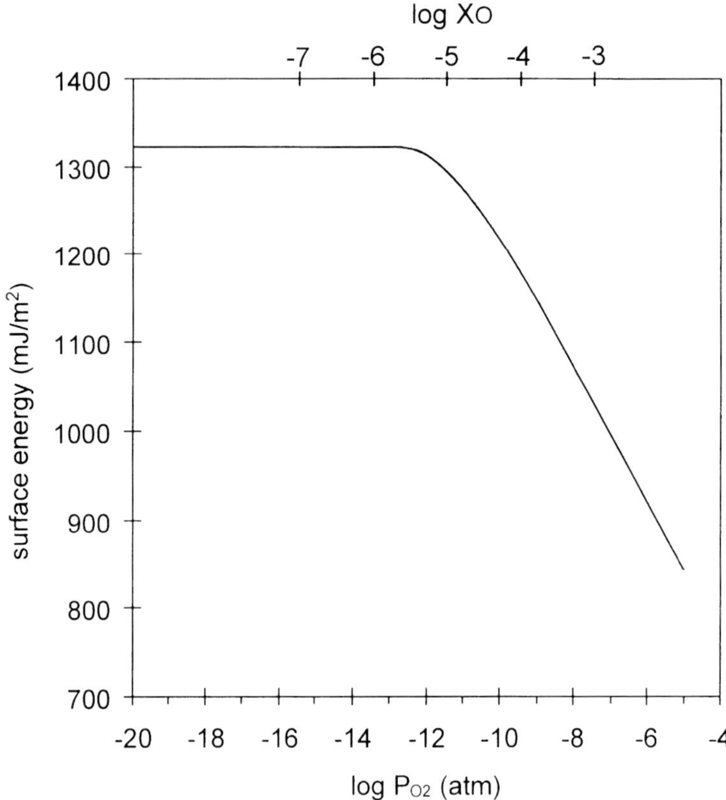

Figure 4.3. Experimental values of surface energy of liquid Cu-O mixtures as a function of partial pressure of oxygen P_{O2} or molar fraction of dissolved oxygen X_O at 1108°C. Data from work reported in (Gallois and Lupis 1981) [13].

the area of the adsorbed species, as calculated from the adsorption isotherm (equation (6.16)), with that of neutral atoms or with the area of oxygen ions at an appropriate ionic compound surface. For instance, in the case of oxygen adsorbed at the liquid Fe surface, the area of ≈ 9 Å^2 is much closer to that occupied by oxygen ions in the plane (111) of FeO (8.2 Å^2) than to the 1.2 Å^2 of neutral oxygen atoms in a close-packed structure. Kozakevitch and Urbain concluded that the adsorption layer was an ionic double-layer, the vapour side of the surface being occupied by anions and the liquid side by positively charged metal atoms. The formation of such a layer was assumed to produce a substantial decrease of the surface energy.

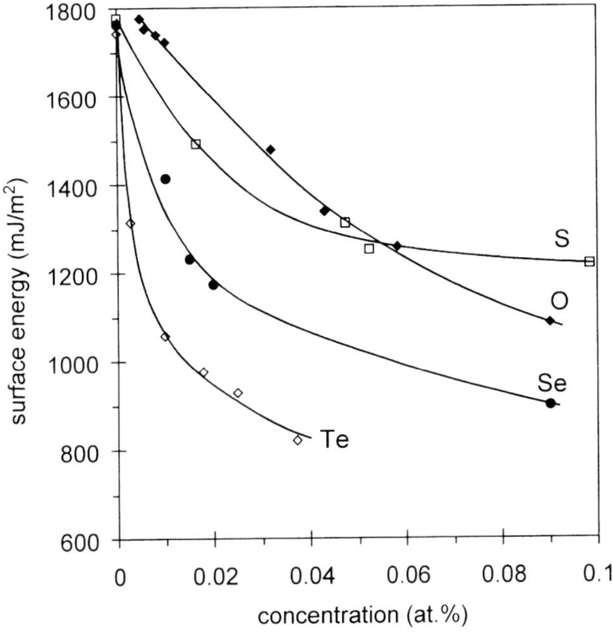

Figure 4.4. Surface energies of solutions of O, S, Se and Te in liquid Fe at 1600C (Ogino et al. 1983) [14].

4.1.2 Liquid alloys

The surface energy of a metal can be affected also by the adsorption of a metallic solute. The surface energy of a binary A-B alloy has been calculated by Guggenheim (1945) from a regular solution model using mean-field approximation statistics (in which the atoms A and B are supposed to be randomly distributed on the sites of the "quasi-crystalline liquid" lattice) and a single parameter λ (*the exchange energy*) which accounts for the change of energy of the system when heteroatomic A-B pairs are formed from homoatomic A-A and B-B pairs:

$$\lambda = ZN_a\left[\varepsilon_{AB} - \frac{\varepsilon_{AA} + \varepsilon_{BB}}{2}\right] \tag{4.3}$$

In (4.3), Z is the mean coordination number in the liquid, N_a is the Avogadro's number and ε_{ij} is the interaction energy of an i-j pair. Recalling that bond energies ε_{ij} have negative values, a high positive value of λ is due to A-B interactions being less strong than the average value of homoatomic A-A and B-B interactions. In the regular solution model, the enthalpy of mixing of A-B alloy, ΔH_m, is simply given by

$$\Delta H_m = \lambda X_B(1 - X_B) \tag{4.4}$$

where X_B is the molar fraction of B in the A-B alloy.

The interface between the liquid and the vapour phase is assumed to be a liquid monolayer in which the molar fraction of B is Y_B (see also Figure 6.27). The use of the monolayer approximation in describing equilibrium adsorption in binary liquids is satisfactory only when the temperature is not close to the critical temperature of the liquid mixture i.e., the temperature below which the A-B solution consists of a mixture of two solutions (Defay et al. 1966, Eustathopoulos and Joud 1980).

Using the symbols introduced in Section 1.1, an expression for σ_{LV} can be derived from the regular solution model (Guggenheim 1945):

$$\begin{aligned}
\sigma_{LV} &= \sigma_{LV}^A + \frac{RT}{\Omega_m}\ln\frac{1 - Y_B}{1 - X_B} + \frac{m_2\lambda}{\Omega_m}(Y_B^2 - X_B^2) - \frac{m_1\lambda}{\Omega_m}X_B^2 \\
&= \sigma_{LV}^B + \frac{RT}{\Omega_m}\ln\frac{Y_B}{X_B} + \frac{m_2\lambda}{\Omega_m}[(1 - Y_B)^2 - (1 - X_B)^2] - \frac{m_1\lambda}{\Omega_m}(1 - X_B)^2
\end{aligned} \tag{4.5}$$

where m_1 and m_2 are structural parameters defined in Fig. 1.3 and the molar area Ω_m is assumed to be the same for A and B. Solving this equation allows calculation of Y_B and σ_{LV} as a function of X_B:

$$\frac{Y_B}{1 - Y_B} = \frac{X_B}{1 - X_B}\exp\left[-\frac{1}{RT}\{(\sigma_{LV}^B - \sigma_{LV}^A)\Omega_m - m_1\lambda + 2\lambda((m_1 + m_2)X_B - m_2 Y_B)\}\right] \tag{4.6}$$

For a dilute solution, when $X_B \to 0$ then $Y_B \to 0$, and equation (4.6) reduces to:

$$\left(\frac{Y_B}{X_B}\right)_{X_B \to 0} = \exp\left(-\frac{E_{B(A)}^{\infty,LV}}{RT}\right) \tag{4.7}$$

where the adsorption energy of B at infinite dilution, $E_{B(A)}^{\infty,LV}$, is given by:

$$E_{B(A)}^{\infty,LV} = (\sigma_{LV}^B - \sigma_{LV}^A)\Omega_m - m_1\lambda \tag{4.8}$$

A negative value of adsorption energy leads to a high value of the "enrichment factor" Y_B/X_B and to a sharp decrease of σ_{LV} with the first additions of B in A. Equation (4.8) shows that factors favoring adsorption of B are (i) a low value of σ_{LV}^B compared to σ_{LV}^A and (ii) weak A-B interactions i.e., a tendency of A-B alloy to form a miscibility gap ($\lambda \gg 0$).

Figure 4.5 shows experimental and calculated curves of σ_{LV} and Y_{Cu} vs X_{Cu} for Fe-Cu alloys. The "experimental" Y_{Cu} vs X_{Cu} curve was deduced from the experimental σ_{LV}-X_{Cu} curve applying the Gibbs adsorption equation (6.16) (Nogi et al. 1991). The thermodynamic properties of mixing of Fe-Cu alloys can be described satisfactorily by the regular solution model. Moreover, the molar volumes of Fe and Cu, from which the molar area is calculated, are close, as assumed when deriving equation (4.5). The experimental curve of Figure 4.5.a shows a marked decrease of σ_{LV} with the first additions of low-surface-energy Cu, but the additions of Fe in Cu have a comparatively little effect on σ_{LV}^{Cu}. This behaviour, which is a common feature, is discussed in Section 6.5.1 (see also Figure 6.28). The σ_{LV} curve calculated supposing the Fe-Cu alloys to be ideal solutions i.e., taking $\lambda = 0$, reproduces this behaviour very imperfectly, but taking into account enthalpy of mixing effects (the parameter λ for this system is positive) leads to a better agreement with experimental results. Both calculated and experimental values of surface composition (Figure 4.5.b) indicate marked enrichment of the surface in Cu, with small Cu additions to Fe.

It should be noted that the regular solution model has been extended to ternary alloys and applied successfully for systems for which interactions cause positive deviations from ideality (Joud et al. 1974).

These treatments suppose the surface molar areas of A and B atoms to be equal. However, for alloys with $\lambda \gg 0$, the low σ_{LV} component, B, can have a molar volume much greater than that of A (up to 100%). Typical examples are the Cu-Pb, Cu-Bi and Al-Pb systems, for which the experimental decreases in σ_{LV} caused by adding small quantities of B are significantly more than those predicted by the regular solution model. These differences have been attributed to the fact that not all the sites of the "quasicrystalline lattice" on which A and B atoms are distributed are equivalent (Goumiri et al. 1979).

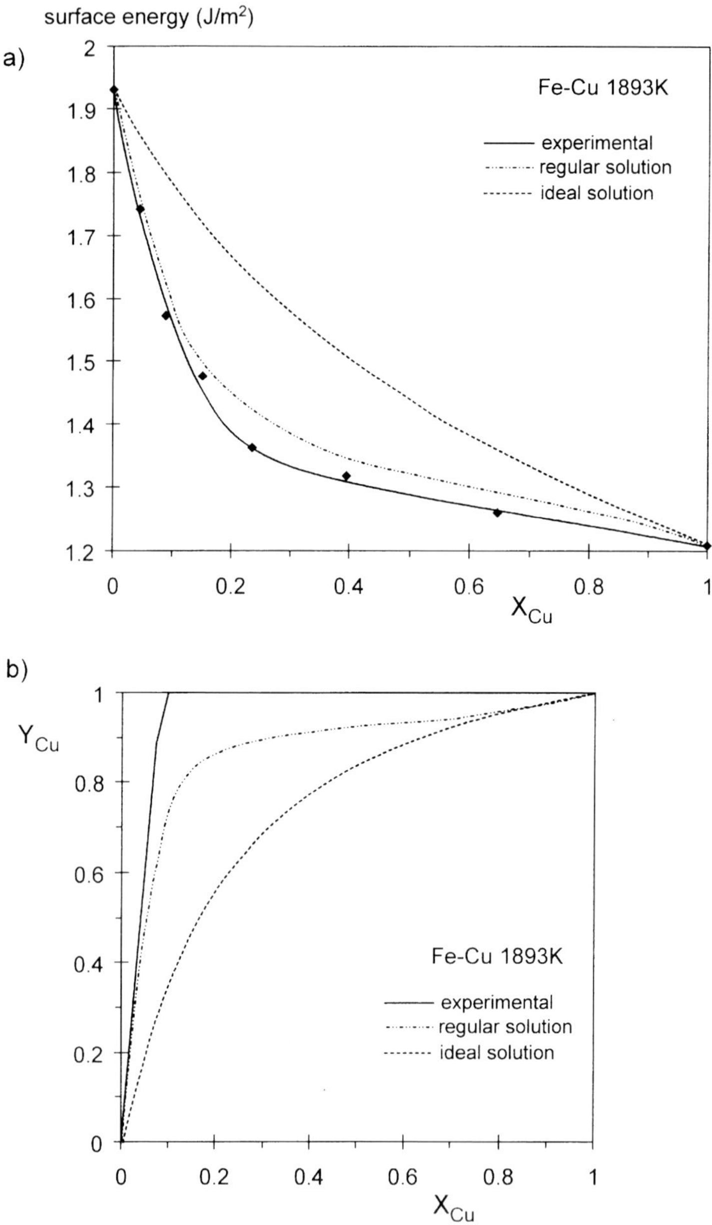

Figure 4.5. (a) Experimental (Nogi et al. 1991) and calculated surface energy and (b) surface composition curves for liquid Fe-Cu alloys at 1893K. The calculated curves were obtained taking $\sigma_{LV}^{Fe} = 1.93\,\mathrm{J/m^2}$, $\sigma_{LV}^{Cu} = 1.21\,\mathrm{J/m^2}$, $\Omega_m = 3.6 \times 10^4\,\mathrm{m^2/mole}$ and $\lambda = 30700\,\mathrm{J/mole}$.

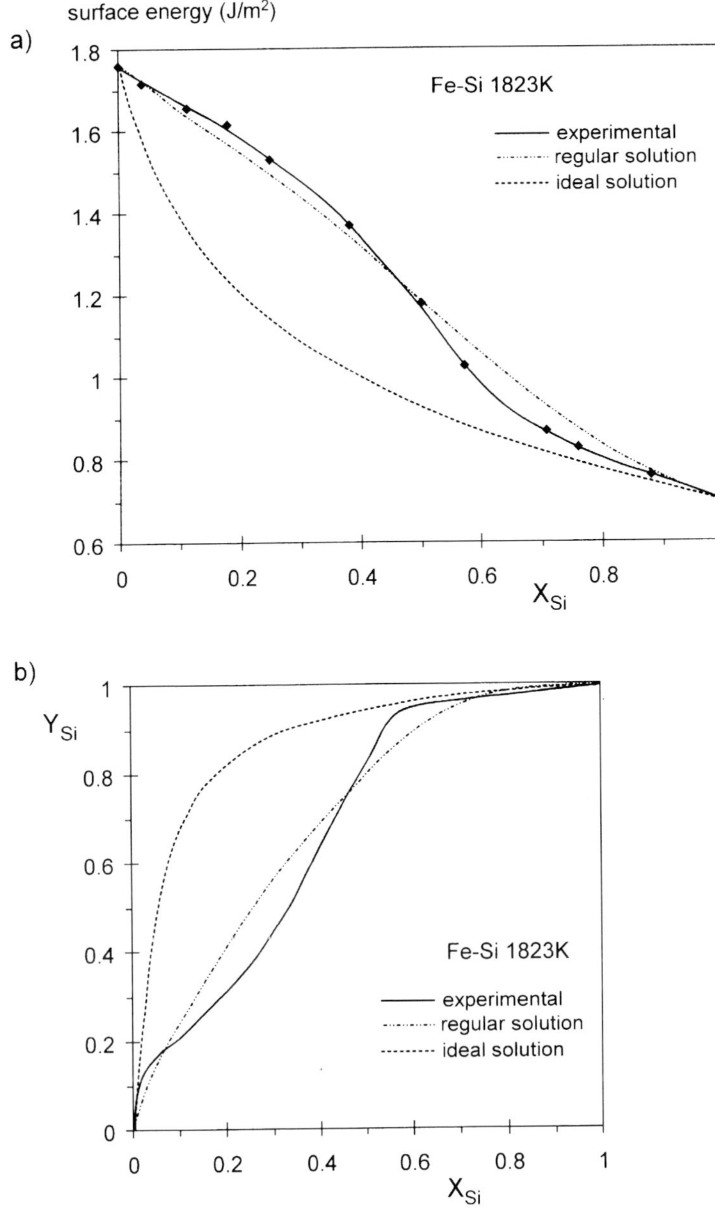

Figure 4.6. (a) Experimental (Popel et al. 1970) and calculated surface energy and (b) surface composition curves for liquid Fe-Si alloys at 1823K. The calculated curves were obtained taking $\sigma_{LV}^{Fe} = 1.76\,\mathrm{J/m^2}$, $\sigma_{LV}^{Si} = 0.70\,\mathrm{J/m^2}$, $\Omega_m = 4.2 \times 10^4\,\mathrm{m^2/mole}$ and $\lambda = -79000\,\mathrm{J/mole}$.

 Wettability at high temperatures

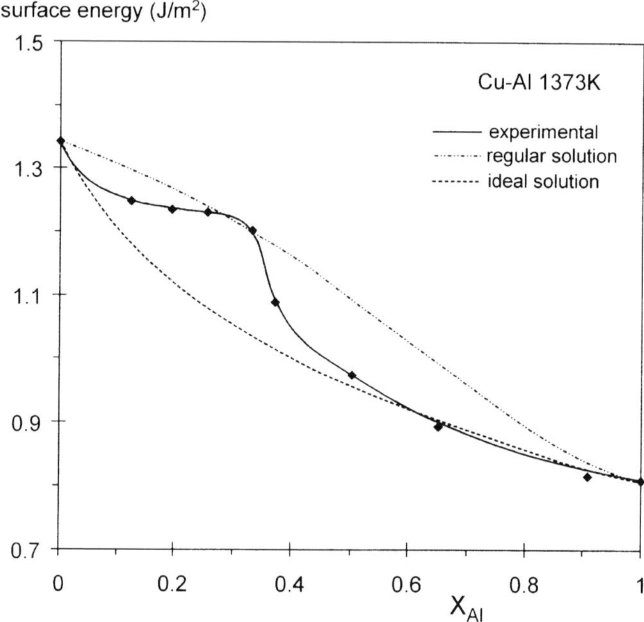

Figure 4.7. Experimental (Eremenko et al. 1961) and calculated surface energy curves for liquid Cu-Al alloys at 1373K. The calculated curves were obtained taking $\sigma_{LV}^{Cu} = 1.343\,\text{J}/\text{m}^2$, $\sigma_{LV}^{Al} = 0.810\,\text{J}/\text{m}^2$, $\Omega_m = 4.3 \times 10^4\,\text{m}^2/\text{mole}$ and $\lambda = -55000\,\text{J}/\text{mole}$.

Figure 4.6.a shows both experimental (Popel et al. 1970) and calculated σ_{LV} curves for Fe-Si alloys. This system exhibits strong heteroatomic interactions, ($\lambda \ll 0$), as shown by the negative deviation of activities from ideality in the liquid alloys and by the formation of high-melting-point intermetallics. The shape of the experimental σ_{LV}-X_{Si} curve is typical of systems with $\lambda \ll 0$ and is significantly different from that observed for the Fe-Cu system, especially in the sign of its curvature at low B solute concentrations. This is in agreement with the simple regular solution model which predicts that strong bulk interactions (i.e. $\lambda \ll 0$) decrease the tendency for adsorption (Figure 4.6.b). However, in some alloys with $\lambda \ll 0$, the experimental σ_{LV} curve exhibits a plateau at low B solute concentrations. This feature is not reproduced satisfactorily by the regular solution model (Figure 4.7) and is due to the fact that for alloys with $\lambda \ll 0$, short-range chemical ordering (i.e., clustering) effects cannot be ignored. Such effects are taken into account by more sophisticated statistical models allowing a better representation of experimental data (Laty et al. 1976).

To sum up, the factors favouring adsorption of a solute B in A, leading to a strong decrease of σ_{LV}^A, are $\sigma_{LV}^B \ll \sigma_{LV}^A$, $\lambda \gg 0$ and $v_m^B \gg v_m^A$.

4.1.3 Solid metals

Our experimental knowledge of the surface energies of metallic solids is limited compared to that of liquid metals and consists of values of σ_{SV} for the most common pure metals determined mainly by the "zero-creep" method (Hondros 1970). This technique is based on the observation that thin wires or foils, when heated to a temperature close to the melting point (usually $T > 0.9T_F$), contract to minimise the total surface energy of the system. Thus, σ_{SV} can be derived by measuring the applied load needed to suppress contraction, i.e., to produce a zero-creep. At these high temperatures, when small loads are applied the surface area changes occur by diffusion without any elastic deformation. Therefore, this kind of measurements leads to the determination of the surface energy rather than the surface tension of the solid (see Section 1.1). Compilations of σ_{SV} data are given in (Eustathopoulos and Joud 1980, Kumikov and Khokonov 1983) (Table 4.2).

Table 4.2. Solid/vapour surface energies of pure metals measured by the zero-creep technique (Eustathopoulos and Joud 1980).

Metal	T_{exp} (K)	T_{exp}/T_F	σ_{SV} (mJ/m^2)
Ag	1183	0.86	1140 ± 90
Au	1273	0.95	1400 ± 50
Co	1678	0.95	2282 ± 300
Cu	1243	0.92	1650 ± 100
γ-Fe	1648	0.91	2150 ± 325
δ-Fe	1723	0.95	2220 ± 250
In	420	0.98	674 ± 74
Nb	2523	0.92	2100 ± 100
Ni	1488	0.86	2385 ± 100
Pb	590	0.98	620 ± 20
Sn	488	0.97	685
Ti	1873	0.96	1700
Zn	653	0.94	830

For cubic metals, the value of σ_{SV} generally exceeds that of σ_{LV} at T_F by 10 to 20% as shown by the data presented in Table 4.3. There is some anisotropy of σ_{SV} values

 Wettability at high temperatures

but for metals with cubic structures (fcc or bcc) it is only a few percent when T is close to T_F (Eustathopoulos and Joud 1980). However, anisotropy is much higher (several tens of percent) for hexagonal structure metals like Zn and Cd whose axial ratio deviates strongly from ideality.

As for liquid metals, oxygen adsorbs strongly on solid surfaces, markedly decreasing σ_{SV} values, but this effect is very anisotropic (Hondros and McLean 1970b). Substantial reductions in σ_{SV} have also been observed with very small additions (less than 10^{-1} at.%) of low-surface-energy metals like Sb or Bi in Cu, and Sn in δ-Fe (Hondros 1978). Adsorption at solid surfaces (as well as at grain boundaries) is strongly reinforced by the relaxation of the lattice strain energy caused by solutes with atomic sizes much larger than that of the solvent. This contribution to the adsorption energy is always important and may even be predominant compared to other contributions to the adsorption energy, i.e., the cohesion energy of pure metals ($(\sigma_{SV}^B - \sigma_{SV}^A)\Omega_m$) and chemical A-B interactions ($m_1\lambda$) cited in equation (4.8) (Joud and Passerone 1995).

Table 4.3. Surface and interface energies of pure Cu at temperatures close to the melting point.

σ_{LV}	$1355 \pm 60 \, \mathrm{mJ/m^2}$	From Table 4.1
σ_{SV}	$1650 \pm 100 \, \mathrm{mJ/m^2}$	(Eustathopoulos and Joud 1980)
σ_{SL}	$235 \pm 35 \, \mathrm{mJ/m^2}$	(Eustathopoulos 1983)

Absolute values of surface energies of high-melting-point metals A in equilibrium with saturated vapours of a metal B have been determined at $T \ll T_F^A$ by the "multiphase equilibrium technique" (MPE). This consists of placing a liquid drop of a metal B on a coarse-grained solid substrate of metal A and measuring both the contact angle θ and the dihedral angles Φ and ψ formed at the grain-boundary grooves in contact respectively with the liquid and the vapour (Figure 4.8). Then, combining the Young equation (1.16) with the Smith equation (1.30) applied to grain boundary grooves ($\sigma_{gb} = 2\sigma_{SV}\cos(\psi/2) = 2\sigma_{SL}\cos(\Phi/2)$ where σ_{gb} is the grain boundary energy), the surface energy of the solid in equilibrium with a saturated vapour of metal B can be expressed in terms of measurable quantities:

$$\sigma_{SV} = \frac{\sigma_{LV}\cos(\theta)\cos(\Phi/2)}{\cos(\Phi/2) - \cos(\psi/2)} \tag{4.9}$$

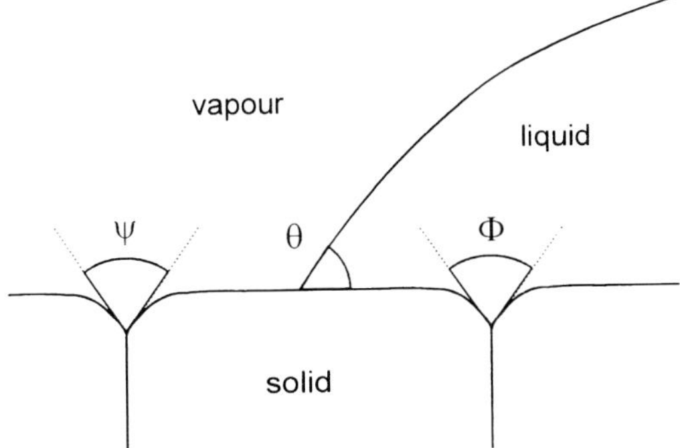

Figure 4.8. Definition of the dihedral angles ψ and Φ formed at grain-boundary grooves in the multiphase equilibrium technique.

To determine the surface energy of the solid A in equilibrium with its own vapour, σ_{SV}^0, a further experiment is needed to measure the dihedral angle ψ^0 of grain boundary grooves in vacuum or a neutral gas. Then, assuming that the grain boundary energy of solid A in contact with liquid B is not affected by adsorption of B, σ_{SV}^0 is obtained from the equation:

$$\sigma_{SV}^0 = \sigma_{SV}\frac{\cos(\psi/2)}{\cos(\psi^0/2)} \tag{4.10}$$

In practice, θ is measured in short-time sessile drop experiments to avoid distortion of the triple line (see Section 1.2.4). The dihedral angles Φ, ψ and ψ^0 are obtained after prolonged experiments, that allow the grooves to grow deep enough for measurements by metallographic or interferometric techniques (Eustathopoulos 1983). To obtain an acceptable accuracy for dihedral angles, several tens or even hundreds of grooves must be measured. If the value of σ_{LV} is known, σ_{SV}^0 (and σ_{SV}, σ_{SL} and σ_{gb}) can be calculated as indicated for W in Table 4.4.

In the example of Table 4.4, the surface energy of solid W in equilibrium with a saturated vapour of Cu is lower than σ_{SV}^0 due to adsorption of Cu atoms on the W surface. This is generally characteristic of metallic A-B pairs having a low mutual miscibility (Eustathopoulos and Joud 1980). For this reason, results of sessile drop experiments for such systems cannot be interpreted by taking for the surface energy of the solid metal the value of σ_{SV} of pure A in equilibrium with its own vapour (see Sections 1.4.2 and 5.2).

Table 4.4. Interfacial energies at 1500°C in Ar/liquid Cu/solid W system, calculated using data on dihedral angles and contact angle of Hodkin et al. (1970) and taking $\sigma_{LV}^{Cu} = 1276$ mJ/m^2 (see Table 4.1). Note that the contact angle is close to but higher than zero.

Angles (deg)	Interfacial energies (mJ/m^2)
$\theta \cong 0$	$\sigma_{SV}^0 = 3200$
$\Phi = 113.1$	$\sigma_{SV} = 2400$
$\psi = 150.1$	$\sigma_{SL} = 1120$
$\psi^0 = 157.7$	$\sigma_{gb} = 1240$

4.1.4 Solid/liquid interfaces

At S/V and L/V surfaces, the atomic density decreases sharply within a few layers from its value in the bulk phase to zero. This leads to a high excess energy of solid atoms at the surface and hence to a high surface energy. At S/L interfaces of pure metals, the change of atomic density is comparatively small, a few percent, so the value of σ_{SL} for a given metal is generally much lower than those of σ_{SV} and σ_{LV} (Table 4.3).

For pure bcc or fcc metals, experimental values of σ_{SV} and σ_{SL} reveal a relation $\sigma_{SL}/\sigma_{SV} \cong 0.15 \pm 0.05$ (Eustathopoulos 1983). The constancy of this ratio is compatible with the simple nearest-neighbour interaction model of Skapski (1956) (see Section 5.1). Usually, σ_{SL} for pure metals and A-B alloys is of the order of 10^2 mJ/m^2. However, for some highly immiscible solid A/liquid B systems, values of σ_{SL} up to about 10^3 mJ/m^2 can be observed (Table 4.4) (Eustathopoulos 1983), due at least partially to the contribution of heteroatomic interactions. This effect is discussed in more detail in Section 5.2.

4.2. DATA FOR NON-METALLIC COMPOUNDS

4.2.1 Liquid oxides and halides

The surface energy of molten oxides has been studied less extensively than that of pure liquid metals. For instance, the surface energy of molten Al_2O_3, which is the most widely studied oxide, has been measured by 12 teams (Ikemiya et al. 1993) while that of Fe has been measured by 28 (Keene 1993). One reason for this difference is the experimental difficulties arising from the high melting point of many oxides but nevertheless experimental data for surface energies exist for about 30 oxides (Table 4.5) (Ikemiya et al. 1993).

Table 4.5. Surface energy of molten pure oxides at their melting point. From the compilation of (Ikemiya et al. 1993) except for Al_2O_3 (see the text) and Cr_2O_3 (Rasmussen 1972).

Oxide	T_F (K)	σ_{LV} (mJ/m^2)	Oxide	T_F (K)	σ_{LV} (mJ/m^2)
Al_2O_3	2320	630	MnO	2058	630
B_2O_3	723	80 (1173K)	MoO_3	1068	70
BaO	2196	520	Na_2O	1193	308 (1673K)
BeO	2843	415	Nb_2O_5	1773	279
Bi_2O_3	1098	213	P_2O_5	836	60
CaO	2860	670	PbO	834	132 (1173K)
CoO	2078	550	SiO_2	1993	307
Cr_2O_3	2573	812	Sm_2O_3	2593	815
FeO	1641	545	Ta_2O_5	2150	280
GeO_2	1389	250	TiO_2	2143	380
K_2O	980	156 (1673K)	Ti_2O_3	2090	584
La_2O_3	2573	560	V_2O_5	943	80
MgO	3073	660	WO_3	1743	100

There is insufficient information to allow classification of liquid oxide data into two types according to their accuracy, as was done for liquid metals. In the case of Al_2O_3, the determinations of σ_{LV} values by mainly the drop weight, sessile drop and maximum bubble pressure techniques converge to a value of 630 ± 70 mJ/m^2 near the melting temperature (Ikemiya et al. 1993). Moreover, the possible effect of P_{O2} on the σ_{LV} value of Al_2O_3 appears to be weak since similar values have been obtained when working in vacuum, in neutral or reducing atmosphere or air. Like liquid metals, the dispersion of values of temperature coefficient, $d\sigma_{LV}/dT$, is high and values lying between -0.15 and -0.5 mJ.m^{-2}.K^{-1} have been reported for Al_2O_3 (Ikemiya et al. 1993).

Because of the success of correlations between the molar surface energy of pure metals ($\cong \sigma_{LV}.V_m^{2/3}$) and various quantities reflecting the cohesion energy of the metal (heat of evaporation L_e, melting point T_F, etc), similar correlations have been assessed for molten oxides. An example is shown in Figure 4.9 in which the experimental values of σ_{LV} are plotted versus $C_3.T_F.V_m^{-2/3}$ where C_3 is a constant equal to 1.18×10^{-8} (Tanaka et al. 1996) (in this correlation, the melting temperature was preferred to the heat of evaporation because this last quantity is not available for many oxides). Except for BeO and Sm_2O_3, the quality of the

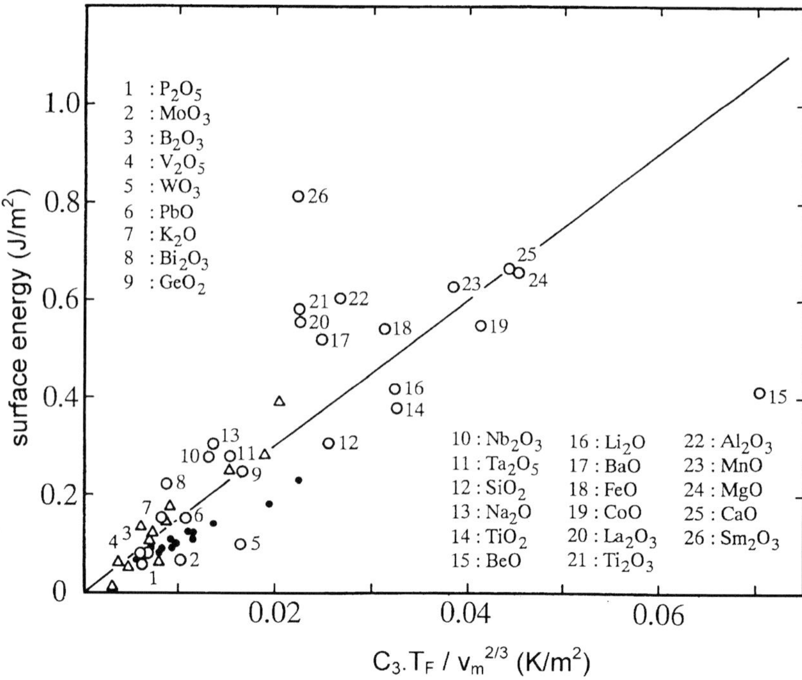

Figure 4.9. Correlation between surface energy of molten oxides and $C_3.T_F.v_m^{-2/3}$ (with $C_3 = 1.18 \times 10^{-8}$). Non-numbered symbols (full circles and hollow triangles) relate to the ionic melts of Figure 4.10. Reprinted from (Tanaka et al. 1996) with kind permission of the authors.

correlation is similar to that observed for liquid metals (Figure 4.1), indicating that the molar surface energy of oxides is also nearly proportional to their cohesion energy. However, the proportionality coefficient is found to be significantly lower than for pure metals. This suggests that the atomic and electronic rearrangements accompanying the creation of a surface is particularly important in oxides and hence results in a marked surface energy decrease (Kruse et al. 1996).

In Figure 4.10, values of σ_{LV} for molten halides at T_F are plotted versus the quantity $C_3.L_e.v_m^{-2/3}$ (Tanaka et al. 1996) (it should be noted that the constant $C_3 = 1.18 \times 10^{-8}$ is similar but not equal to the constant C_1 of equation (4.1)). Fused halides have comparatively low surface energies (lying between 0.05 and 0.4 J/m^2) for two main reasons: (i) their heat of evaporation is low compared to that of metals (typical values being 30 as compared to 300 kJ/mole) and (ii) the coefficient of proportionality between σ_{LV} and $L_e.v_m^{-2/3}$ for halides is (as for oxides) only half of that for metals.

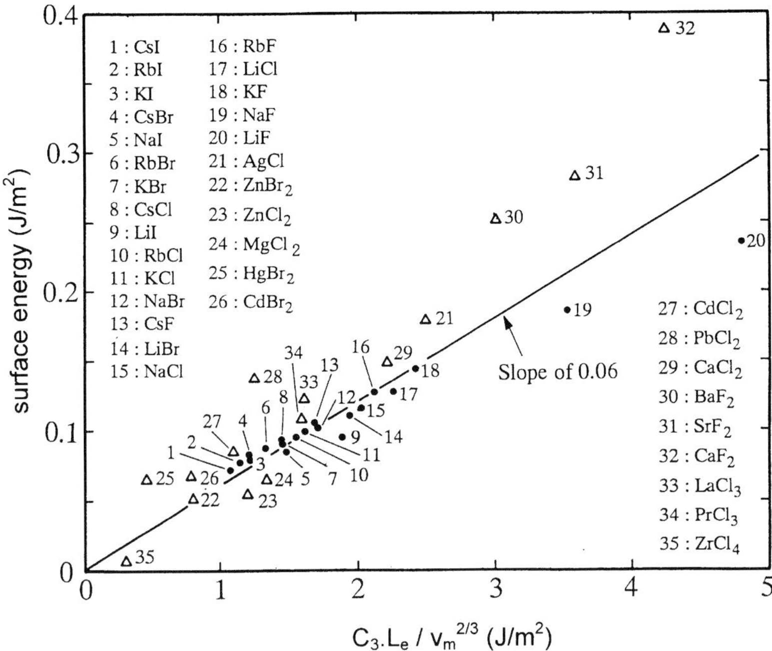

Figure 4.10. Correlation between surface energy of molten halides and $C_3.L_e.V_m^{-2/3}$ (with $C_3 = 1.18 \times 10^{-8}$). Reprinted from (Tanaka et al. 1996) with kind permission of the authors.

As the surface energy varies by a factor 10 from one oxide or halide to another, tensio-active effects, such as those observed in liquid alloys, are to be expected also for mixtures of fused oxides or fused halides. One example is shown in Figure 4.11 which shows the effect of additions of various oxides on the surface energy of molten FeO. A sharp decrease of σ_{LV} is observed in the case of additions of low-surface-energy P_2O_5 or Na_2O while additions of high surface energy Al_2O_3, MnO or CaO have only weak effects on σ_{LV}. Because of the complexity of the structure of molten oxides, there is no statistical thermodynamic model that allows calculation of σ_{LV} values for binary oxide mixtures. However, useful semi-empirical techniques for estimating these quantities were proposed in (Tanaka et al. 1996).

4.2.2 Solid oxides

In the literature, there is a very limited number of measurements of the surface energy of non-metallic solids. Most of them are for Al_2O_3 and were derived using

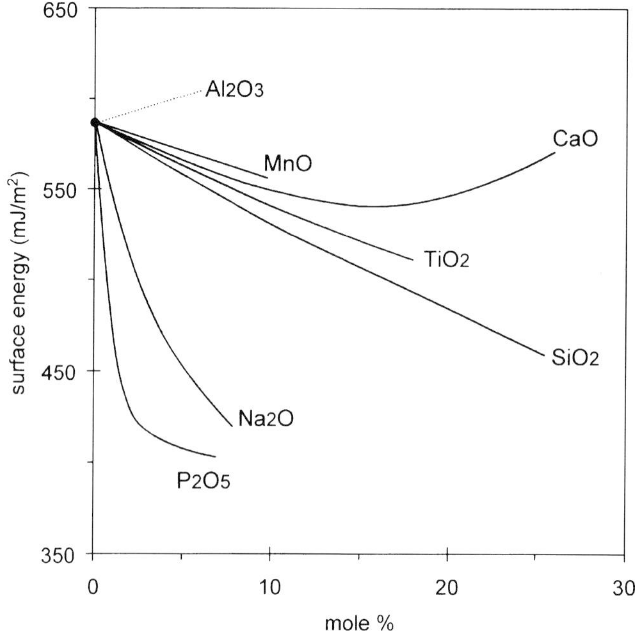

Figure 4.11. Surface energies of binary mixtures based on FeO at 1400–1420°C. Data from work reported in (Kozakevitch 1949) [15].

the MPE technique (see Figure 4.8) with a non-reactive molten metal as the auxiliary liquid. Deriving meaningful data from these measurements requires care. As when applied to metals, the method needs the measurement of a great number of parameters and a small error on dihedral angles often produces a large uncertainty in the value of σ_{SV}. Additionally, by definition polycrystalline ceramics must be used but for these materials the impurity content is usually high. Because the growth of measurable dihedral angle grooves needs high temperatures and long equilibration times, impurities may easily migrate to surfaces and interfaces and affect results.

An example of the application of the MPE technique is given in Table 4.6 for Al_2O_3, which was studied using Sn and Co as the auxiliary liquids. The dihedral angles ψ, ψ^0 and Φ in these systems were measured by Nikolopoulos (1985). For both systems, at each temperature, it was found that the difference between ψ and ψ^0 is of the order of the experimental error, meaning that $\sigma_{SV} \cong \sigma_{SV}^0$ (equation (4.10)). In other words, adsorption of metallic vapours on Al_2O_3 surfaces was negligible. This result is to be expected for a non-reactive metal/oxide system in which metal/oxide interactions are weak, as will be discussed in Section 6.2.2. Combining values of dihedral angles with θ values derived from short time sessile drop experiments on monocrystalline Al_2O_3 (as discussed in Section 3.3, the use of monocrystalline substrates leads to more reliable values of contact angles than polycrystalline ones), σ_{SV} can be calculated by substitution in equation (4.9) as was done for the data presented in Table 4.6. The uncertainty of the values for σ_{SV} is as high as 30–40% and is typical of this kind of measurements. Note that if the ratio of σ_{SV}/σ_{LV} for an oxide is as close to unity as for pure metals, the value of σ_{SV} of $750\ mJ/m^2$ obtained at the highest temperature is compatible with the value of σ_{LV} of $630 \pm 70\ mJ/m^2$ for molten Al_2O_3. Therefore, when a rough estimate of the surface energy of a solid oxide is needed at a temperature not too far from its melting point, the value of σ_{LV} of this oxide can be used. Finally it should be noted that values of σ_{SL} calculated from data of Table 4.6 lie in the range $1200–1500\ mJ/m^2$.

Table 4.6. Experimental determination of σ_{SV} of Al_2O_3 by the MPE technique.

T (K)	auxiliary liquid	σ_{LV} [a] (mJ/m^2)	$\bar{\psi}$ [b] (deg)	$\bar{\Phi}$ [b] (deg)	θ (deg)	σ_{SV} (mJ/m^2)
1473	Sn	463	139	146.9	121 [c]	1040 ± 375
1623	Sn	447	137	145.5	120 [c]	947 ± 330
1783	Co	1878	139	159.6	113 [d]	750 ± 220
1923	Co	1826	138	157.9	111 [d]	752 ± 210

[a] from Table 4.1, [b] from (Nikolopoulos 1985), [c] from (Rivollet et al. 1987), [d] from (Chatain et al. 1986)

A similar study was done for UO_2 using Cu (Hodkin and Nicholas 1973) and Ni (Nikolopoulos et al. 1977) as auxiliary liquids. As in Sn/Al_2O_3 and Co/Al_2O_3 couples, the effect of adsorption of metal atoms on UO_2 surface at 1500°C was found to be negligible, i.e. $\sigma_{SV} \cong \sigma_{SV}^0$. The surface energy of UO_2 was calculated using values of σ_{LV} of pure metals given in Table 4.1 and contact angle and dihedral angle data. The value of σ_{SV}^0 when using liquid Cu was $340\ mJ/m^2$ and $850\ mJ/m^2$ when using liquid Ni. This difference in values for UO_2 is much higher than the experimental error and underlines the difficulties in measuring surface energies of polycrystalline ceramics by the MPE technique. Complications may have been caused also by minor differences in the stoichiometry of the UO_2 used by the two

teams, as shown by results obtained by Hodkin and Nicholas (1977) by varying the
O/U ratio in UO_2.

4.2.3 Carbon and solid carbides

Due to obvious technical difficulties, neither the zero-creep nor the multiphase
equilibrium techniques have been used to measure the surface energy of *graphite*.
However, a 150 mJ/m^2 value (Donnet et al. 1982) for the basal plane of graphite
was measured at room temperature by wetting experiments using the so-called
"two liquids" method and alkane/water/graphite systems (Donnet et al. 1977).
This technique does not allow measurement of the total surface energy but only the
van der Waals component of σ_{SV}, σ_{SV}^d. However, because the graphite basal planes
are bonded by van der Waals dispersion force interactions,
$\sigma_{SV}(\text{basal}) \cong \sigma_{SV}^d(\text{basal}) = 150\,\text{mJ}/\text{m}^2$.

In wetting experiments of a pure metal or alloy on graphite, adsorption of the
liquid components on to the solid surface from the vapour phase can occur, to
decrease σ_{SV}. However, for inert non-wetting metal/graphite systems, this effect is
unlikely, as shown by results of Gangopadhyay and Wynblatt (1994). Indeed, at
temperatures close to (but slightly lower than) the melting point of Pb, adsorption
of Pb and Ni from Pb-Ni alloys on graphite surface, monitored by in-situ Auger
analyses, was negligible. Negligible adsorption is expected *a fortiori* at higher
temperatures.

The formation of a surface orientated perpendicularly to the basal plane needs the
rupture of chemical carbon-carbon bonds. The surface energy for this prismatic
orientation has not been determined experimentally, but is expected to be an order
of magnitude higher than that of the basal plane (150 mJ/m^2).

In the case of *vitreous carbon*, the component σ_{SV}^d of σ_{SV} was measured by the
technique used for graphite (Donnet et al. 1982) and found to be 32 mJ/m^2. This is
five times lower than that of the graphite basal plane and may be explained by the
low density of vitreous carbon (1.5×10^3 kg/m^3) compared to that of graphite
(2.26×10^3 kg/m^3) (see also Section 8.1). Because the atomic structure of polished
surfaces of vitreous carbon is not known, it is not possible to evaluate the
percentage of σ_{SV} that is due to σ_{SV}^d.

The formation of the low-energy (111) surface planes of *diamond* requires
breaking of chemical carbon-carbon bonds. Although surface reconstruction
occurring at high temperatures can decrease the surface energy, this is still
expected to be much higher than that of the basal plane of graphite. A rough
estimation of the surface energy made by a simple nearest-neighbour model leads
to a value close to 4000 mJ/m^2 (Appendix F).

Carbides can be covalent or metal-like, the most important of the covalent carbides being SiC which like carbon crystallizes in both hexagonal and cubic structures. However, contrary to carbon, the basal planes in the hexagonal structure and the (111) faces of the cubic variant are linked by chemical bonds, so the corresponding surface energies should be about 10^3 rather than 10^2 mJ/m^2. Estimated values for the surface energy of both faces are close to 1500 mJ/m^2 (Appendix F).

The only measurement of surface energy of a metal-like carbide was performed by the multiphase equilibrium technique for the molten U/UC system (Hodkin et al. 1971). In this system, because of the metallic character of UC, good wetting is observed, i.e., $\theta < 90°$ (see also Section 7.2). The surface energy of UC at 1325°C was found to be 730 mJ/m^2 while the interfacial energy between U (saturated in C) and UC was about 140 mJ/m^2. Thus the σ_{SL}/σ_{SV} ratio is closer to those for metal/metal rather than metal/oxide systems. It must be emphasized that at high temperature UC exhibits a range of non-stoichiometry which can affect its surface and interfacial energies. Therefore, the value $\sigma_{SV} = 730$ mJ/m^2 corresponds to the particular stoichiometry of UC in equilibrium with molten U at 1325°C.

For other metal-like carbides (TiC, ZrC, VC, TaC), an estimation of the *minimum* value of σ_{SV} can be made using contact angle data of ferrous metals (see Table 7.12) and assuming that $\sigma_{SL} = 0$. This leads to values close to 1500 mJ/m^2. However, the calculations of Warren (1980) suggest that σ_{SL} in ferrous metal/metallic carbide systems is close to 500 mJ/m^2, so a more plausible value for σ_{SV} is 2000 mJ/m^2.

Returning to the U/UC system in which contact angles close to 40°–60° were observed in the range 1300–1600°C, an interesting observation was that the dihedral angle ψ was nearly equal to ψ^0 as for non-wetting metal/oxide systems. Thus adsorption of metal atoms on the ceramic surface was negligible. This behaviour differs from that observed for several low melting point metals B on high melting point metals A systems (see for instance Table 4.4), where adsorption of B atoms on solid A surfaces reduces σ_{SV}^A (see Figure 5.2). The condition for adsorption is $\sigma_{LV}^B \ll \sigma_{SV}^A$ (Section 5.2) but this condition is not met by the pseudo-metallic U/UC system because σ_{LV} of liquid U is about 1500 mJ/m^2, higher than the surface energy of UC by a factor two.

4.3. CONCLUDING REMARKS

Measurements of the surface energy of pure liquid metals performed in the last decades by different investigators and different techniques have led to values of

σ_{LV} differing by less than 5% for the most common metals. However, the dispersion is much more marked for the temperature coefficient of σ_{LV} (30–100%). The influence of dissolved oxygen on the surface energy of liquid metals can be large and has been studied formally for about ten elements. Surface energy variations with alloy composition have been determined for several binary systems and compilations of data are given in (Khilya 1980, Keene 1987). The experimental curves can be predicted satisfactorily using the monolayer-regular solution model or other more sophisticated approaches. For alloys containing more than two components, experimental measurements of σ_{LV} are essential, except for some simple ternary systems for which modelling is possible.

Although fused oxides and halides have been less extensively studied than liquid metals, surface energies have been determined for a number of such compounds. In the absence of models for estimating the surface energy of oxide or halide mixtures, this quantity must be determined experimentally.

Experimental data for the surface energy of solid compounds exist only for some pure metals and for a few non-metallic solids (Al_2O_3, UO_2, UC, graphite). Even for these materials, the scatter of surface energy values is as high as 50%, owing to impurity effects and difficulties inherent in the experimental methods that have to be used. Thus it is not possible to analyse wetting data for high-temperature solids by an approach needing the accurate knowledge of surface energy of these solids.

REFERENCES FOR CHAPTER 4

Chatain, D., Rivollet, I. and Eustathopoulos, N. (1986) *J. Chim. Phys.*, **83**, 561

Chung, W. B., Nogi, K., Miller, W. A. and McLean, A. (1992) *Mater. Trans. JIM*, **33**, 753

Defay, R., Prigogine, I., Bellemans, A. and Everett, D. H. (1966) in *Surface Tension and Adsorption*, Longmans, London, p. 161

Donnet, J. B., Schultz, J. and Tsutsumi, K. (1977) *J. Colloid Interf. Sci.*, **59**, 272 (part I) and 277 (part II)

Donnet, J. B., Schultz, J. and Shanahan, M. E. R. (1982) in *Proc. Intern. Symp. on Carbon: New Processing and New Applications*, Toyohashi (Japan), p. 57

Eremenko, V. N., Nizhenko, V. I. and Naidich, Y. V. (1961) *Izv. Akad. Nauk. SSSR*, **3**, 150

Eustathopoulos, N. and Joud, J. C. (1980) in *Current Topics in Materials Science*, Ed. E. Kaldis, Vol. 4, p. 281, North Holland, Amsterdam

Eustathopoulos, N. (1983) *International Metals Reviews*, **28**, 189

Eustathopoulos, N., Drevet, B. and Ricci, E. (1998) *J. Crystal Growth*, **191**, 268

Eustathopoulos, N., Ricci, E. and Drevet, B. (1999) in "Tension Superficielle", Traité Matériaux Métalliques M67, Techniques de l'Ingénieur, Paris.

Gallois, B. and Lupis, C. H. P. (1981) *Metall. Trans. B*, **12**, 549

Gangopadhyay, U. and Wynblatt, P. (1994) *Metall. Mater. Trans. A*, **25**, 607

Goumiri, L., Joud, J. C. and Desré, P. (1979) *Surf. Sci.*, **88**, 461

Goumiri, L. and Joud, J. C. (1982) *Acta Metall.*, **30**, 1397

Guggenheim, E. A. (1945) *Trans. Faraday Soc.*, **41**, 150

Hasouna, A. T., Nogi, K. and Ogino, K. (1991) *Mater. Trans. JIM*, **32**, 74

Hodkin, E. N., Nicholas, M. G. and Poole, D. M. (1970) *J. Less-Common Metals*, **20**, 93

Hodkin, E. N., Mortimer, D. A., Nicholas, M. G. and Poole, D. M. (1971) *J. Nucl. Mater.*, **39**, 59

Hodkin, E. N. and Nicholas, M. G. (1973) *J. Nucl. Mater.*, **47**, 23

Hodkin, E. N. and Nicholas, M. G. (1977) *J. Nucl. Mater.*, **67**, 171

Hondros, E. D. (1970) in *Techniques and Properties of Solid Surfaces*, ed. R. F. Bunshah, Wiley, NY, vol. IV(2), ch. 8A

Hondros, E. D. and McLean, M. (1970b) in *Structure et Propriétés des Surfaces*, CNRS Colloque N.187, Paris, p. 219

Hondros, E. D. (1978) in *Precipitation Processes in Solids*, Warrendale, PA, The Metallurgical Society of AIME, p. 1

Ikemiya, N., Umemoto, J., Hara, S. and Ogino, K. (1993) *ISIJ International*, **33**, 156

Joud, J. C., Eustathopoulos, N., Bricard, A. and Desré, P. (1974) *J. Chim. Phys.*, **7-8**, 1113

Joud, J. C. and Passerone, A. (1995) *Heterogeneous Chemistry Reviews*, **2**, 173

Keene, B. J. (1987) *Surface and Interface Analysis*, **10**, 367

Keene, B. J. (1993) *International Materials Reviews*, **38**, 157

Khilya, G. P. (1980) *Adgezija Rasplavov i Pajka Materialov*, **5**, 11 (in Russian)

Kozakevitch, P. (1949) *Revue Métall.*, **46**, 505

Kozakevitch, P. and Urbain, G. (1961) *Mem. Sci. Rev. Met.*, **58**, 517

Kozakevitch, P. (1968) in *Surface Phenomena of Metals*, Society of Chemical Industry Monograph 28, London, p. 223

Kruse, C., Finnis, M. W., Lin, J. S., Payne, M. C., Milman, V. Y., De Vita, A. and Gillan, M. J. (1996) *Phil. Mag. Lett.*, **73**, 377

Kumikov, V. K. and Khokonov, K. B. (1983) *J. Appl. Phys.*, **54**, 1346

Laty, P., Joud, J. C. and Desré, P. (1976) *Surf. Sci.*, **60**, 109

Nakashima, K., Takihira, K., Mori, K. and Shinozaki, N. (1992) *Mater. Trans. JIM*, **33**, 918

Nicholas, M. E., Joyner, P. A., Tessem, B. M. and Olson, M. D. (1961) *J. Phys. Chem.*, **65**, 1373

Nikolopoulos, P., Nazare, S. and Thümmler, F. (1977) *J. Nuclear Materials*, **71**, 89

Nikolopoulos, P. (1985) *J. Mater. Sci.*, **20**, 3993

Niu, Z., Mukai, K., Shiraishi, Y., Hibiya, T., Kakimoto, K. and Koyama, M. (1998) in *Proc. 2nd Int. Conf. on High Temperature Capillarity*, Cracow (Poland), 29 June-2 July 1997, ed. N. Eustathopoulos and N. Sobczak, published by Foundry Research Institute (Cracow), p. 182

Nogi, K., Chung, W. B., McLean, A. and Miller, W. A. (1991) *Mater. Trans. JIM*, **32**, 164

Ogino, K. Taimatsu, H. and Nakatani, F. (1982) *J. Japan Inst. Metals*, **46**, 957

Ogino, K., Nogi, K. and Yamase, O. (1983) *Transactions ISIJ*, **23**, 234

Passerone, A., Sangiorgi, R. and Caracciolo, G. (1983) *J. Chem. Thermodyn.*, **15**, 971

Rasmussen, J. J. (1972) *J. Amer. Ceram. Soc.*, **55**, 326

Popel, S. I., Shergin, L. M. and Zarevski, B. V. (1970) *Zh. Fiz. Khim.*, **44**, 260

Rivollet, I., Chatain, D. and Eustathopoulos, N. (1987) *Acta Metall.*, **35**, 835

Sangiorgi, R., Passerone, A. and Muolo, M. L. (1982) *Acta Met.*, **30**, 1597

Sharan, A. and Cramb, A. W. (1997) *Metall. Mater. Trans. B*, **28**, 465

Skapski, A. (1948) *J. Chem. Phys.*, **16**, 386

Skapski, A. (1956) *Acta Met.*, **4**, 576

Taimatsu, H. and Sangiorgi, R. (1992) *Surface Science*, **261**, 375

Tanaka, T., Hack, K., Iida, T. and Hara, S. (1996) *Z. Metallkd.*, **87**, 380

Warren, R. (1980) *J. Mater. Sci.*, **15**, 2489

Chapter 5
Wetting properties of metal/metal systems

In the absence of barriers to wetting such as oxide films on the liquid or the solid, molten metals wet metallic substrates ($\theta < 90°$) whatever the intensity of interaction between the liquid and the solid. Wetting in this kind of systems can be discussed using the Young-Dupré equation and a simple nearest-neighbour interaction model described in Section 1.1. This model contains several approximations which make it only a rough description of the actual system. For instance, the solid and liquid metals are assumed to have the same structure and coordination, Z. However, for solid fcc metals $Z = 12$, while the average value of Z for liquid metals is in the range 9–11. Moreover, in this description, the interfacial energies are assumed to result only from bond energies and possible entropy contributions to these quantities are neglected. This approximation is particularly questionable for metal/metal interfaces because the excess entropy due to the interface can be substantial. This entropy results both from the variation of atomic order between the bulk liquid and liquid atoms at the interface (Ewing 1971, Spaepen and Meyer 1976) and from the atomic roughness of the solid in contact with the liquid (Eustathopoulos 1983). For metal/ceramic systems for which σ_{SL} is of the order of 1–2 J/m^2, entropic effects are comparatively weak and may be neglected. However, in metal/metal systems for which σ_{SL} is usually in the range 0.05 to 0.5 J/m^2, the contribution of interfacial entropy to σ_{SL} may become important. Despite drawbacks, the model is useful for qualitative and sometimes semi-quantitative discussion of the effects of variables such as the orientation of the solid surface and the intensity of solid-liquid interactions.

5.1. A PURE LIQUID METAL ON ITS OWN SOLID

As shown in Section 1.4.1, the Young-Dupré equation can be written:

$$\cos \theta = 2 \frac{\varepsilon_{SL}}{\varepsilon_{LL}} - 1 \tag{5.1}$$

Assuming that, for a pure metal on its own solid, the solid-liquid and liquid-liquid bond energies, ε_{SL} and ε_{LL}, are equal, which is the approximation used in the classical work of Skapski (1956), and knowing that the molar heat of melting of the pure metal L_m is related to pair energies by

$$L_m = ZN_a \frac{\varepsilon_{LL} - \varepsilon_{SS}}{2} \tag{5.2}$$

where Z is the coordination number and N_a is Avogadro's number, the expressions (1.11) for interfacial energy (written for a pure metal with $A = S$ and $B = L$) and (5.1) for contact angle become:

$$\sigma_{SL} = \frac{m_1 L_m}{\Omega_m} \tag{5.3}$$

$$\cos \theta = 1 \tag{5.4}$$

where Ω_m is the molar area and m_1 denotes the fraction of bonds which are of S/L type for a solid atom at the interface (see Figure 1.3). Equation (5.4) is in good agreement with experimental results showing contact angles close to $0°$ for fcc metals with low-anisotropy surface energy. Thus, the contact angle for Au/Au is $7°$ (Naidich et al. 1975) and for Cu/Cu close to $0°$ (Wenzl et al. 1976). For the semi-metal Ga/Ga, the contact angle is $6°$ (Dokhov et al. 1971) and for the semi-conductors Si/Si the angle is $12° \pm 2°$ (Naidich et al. 1975, Surek 1976) and for Ge/Ge is $9°$ to $30°$ depending on the orientation of the solid surface (Naidich et al. 1981). In view of the crude model, this quality of agreeement between experiments and model predictions is almost certainly due to compensating errors made in the model.

The variation of contact angle with *orientation* of surface planes can be understood by considering the Young-Dupré equation and the definition of W_a based on the simple model of Section 1.1:

$$W_a = -\frac{Zm_1}{\omega} \varepsilon_{SL} \tag{5.5}$$

where ω is the surface area per atom. When the atomic density of a surface plane increases, both m_1 and ω decrease but as the decrease of m_1 is more pronounced, the work of adhesion of the liquid on this plane decreases. Therefore, in systems developing strong interactions at the interface, the wetting of the less close-packed (high-index) planes should be better than that of the closest-packed (low-index) planes. This prediction is confirmed by the results of Naidich et al. (1981) showing that contact angles of liquid Ge on (111), (110) and (100) faces were found to be $30°$, $17°$ and $9°$ respectively, i.e., they decreased in the same order as the atomic density of these planes. Note that the opposite effect of atomic density is expected in liquid/solid systems with high cohesion energies but developing weak, van der Waals, interactions at the interface (see Section 8.1).

5.2. SYSTEMS WITH NEGLIGIBLE MUTUAL SOLUBILITY

Often the liquid and solid are not the same metal and sometimes they are mutually insoluble. When the equilibrium value of the molar fraction of solid component A in a liquid B $X_A^L \ll 1$, by equalling the chemical potentials of A in the solid and liquid phases, X_A^L is related to the regular solution parameter λ given by equation (4.3) by:

$$X_A^L \cong K_4 \exp\left[\frac{-\lambda}{RT}\right] \quad \text{with} \quad K_4 = \exp\left[-L_m^A\left(1 - \frac{T}{T_F^A}\right)/(RT)\right] \tag{5.6}$$

where L_m^A and T_F^A are the heat of melting and the melting temperature of pure A, respectively. Thus, a very low solubility of A in liquid B is expected for temperatures $T \ll T_F^A$ and high positive values of λ.

Typical examples are the Pb/Fe and Cu/W couples for which the solubility of Fe in molten Pb and W in molten Cu at temperatures close to the melting point of the liquid is as low as a few ppm. In this type of system for which spreading times are of the order of 10^{-2} second for millimetre-size droplets (see Section 2.1.1), the final contact angles can be as low as $10°$ (Table 5.1). This indicates that strong A-B interactions can be established *at the interface*, even for nearly insoluble metallic couples, i.e., for systems having weak A-B interactions in the *bulk liquid* ($\lambda \gg 0$).

To understand this apparent paradox, consider a Cu/W couple. For this system, $\lambda = 90$ kJ/mole (Appendix G) and the heats of evaporation are 316 kJ/mole for Cu ($= -ZN_a\varepsilon_{CuCu}/2$) and 821 kJ/mole for W ($= -ZN_a\varepsilon_{WW}/2$) (Hultgren et al. 1973). Thus the value of $ZN_a\varepsilon_{CuW}$ can be calculated from equation (4.3) and the various interaction values are presented in the bond energy diagram of Figure 5.1 in terms of molar units ($\varepsilon^* = ZN_a\varepsilon$). This diagram shows that even in the very insoluble Cu/W couple, the exchange energy λ is only a small quantity compared to the average cohesion energy of the two metals (λ is less than 10% of $(\varepsilon_{CuCu}^* + \varepsilon_{WW}^*)/2$). As a consequence, ε_{CuW}^* has a very negative value ($|\varepsilon_{CuW}^*|$ is

Table 5.1. Contact angles for nearly insoluble metal/metal systems. x_A^L is the molar fraction of A in liquid B at equilibrium. For all systems, the corresponding molar fraction of B in solid A is negligible ($x_B^S \approx 0$).

Liquid B /solid A	T (°C)	$x_A^{L \ (a)}$	Atmosphere	θ (deg)	Reference for θ values
Pb/Fe	350	$\sim 10^{-5}$	H_2	57	(Popel et al. 1971)
	750	3×10^{-5}	H_2	53	
	750	3×10^{-5}	Ar-5% H_2	53	(Gomez-Moreno et al.
	1000	4×10^{-4}	Ar-5% H_2	32	1982)
	1000	4×10^{-4}	H_2	27	(Eremenko and Lesnik 1963)
Ag/Fe	980	6×10^{-4}	Ar+ H_2	40	(Pique et al. 1981)
	1000	6×10^{-4}	He ($P_{O2} = 5 \times 10^{-19}$ atm)	57-45-36 [b]	(Tomsia et al. 1982)
Ag/W	960	6×10^{-9}	high vacuum	78	(Sugita et al. 1970)
	1000	6×10^{-9}	high vacuum	50	(Naidich 1981b)
Ga/W	975	2×10^{-3}	N_2-10% H_2	11	(Weirauch 1998)
Au/W	1063	1.37×10^{-3}	high vacuum	66	(Sugita et al. 1970)
Cu/W	1100	2×10^{-5}		10-55	see Table 5.2
Cu/Mo	1100	6.7×10^{-4}	high vacuum	30	(Yupko et al. 1991)
	1100	6.7×10^{-4}	He-18% H_2	10	(Yupko et al. 1986)

[a] estimated values using data of enthalpy of mixing of Appendix G except values for Pb/Fe (Stevenson and Wulff 1961) and Ag/Fe, Au/W, Cu/Mo (Massalski 1990) which are experimental ones.

[b] these values correspond to three kinds of Fe substrate containing 1200, 60 and 5 ppm of oxygen respectively.

even higher than $|\varepsilon^*_{CuCu}|$), which indicates that strong interactions can be established at the Cu/W interface.

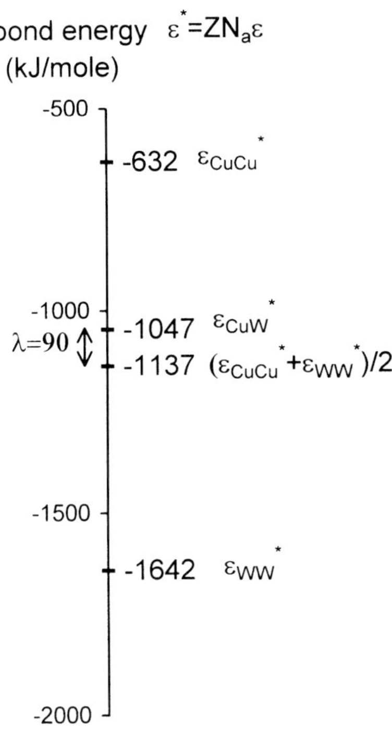

Figure 5.1. Bond energy diagram for the Cu-W system.

Writing equation (1.11) for a liquid B in contact with a solid A (Figure 5.2.a), and assuming that the bond energy between solid A and liquid B, ε^{SL}_{AB}, is equal to the bond energy between liquid A and liquid B, ε^{LL}_{AB}, as for an interface between a liquid and its own solid, we derive:

$$\sigma_{SL} = \frac{ZN_a m_l}{\Omega_m}\left(\varepsilon^{LL}_{AB} - \frac{\varepsilon^{SS}_{AA} + \varepsilon^{LL}_{BB}}{2}\right) \tag{5.7}$$

Introducing equations (4.3), (5.2) and (5.3) into (5.7), one gets:

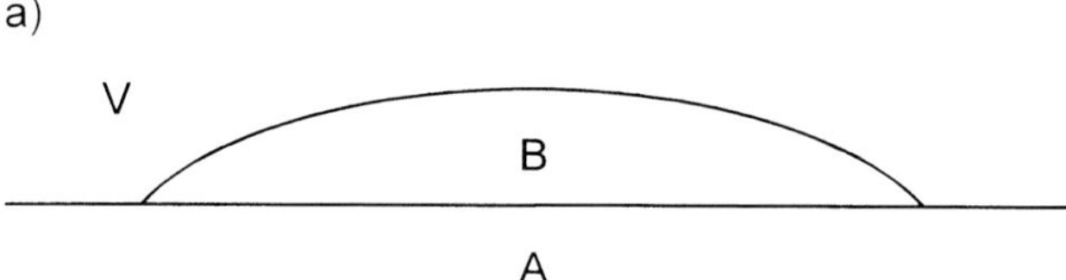

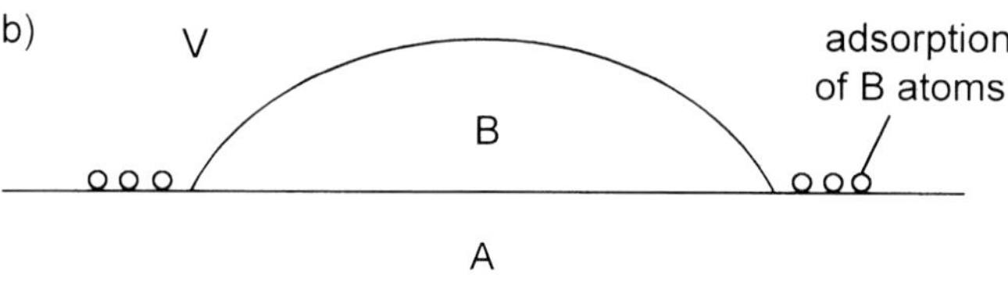

Figure 5.2. a) Wetting of liquid B on a "dry" A surface. b) The partial saturation of A broken bonds by adsorption of B decreases σ_{SV}^A and increases θ.

$$\sigma_{SL} = \sigma_{SL}^A + \frac{m_l \lambda}{\Omega_m} \tag{5.8}$$

The first term of the right hand side of equation (5.8) is the interfacial energy of pure solid A in contact with pure liquid A, while the second term takes into account the fact that the liquid is not pure A but pure B.

Introducing equation (5.8) in the Young equation $\cos\theta = (\sigma_{SV}^A - \sigma_{SL})/\sigma_{LV}^B$ and replacing σ_{LV}^B by $m_l L_e^B / \Omega_m$ where L_e^B is the molar heat of evaporation of B (see the similar equation (1.15)), one obtains:

$$\cos\theta = \frac{\sigma_{SV}^A - \sigma_{SL}^A}{\sigma_{LV}^B} - \frac{\lambda}{L_e^B} \tag{5.9}$$

Assuming perfect wetting of pure molten A on pure solid A at $T = T_F^A$ is also valid for supercooled liquid A, i.e., at $T < T_F^A$, then

$$\sigma_{SV}^A - \sigma_{SL}^A = \sigma_{LV}^A \tag{5.10}$$

and equation (5.9) becomes:

$$\cos\theta = \frac{\sigma_{LV}^{A}}{\sigma_{LV}^{B}} - \frac{\lambda}{L_{e}^{B}} \qquad (5.11)$$

As a general rule, $\sigma_{LV}^{A}/\sigma_{LV}^{B}$ in equation (5.11) is much higher than λ/L_{e}^{B}. For the Cu/W system, these terms are 1.9 and 0.3, and for the Pb/Fe system are 4 and 0.7. Therefore, the condition $\sigma_{LV}^{A} \gg \sigma_{LV}^{B}$, which is fulfilled for any combination of a low-melting point metal B and a high-melting point metal A, implies generally perfect wettability of A by B. Actually, wetting is good but not perfect (Table 5.1).

The main reasons for this lack of perfection are that (i) a drop of A supercooled at $T \ll T_{F}^{A}$ does not probably wet perfectly its own solid so that $\sigma_{SV}^{A} - \sigma_{SL}^{A}$ is lower than σ_{LV}^{A} (see (5.10)) and (ii) the solid surface is not pure A but is modified by adsorption of B atoms, Figure 5.2.b, which decreases the surface energy of the solid σ_{SV}^{A} and hence increases θ. The larger the difference $(\sigma_{LV}^{A} - \sigma_{LV}^{B})$, the greater the adsorption of B on A and the consequent decrease of the surface energy of the solid. This is confirmed by experimental results obtained by the multi-phase equilibria technique for this type of system showing that the ratio of the surface energy of solid A in equilibrium with a saturated vapour of B to the surface energy of pure solid A at the same temperature, $\sigma_{SV}^{A}/\sigma_{SV}^{0,A}$, lies in the range 0.6 to 0.8 (Eustathopoulos and Joud 1980).

In metal B/metal A systems, an increase in temperature is expected to favour desorption of B from the A surface, thus increasing σ_{SV}. Therefore, contact angles should decrease with increasing temperature as in Figure 5.3 not only because both σ_{LV} and σ_{SL} decrease with this parameter but also because σ_{SV} increases (Pique et al. 1981).

Contact angle values are sensitive to impurities, in particular to oxygen because this can modify not only the surface properties of liquid metals (as discussed in Section 4.1.1), but also the surface properties of solid metals and semi-conductors. Two different cases can occur, depending on the value of P_{O2} in the furnace and on the oxygen content of the solid metal. In the first case, a three dimensional oxide layer covers the solid substrate and a non-wetting contact angle is observed. In the second case, the substrate surface has adsorbed oxygen but remains metallic in character so that good wetting ($\theta \ll 90°$) is observed but the contact angles are very sensitive to environmental conditions. This effect is illustrated by the differing contact angles of Cu on polycrystalline W, determined by several teams (Table

5.2). In all cases, good wetting is observed but θ varies between $10°$ and $55°$. Low contact angles were observed when using a reducing atmosphere or *in situ* cleaning of the substrate surface with an electron beam, while the higher contact angles were observed using high vacuum or a neutral gas atmosphere. Similar variations in θ were observed for the Ag/Fe system using Fe substrates of different purity (Table 5.1). Pollution of W surface by oxygen is confirmed by the wetting kinetics. While spreading times as low as 10^{-2} second are expected in non-reactive systems (Section 2.1.1), a stationary contact angle was obtained only after 20 minutes at $1100°C$ for Cu on W (Lorrain 1996). In this case, spreading kinetics are controlled by deoxidation of W, occurring by evaporation of W suboxides. At a lower temperature ($900°C$), deoxidation kinetics slow down and in some cases no deoxidation occurs, resulting in contact angles as high as $130°$, typical of oxidized surfaces (Figure 5.4).

Taking into account the fact that deoxidation of solid metals is very difficult at low temperatures, it is usual to observe non-wetting contact angles for insoluble metal/metal systems.

Table 5.2. Contact angles measured by different authors for the Cu/W system at $1100°C$.

Substrate characteristics	Atmosphere	θ (deg)	Reference
surface purified by electron beam	high vacuum ($\approx 10^{-6}$ mbar)	10	(Doussin and Omnes 1967)
prior heat treatment for 1h at $1100°C$ in H_2	H_2	30	(Ligachev et al. 1979)
	He	55	
–	H_2	10	(Lesnik et al. 1984)
prior heat treatment for 30 min at $900°C$ under vacuum	high vacuum ($<10^{-5}$ mbar)	50	(Nicholas and Poole 1967)
–	high vacuum ($<10^{-5}$ mbar)	40	(Gorskii et al. 1974)
prior heat treatment for 2h at $1100°C$ under vacuum ($R_a = 15$ nm)	high vacuum ($< 5 \times 10^{-5}$ mbar)	47 ± 5	(Lorrain 1996)

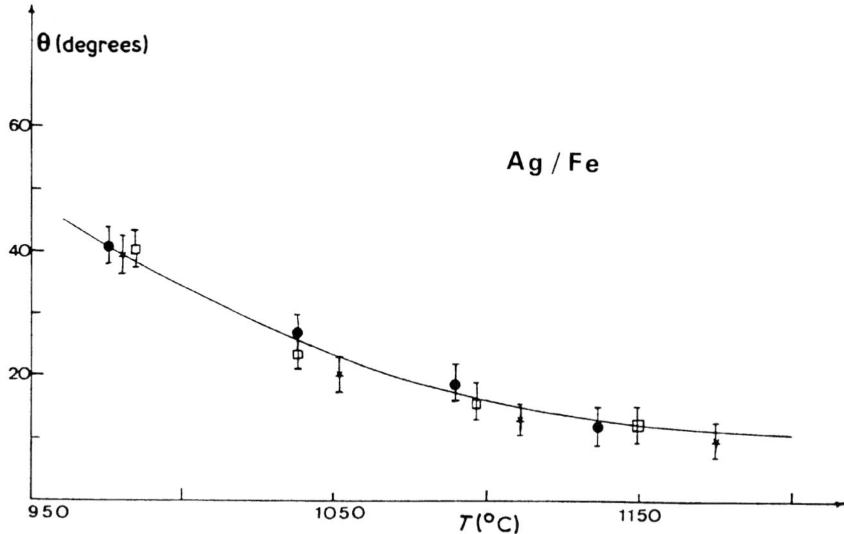

Figure 5.3. Stationary contact angle versus temperature data for the Ag/Fe system obtained from three experiments (Pique et al. 1981) [16].

5.3. SYSTEMS WITH SIGNIFICANT MUTUAL SOLUBILITY

When the solubility of A in B or B in A is not negligible, definition of a contact angle using the Young equation is problematic (see Figure 2.18) unless the two phases are presaturated. Dissolution of A in B, or B in A, or both, will produce a continuous change in drop volume and a non-planar solid/liquid interface. Therefore only data obtained with presaturated phases are given in Table 5.3. For all systems, except Si/Au, the regular solution parameter λ is positive.

When comparing results in Table 5.3 with those for insoluble systems presented in Table 5.1, the contact angles for systems with mutual solubility are usually lower. (Note, however, that the difference is small, even less than that caused by environment effects for the Cu/W system (Table 5.2)). For instance, the contact angle of Pb on Ni at 740°C is 25° i.e., significantly lower than the contact angle of the insoluble Pb/Fe couple ($\theta = 53°$, Table 5.1). Because Ni is not tensio-active in molten Pb, the surface energy of Pb-4 at.% Ni is nearly equal to that of pure Pb. Moreover, the surface energies of solid Ni and Fe are very close (see Table 4.2). Therefore, the different contact angles in these systems must be due to significantly different solid-liquid interfacial energies. Similar reasoning accounts for the different contact angles observed at 350°C for Pb on Ag (19°) and on Fe (57°).

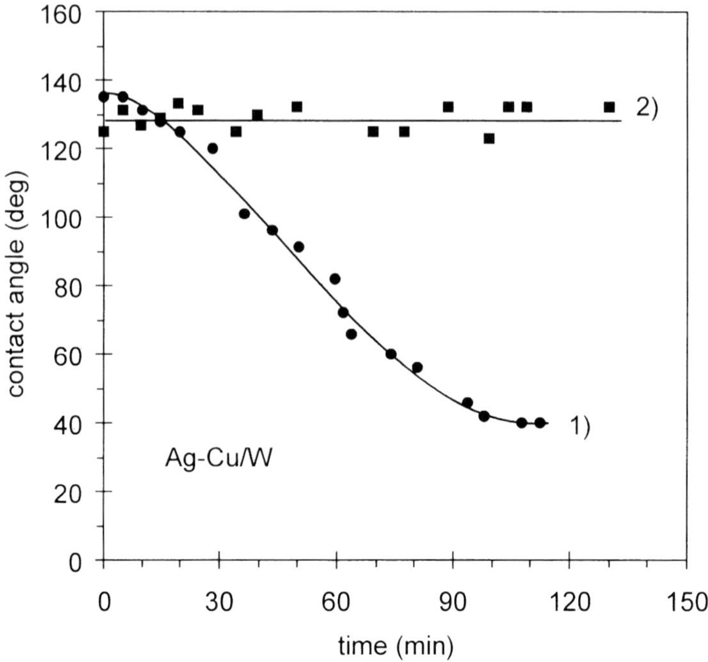

Figure 5.4. 1) Contact angle versus time for a eutectic (Ag-Cu) drop on polycrystalline W at 900°C in a high vacuum. Before the experiment, the W substrate was heat-treated in high vacuum at 1100°C for 2 h. Despite this treatment, the surface remained oxidised and a slow spreading, controlled by W deoxidation, was observed. 2) The same without prior heat treatment of W. In this case segregation of O at the W surface, by fast grain-boundary diffusion, prevents deoxidation of the substrate, resulting in non-wetting behaviour. From Lorrain (1996).

By neglecting segregation at the S/L interface, a simple expression can be obtained relating the solid-liquid interfacial energy σ_{SL} of a system displaying mutual solubility to the energies of the pure components, σ_{SL}^i (i = A, B), the molar fractions of components i in phases j, X_i^j, and the regular solution parameters in the solid and liquid phases, λ^S and λ^L (Gressin et al. 1982):

$$\sigma_{SL} = \sigma_{SL}^A X_A^S + \sigma_{SL}^B X_B^S + \frac{m_l \lambda^L}{\Omega_m}(X_A^L X_B^S + X_A^S X_B^L) - \frac{m_l \lambda^L}{\Omega_m} X_A^L X_B^L - \frac{m_l \lambda^S}{\Omega_m} X_A^S X_B^S$$

$$(5.12)$$

Table 5.3. Contact angles for systems with mutual solubility.

Liquid B /solid A	T (°C)	X_A^{L} [b]	X_B^{S} [b]	Atm.	θ (deg)	Reference for θ values
$Pb^{(a)}/Ag^{(a)}$	350	0.065	$\approx 10^{-2}$	$Ar+H_2$	19	(Passerone et al. 1982)
$Pb^{(a)}/Cu$	400	0.007	≈ 0	H_2	43	(Bailey and Watkins
	700	0.06	≈ 0	H_2	25	1951–52)
	700	0.06	≈ 0	$Ar+H_2$	21	(Coudurier et al. 1977)
$Pb^{(a)}/Ni$	740	0.04	5×10^{-4}	$Ar+H_2$	25	(Foucher 1998)
$Ag^{(a)}/Cu^{(a)}$	782	0.40	0.05	He	19	(Sharps et al. 1981)
$Si^{(a)}/Au$	370	0.81	≈ 0	high vacuum	34	(Naidich et al. 1975)

[a] presaturated in the other element
[b] from (Massalski 1990) except for Pb/Ni (Foucher 1998)

When the solubility of B in the solid A is negligible, this equation reduces to:

$$\sigma_{SL} = \sigma_{SL}^{A} + \frac{m_1 \lambda^{L}}{\Omega_m}(1 - X_A^{L})^2 \tag{5.13}$$

and for $X_A^{L} \ll 1$, equation (5.13) reduces to (5.8). For example, λ^{L} values are very different for the Pb/Fe and Pb/Ni systems (the enthalpy of mixing of Fe in molten Pb is 91 kJ/mole, but for Ni is 40 kJ/mole (Appendix G)), resulting in a substantial difference in σ_{SL}. For both Ni and Fe, the first term of equation (5.13) is about 0.3 J/m^2 (Eustathopoulos 1983), but the second term is equal to 0.28 J/m^2 for Pb/Ni and 0.65 J/m^2 for Pb/Fe.

Equations (5.6) and (5.13) clarify the relation between solubility and wetting. For a given solid metal A, at fixed temperature, a comparatively high value of X_A^{L} results from a low value of λ, which in turn leads to a low value of σ_{SL} (equation (5.13)). This relation between wettability and solubility in metallic systems, through their dependence upon the same energy parameter (i.e., λ), was identified long ago by Wassink (1967).

5.4. EFFECTS OF ALLOYING ELEMENTS

The wetting of a liquid B on a solid A may be improved by additions to B of an element M. Thus, while Figure 5.5 shows that additions of Sn and Pb in Ag have little effect on wetting of Fe, that of Cu is moderate and of Ni is strong. Because the surface energies of Pb (390 mJ/m^2) and Sn (490 mJ/m^2) are much lower than that of Ag (910 mJ/m^2), additions of 10 at.% of Pb and Sn into Ag produce a decrease of the surface energy of as much as 30% (Eustathopoulos and Joud 1980). According to the Young equation, this decrease of σ_{LV} would lead to perfect wetting on Fe if σ_{SV} and σ_{SL} are constant. This is not observed experimentally, so σ_{SV} and σ_{SL} cannot be constant. Because the addition of these elements cannot increase σ_{SL} significantly (see Sections 6.5.1 and Figure 6.28) one must conclude that adsorption of these elements decreases not only σ_{LV} for molten Ag but also σ_{SV} for solid Fe. This suggests that an increase in wetting can seldom be produced by low-melting-point, low-surface-energy, additives because the resultant decrease of σ_{LV} may be compensated by that of σ_{SV} so that the net effect on the contact angle is weak.

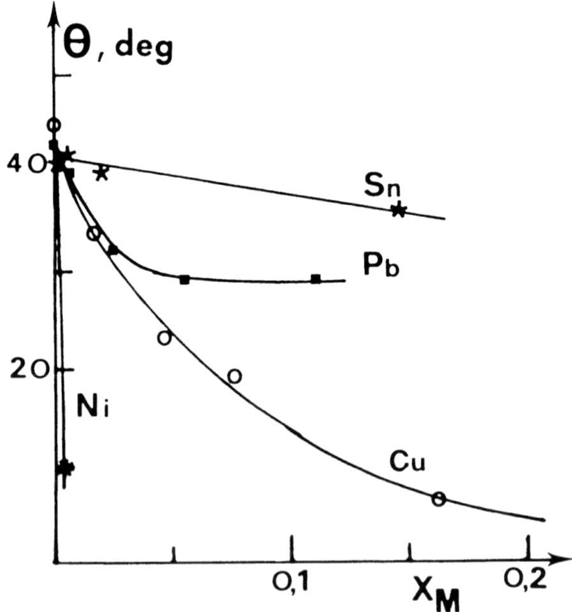

Figure 5.5. Influence of Sn, Pb, Cu and Ni dissolved in Ag on the contact angles of the liquid Ag/solid Fe system at 965°C in an Ar-5% H$_2$ atmosphere (Coudurier et al. 1987) [3].

Compared to Sn and Pb, molten Cu has a high liquid/vapour surface energy (1350 mJ/m^2) that is higher than the surface energy of Ag and closer to the surface energy of Fe. For Ni, the value of σ_{LV} is even higher than that for Ag and does not differ significantly from that of Fe. Thus the beneficial influence of Cu and Ni on θ must be due to a decrease in the solid Fe/liquid Ag interfacial energy. For an element M at infinite dilution in a liquid matrix B, the adsorption energy of an atom M at the liquid side of the interface of a liquid B/solid A system with a negligible mutual solubility is given by (Gomez-Moreno et al. 1982):

$$E_{M(B)}^{\infty,SL} = m_l(\lambda_{AM} - \lambda_{BM} - \lambda_{AB}) \tag{5.14}$$

The more negative the value of $E_{M(B)}^{\infty,SL}$, the stronger the decrease in σ_{SL} and θ (see equation 6.28). For A = Fe, B = Ag and M = Cu, $E_{Cu(Ag)}^{\infty,SL}/m_l$, calculated using λ values of Appendix G, is -70 kJ/mole, indicating a strong adsorption of Cu. For M = Ni, this adsorption is even stronger, with $E_{Ni(Ag)}^{\infty,SL}/m_l$ equal to -180 kJ/mole. Thus there is a much stronger effect on the contact angle of Ag by additions of Ni than of Cu (Figure 5.5). Note that a complete model, applicable also to concentrated B-M solutions and taking into account both adsorptions of M at the liquid side and the solid side of the interface, was proposed by Chatain et al. (1985). However, selection of alloying elements M for improved wetting based only on the value of $E_{M(B)}^{\infty,SL}$ given by (5.14) and on the melting point of M achieved significant improvements with the Pb-M/Fe systems (Gomez-Moreno et al. 1982) and with the Ag-M/Fe systems (Coudurier et al. 1987).

5.5. SYSTEMS THAT FORM INTERMETALLIC COMPOUNDS

Comparison of the results for systems that form compounds (Table 5.4), for which the regular solution parameter λ is negative, with those for systems that do not form compounds (Tables 5.1 and 5.3) does not reveal strong differences in wetting. A more detailed examination of data obtained by individual teams for the same substrate, temperature and working conditions, but using two liquid metals, only one of which forms intermetallic compounds, shows some limited effects on contact angle values. Thus, the contact angle of Sn on Cu at 400°C (Table 5.4) is 20° lower than the contact angle of Pb on the same substrate (Table 5.3). A similar difference exists between contact angle values for Sn/Fe (Table 5.4) and Pb/Fe (Table 5.1). Moreover, the final wetting in systems in which intense reactions occur at the interface is usually less sensitive to environmental factors than in insoluble systems. This is because thin oxide layers on solid or liquid metals are disrupted locally by reaction product formation initiated either by diffusion of reacting components through the thin oxide layer or at oxide defects.

Table 5.4. Contact angles for systems that form intermetallic compounds.

Liquid B / solid A	T (°C)	X_A^{L} [a]	Intermetallic compounds	Atm. [b]	θ (deg)	Ref. for θ values
Sn/Fe	350	≈ 0	$FeSn_2$, FeSn	H_2	43	(Popel et al.
	750	0.02	FeSn, Fe_3Sn_2	H_2	38	1971)
	420	≈ 0	$FeSn_2$, FeSn	HV	42	(Harvey 1965)
Sn/Cu	400	0.12	Cu_6Sn_5, Cu_3Sn, Cu_4Sn	H_2	23	(Bailey and Watkins 1951–52)
Al/Fe	750	0.022	$FeAl_3$, Fe_2Al_5, $FeAl_2$, FeAl	HV	33	(Eremenko et al. 1973)
Al/Fe [c]	700	0.012	$FeAl_3$, Fe_2Al_5, $FeAl_2$, FeAl	H_2	35	(Ebrill 1999)
Al/Fe [d]	680	0.011	$FeAl_3$, Fe_2Al_5, $FeAl_2$, FeAl	H_2	30	(Tarasova et al. 1980)
In/Co	300	≈ 0	$CoIn_3$, $CoIn_2$	HV	32	(Lesnik et al.
	500	0.003	$CoIn_2$	HV	18	1970)

[a] From (Massalski 1990), [b] HV = high vacuum, [c] Fe-0.055 wt.% C-0.24 wt.% Mn, [d] Fe-0.9 wt.% C-2 wt.% Mn

The effect of temperature on contact angle values is similar for systems that do or do not form intermetallic compounds, as can be seen by comparison of data for In/Co which does, Table 5.4, and Ag/Fe which does not, Figure 5.3. For the In/Co system, a final contact angle is observed in less than 10^{-2} second whatever the temperature and thereafter remains constant for up to 1 h (Lesnik et al. 1970), as for non-reactive couples like Pb/Fe (Section 2.1.1). In the highly reactive Al/Fe system, spreading times close to 10^{-2} second have been found, the final contact angle being close to 35°, Figure 5.6. Similar good wetting has been obtained for Al

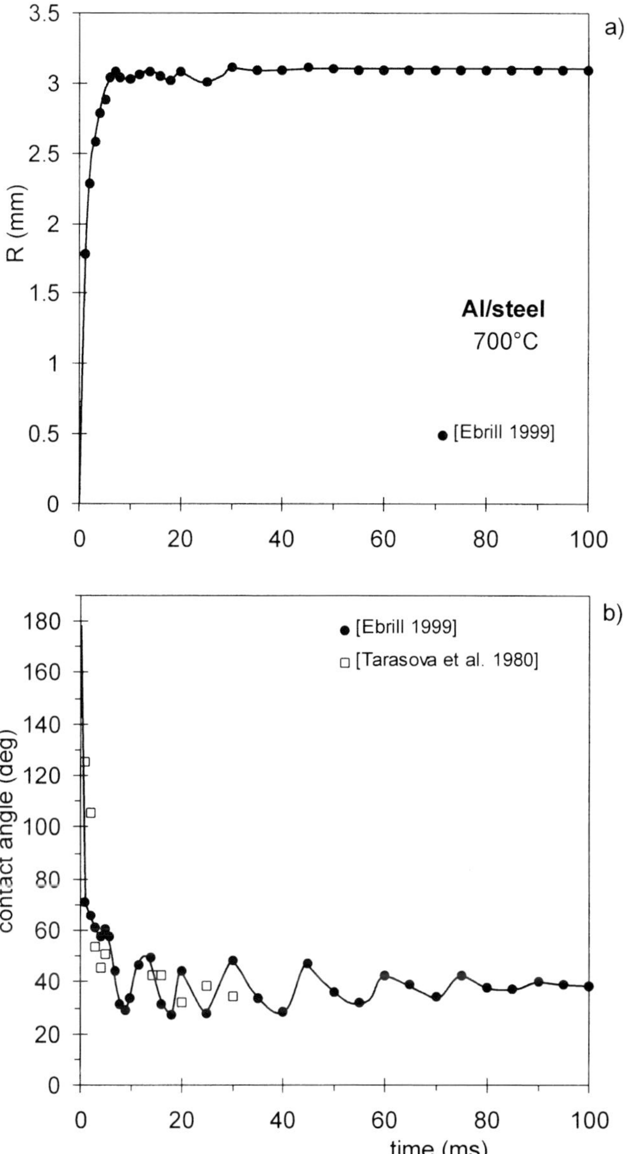

Figure 5.6. Variation of drop base radius (a) and contact angle (b) with time for Al on steel at $T \cong 700°C$ in dry H_2 using the dispensed drop technique (drop mass of about 50 mg). In the experiment of Ebrill, oscillations of the triple line are due to the kinetic energy of the drop falling from a height of 12–13 mm (see Section 2.1.1). Data from works reported in (Tarasova et al. 1980, Ebrill 1999).

on Ni and Co after much longer spreading times (about 20 seconds) owing probably to oxidation of Al (Eremenko et al. 1980). Note that for Al and Al alloys on low-C steel, formation of intermetallics at the melt/solid interface was detected even for contact times as short as 20 milliseconds (Durandet et al. 1998).

5.6. WETTING UNDER TECHNICAL CONDITIONS

The behaviour described in the preceding Sections has been that observed during rigorously controlled laboratory studies using carefully cleaned materials in chemically inert environments. In practice, it may not be possible or economic to adopt such procedures, but nevertheless observations of the wetting behaviour in technical conditions can be interpreted provided it is realised that most solid surfaces of technical materials are covered by oxide films. Thus consideration must be given to how metal-metal contact can be achieved as well as to the implications of such contact for wetting behaviour.

The procedures that have been adopted to achieve adequate wetting behaviour of technically important materials are multitudinous and often very material- and application-specific. Nevertheless, Bailey and Watkins (1951–52) systematically studied the wetting behaviour of a considerable number of metal/metal combinations in an attempt to identify improved ways of soldering, brazing and liquid coating. They assessed wetting behaviour in flowing H_2 by a variety of test methods: by flooding the solid surface with a liquid and observing the behaviour when the melt was decanted, by immersing a solid plate and observing draining during its withdrawal, and by spreading drop tests. In the flooding test, a solid foil, after heat treatment at 600°C in H_2 for 15 minutes, came into contact with a liquid metal to form a continuous film on its surface (Figure 5.7.1). Thereafter, the foil was tilted for 15 seconds, leading either to configuration 2.a of Figure 5.7 in which the film is stable, or to configuration 2.b in which the film has collapsed into drops. A third case was observed and classified as "non-wetting", in which no liquid adhered as a film or drops to the solid surface. Bailey and Watkins found that couples forming stable films had receding contact angles close to 0° and equilibrium contact angles of about 20°. Similarly, in systems forming drops, the equilibrium contact angle was more than about 20° and non-wetting couples had equilibrium contact angles of more than 90°.

Selected results of Bailey and Watkins (1951–52) are given in Table 5.5. These demonstrate that configuration 2.a of Figure 5.7 is achieved with a wide range of couples that form intermetallics compounds or in which the liquid member dissolves in the solid. For Ag/Fe, the drop-forming behaviour accords with the results of sessile drop experiments which identified contact angles of 36°–57° as

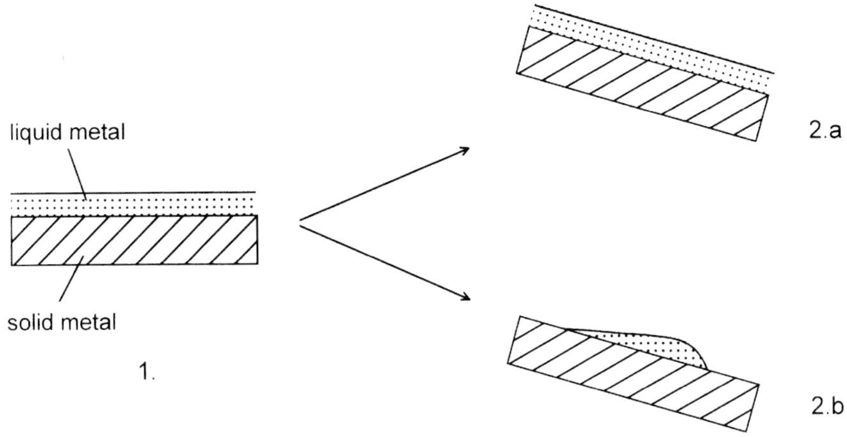

Figure 5.7. Flooding test used in experiments of Bailey and Watkins (1951–52) to determine if a L/S system forms stable films (2.a) or drops (2.b).

equilibrium values at about 1000°C (Table 5.1). Bailey and Watkins identified other couples that were drop-forming or non-wetting (Table 5.5): they all resemble Ag/Fe in being insoluble, the only exception being the Bi/Cu system in which Bi at 400°C dissolves a limited quantity of Cu. The Pb/Fe system was described in Section 5.2 as having a sessile drop contact angle of 57° at 350°C but was classified by Bailey and Watkins as non-wetting at 400°C. However, this possible contradiction can be resolved by reference to the results of their spreading drop tests which yielded contact angles of more than 90° up to 1000°C for both Pb and Bi on the same particularly impure Fe. These results strongly suggest that even in their highly reducing H_2 atmosphere, impure Fe substrates were oxidized.

The Bailey and Watkins study shows that continuous liquid film formation is favoured by interfacial reactivity due to mutual solubility or to formation of intermetallics. This correlation between film stability and reactivity is due to several reasons:

1) As shown in the previous Sections, the stable contact angles for systems displaying mutual solubility or forming intermetallics are somewhat lower than those for insoluble couples.

2) Systems in which there is some reactivity between liquid and solid phases are less sensitive to oxygen pollution than insoluble couples. This is shown by comparing results in Table 5.5 for different liquid metals on Fe. Couples forming intermetallics (Sn/Fe, Zn/Fe) readily established a metal-metal interface and stable liquid film. For the same Fe solid substrate in contact with insoluble metals

such as Pb and Bi, a metal-metal interface is not formed even at temperatures as high as 900°C.

3) Collapse of a liquid film into drops depends on both equilibrium contact angle and the topography of the solid/liquid interface, so film stability is favoured by roughness of the surface. Dissolution of the solid in the liquid, grain-boundary grooving to produce the "secondary wetting" analysed in Appendix H or formation of intermetallics can increase the roughness of the solid surface and cause near-zero receding contact angles.

Table 5.5. Wetting data derived from flooding tests (Bailey and Watkins 1951–52).

System	Temperature (°C)	Phase diagram characteristics	Behaviour
Bi/Au	300	intermetallics	film
Cd/Cu	350	intermetallics	film
Bi/Ni	400	intermetallics	film
Pb/Au	400	intermetallics	film
Sn/Fe	400	intermetallics	film
Zn/Au	450	intermetallics	film
Zn/Cu	500	intermetallics	film
Zn/Fe	500	intermetallics	film
Sb/Ag	550 [*]	intermetallics	film
Sb/Cu	600 [*]	intermetallics	film
Al/Fe	700	intermetallics	film
Sb/Fe	700	intermetallics	film
Pb/Ag	400	soluble	film
Ag/Cu	850 [*]	soluble	film
Ag/Au	1000	soluble	film
Ag/Ni	1000	soluble	film
Bi/Cu	400	2 at.% Cu in liquid Bi	drops
Bi/Fe	400	insoluble	non-wetting
Pb/Cu	400	insoluble	drops
Pb/Fe	400	insoluble	non-wetting
Ag/Fe	1000	insoluble	drops

[*] using eutectic liquid

Bailey and Watkins used solid metal substrates that were only moderately oxidizable and therefore their observations are not directly relevant to many current technical requirements. A similar systematic survey using metals and alloys with high oxygen affinities is not known but nevertheless some more

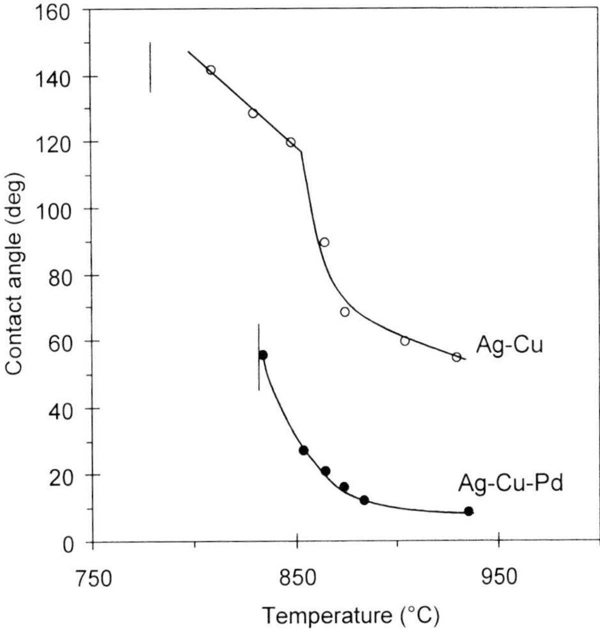

Figure 5.8. Contact angle data for Ag-28 wt.% Cu and Ag-16 wt.% Cu-14 wt.% Pd on AISI 321 steel in a vacuum of 2×10^{-4} Pa during a temperature rise with a heating rate of 5°C/min. The vertical bars identify melting temperatures. From (McGurran and Nicholas 1985).

restricted studies cast light on the influence of metal-metal interactions on wetting behaviour. Thus McGurran and Nicholas (1985) examined the ability of Ag-Cu alloys to wet AISI 321 steel, a Fe-18Cr-8Ni-1Ti (in wt.%) alloy that forms a chemically stable surface of Cr_2O_3. Abraded and degreased coupons of the steel were wetted by sessile drops of Ag-28 wt.% Cu at temperatures of 850°C in a good vacuum. However, sessile drops of an Ag-16 wt.% Cu-14 wt.% Pd alloy wetted much better as soon as they melted at 835°C, as shown in Figure 5.8. Pd had two effects on wetting. First, it improved the "intrinsic" wetting of Ag-Cu eutectic on steel, as seen by the contact angles obtained at the highest temperature in Figure 5.8. This can be easily explained using equation (5.14). Taking A = Fe, B = Ag and M = Pd, we find $E_{Pd(Ag)}^{\infty,SL}/m_1 = -104$ kJ/mole, which indicates a stronger effect of Pd on σ_{SL} and θ than Cu ($E_{Cu(Ag)}^{\infty,SL}/m_1 = -70$ kJ/mole). Second, Pd favoured the disruption of the oxide film on the substrate, because its solubility in Fe is much greater than that of Ag or Cu. Oxide film disruption implies that the surface oxide is flawed in at least some locations. Diffusion of Pd into the substrate in such regions to form a solid solution will cause a volume change and this then could produce decohesion of the oxide/substrate interface to increase the area of liquid metal/solid metal contact and hence decrease the contact angle. Other evidence consistent with such a picture is the detrimental effect of oxidising the steel surface to produce thick oxide films; oxidation for 1 hour at 750°C increased the Ag-Cu-Pd contact angle to 52° while oxidation for 16 hours increased it to 135°. Degrading the vacuum level, and hence presumably increasing the partial pressure of O_2 also caused contact angles to rise, as shown in Figure 5.9. Finally it should be noted that such effects are not specific to surface films of Cr_2O_3, qualitatively similar observations having been made by Wall and Milner (1961–62), for substrates of Al and alloys containing Al whose surfaces were coated by exceptionally tenacious films of Al_2O_3.

The need to deoxidise metallic solid surfaces to make them wettable by liquid metals has been recognized for a long time and various types of liquid fluxes are used to achieve this, as discussed in Section 10.2.

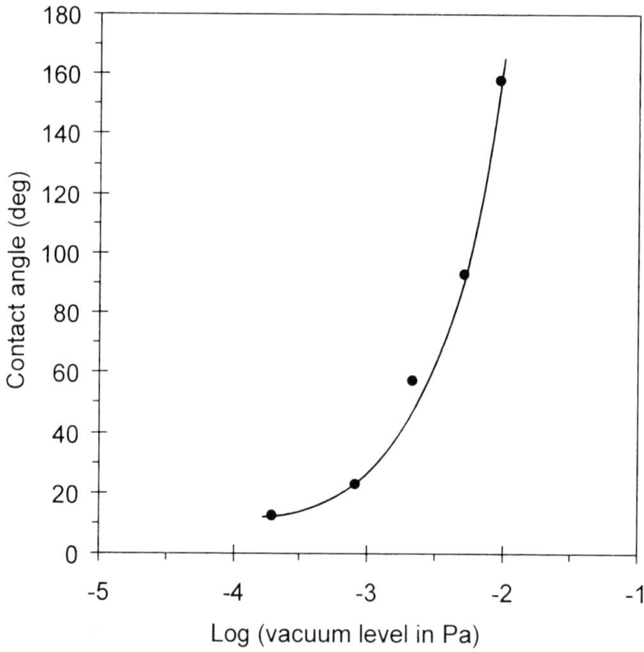

Figure 5.9. The wettability of AISI 321 steel by Ag-16 wt.% Cu-14 wt.% Pd at 880°C as function of vacuum quality (McGurran and Nicholas 1985).

5.7. CONCLUDING REMARKS

Liquid metals wet well metallic substrates ($\theta \ll 90°$) whatever the intensity of interfacial interactions unless the metallic surfaces are oxidized. Adsorption of oxygen on solid substrates can increase contact angles by tens of degrees. Intrinsic wetting seems to be slightly improved in systems with some solubility or which form intermetallics. The main effect of interfacial reactions is the disruption of oxide layers covering the metallic surfaces, allowing the formation of real metal-metal interfaces.

REFERENCES FOR CHAPTER 5

Bailey, G. L. J. and Watkins, H. C. (1951-52) *J. Inst. Metals*, **80**, 57

Chatain, D., Pique, D., Coudurier, L. and Eustathopoulos, N. (1985) *J. Mater. Sci.*, **20**, 2233

Coudurier, L., Eustathopoulos, N., Joud, J. C. and Desré, P. (1977) *J. Chimie Physique*, **3**, 289

Coudurier, L., Pique, D. and Eustathopoulos, N. (1987) *J. Chimie Physique*, **84**, 205

Dokhov, M. P., Zadumkin, S. N. and Karashaev, A. A. (1971) *Russian Journal of Physical Chemistry*, **45**, 1061

Doussin, L. and Omnes, J. (1967) Technical report (report 1/1259M), Office National d'Etudes et de Recherches Aérospatiales, Direction des Matériaux, Chatillon (France)

Durandet, Y., Strezov, L. and Ebrill, N. (1998) in *Proceedings of the 4th International Conference on Zinc and Zinc alloys Coated Steel Sheet (GALVATECH '98)*, Chiba, Japan, The Iron and Steel Institute of Japan, p. 147

Ebrill, N. (1999) Ph.D. Thesis, University of Newcastle, Australia

Eremenko, V. N. and Lesnik, N. D. (1963) in *The Role of Surface Phenomena in Metallurgy*, Ed. V. N. Eremenko, Consultants Bureau, New York, p. 102

Eremenko, V. N., Lesnik, N. D., Pestun, T. S. and Ryabov, V. R. (1973) *Poroshkovaya Metallurgiya*, **7**, 58 (English translation p. 565)

Eremenko, V. N., Ivanova, T. S. and Lesnik, N. D. (1980) *Poroshkovaya Metallurgiya*, **10**, 33 (English translation p. 689)

Eustathopoulos, N. and Joud, J. C. (1980) in *Current Topics in Materials Science*, Vol. 4, Ed. E. Kaldis, North-Holland Publishing Company, p. 281

Eustathopoulos, N. (1983) *International Metals Reviews*, **28**, 189

Ewing, R. H. (1971) *J. Cryst. Growth*, **11**, 221

Foucher, J. (1998) DEA report, INP Grenoble, LTPCM, France

Gomez-Moreno, O., Coudurier, L. and Eustathopoulos, N. (1982) *Acta Metall.*, **30**, 831

Gorskii, D. V., Lesnik, N. D., Teodorovich, O. K. and Flis, A. A. (1974) *Adgez. Rasplavov*, **5**, 95

Gressin, P., Eustathopoulos, N. and Desre, P. (1982) *J. Chimie Physique*, **79**, 545

Harvey, D. J. (1965) in *Liquids : Structure, Properties, Solid Interactions*, ed. T. J. Hughell, Elsevier, Amsterdam, p. 287

Hultgren, R., Desai, P. D., Hawkins, D. T., Gleiser, M. and Kelley, K. K. (1973) *Selected Values of the Thermodynamic Properties of the Elements*, American Society for Metals, Metals Park, Ohio 44073

Lesnik, N. D., Pestun, T. S. and Eremenko, V. N. (1970) *Poroshkovaya Metallurgiya*, **10**, 83 (English translation p. 849)

Lesnik, N. D., Minakova, R. V. and Flis, A. A. (1984) *Adgez. Rasplavov Paika Mater.*, **13**, 54

Ligachev, A. E., Maurakh, M. A., Mitin, B. S. and Shlyapin, S. D. (1979) *Fiz. Obrab. Mater.*, **5**, 155

Lorrain, V. (1996) Ph.D. Thesis, INP Grenoble, France

Massalski, T. B. (1990) *Binary Alloy Phase Diagrams*, 2nd edition, ASM International

McGurran, B. and Nicholas, M. G. (1985) *Brazing and Soldering*, **8**, 43

Naidich, Y. V., Perevertailo, V. M. and Obushchak, L. P. (1975) *Poroshkovaya Metallurgiya*, **7**, 63 (English translation p. 567)

Naidich, Y. V., Grigorenko, N. F. and Perevertailo, V. M. (1981) *J. Cryst. Growth*, **53**, 261

Naidich, Y. V. (1981b) in *Progress in Surface and Membrane Science*, vol. 14, ed. by D. A. Cadenhead and J. F. Danielli, Academic Press, New York, p. 353

Nicholas, M. and Poole, D. M. (1967) *J. Mater. Sci.*, **2**, 269

Passerone, A., Sangiorgi, R. and Eustathopoulos, N. (1982) *Scripta Met.*, **16**, 547

Pique, D., Sangiorgi, R. and Eustathopoulos, N. (1981) *Ann. Chim. (Sci. Mater.)*, **6**, 443

Popel, S. I., Kozhurkov, V. N. and Zakharova, T. V. (1971) *Zashchita Metall.*, **7**, 421 (English translation)

Sharps, P. R., Tomsia, A. P. and Pask, J. A. (1981) *Acta Metall.*, **29**, 855

Skapski, A. S. (1956) *Acta Metall.*, **4**, 576

Spaepen, F. and Meyer, R. B. (1976) *Scripta Met.*, **10**, 37

Stevenson, D. A. and Wulff, J. (1961) *Trans. Met. Soc. AIME*, **221**, 271

Sugita, T., Ebisawa, S. and Kawasaki, K. (1970) *Surface Science*, **20**, 417

Surek, T. (1976) *Scripta Met.*, **10**, 425

Tarasova, A. A., Novozhilova, A. Y. and Filippova, I. A. (1980) *Adhesion of Melts and Soldering of Materials*, **6**, 10 (in Russian)

Tomsia, A. P., Feipeng, Z. and Pask, J. A. (1982) *Acta Metall.*, **30**, 1203

Wall, A. J. and Milner, D. R. (1961-62) *J. Institute of Metals*, **90**, 394

Wassink, R. J. K. (1967) *J. Institute of Metals*, **95**, 38

Weirauch, D. A. (1998) *J. Mater. Res.*, **13**, 3504

Wenzl, H., Fattah, A. and Uelhaff, W. (1976) *J. Cryst. Growth*, **36**, 319

Yupko, V. L., Levchenko, G. V., Luban, R. B. and Kryzhanovskaya, R. I. (1986) *Poroshkovaya Metallurgiya*, **11**, 64 (English translation p. 918)

Yupko, V. L., Garbuz, V. V. and Kryuchkova, N. I. (1991) *Poroshkovaya Metallurgiya*, **10**, 72 (English translation p. 872)

Chapter 6
Wetting properties of metal/oxide systems

Knowledge of wetting and bonding in metal/oxide systems is important in many fields of materials engineering because oxides are the most widely used ceramics. Moreover, metal/oxide interfaces play also a key role in metal/metal and metal/non-oxide ceramic couples in which one of the partners is an oxidisable material, such as stainless steel or SiC. Finally, certain oxides lend themselves well to fundamental studies because they can be obtained easily as high-purity and monocrystalline (Al_2O_3, MgO) or amorphous (SiO_2) solids.

In the first Section, attention is paid to distinguishing between reactive and non-reactive systems from the point of view of wettability. Then, after describing wetting and bonding of non-reactive couples, we discuss the effect on these characteristics of oxygen, which is the most common impurity in solid/liquid/vapour systems, as well as the effect of reactive and non-reactive alloying elements. Finally, in a short Section, we consider some results for the wetting of fluorides which like oxides are very ionic.

6.1. REACTIVE AND NON-REACTIVE SYSTEMS

At high temperatures, there is always some dissolution of the oxide in the liquid metal but to what extent can this reactivity affect wettability? Useful insight can be obtained by studying spreading kinetics: the time for millimetre size droplets to reach capillary equilibrium is less than 10^{-1} second for non-reactive systems (see Section 2.1.1), so much slower spreading kinetics are a strong indication of control by interfacial reactions.

Another method of obtaining the needed insight is to plot equilibrium or stationary contact angles for a series of metal/oxide systems versus a parameter characterising the oxidation-reduction reaction between a liquid metal M and an oxide A_nO_m, leading to the formation of M_pO_q oxide (Naidich 1981, Eustathopoulos and Drevet 1994), or more simply oxide dissolution in the liquid. Consider at a given temperature a pure liquid metal M on an oxide substrate, for instance Al_2O_3. The dissolution reaction of Al_2O_3 in M at the solid/liquid interface can be written as:

$$\frac{1}{2}Al_2O_3 \Leftrightarrow (Al) + \frac{3}{2}(O) \tag{6.1}$$

where the parentheses mean that the elements are dissolved in the matrix M. The equilibrium constant K_D for reaction (6.1) is:

$$K_D(T) = a_{Al}a_O^{3/2} = \exp\left(\frac{\Delta G_{f(Al_2O_3)}^0}{2RT}\right) \tag{6.2}$$

where a_{Al} is the thermodynamic activity of Al in the alloy (reference state: pure liquid Al), a_O is the activity of dissolved oxygen (reference state: one atmosphere of pure O_2) and $\Delta G_{f(Al_2O_3)}^0$ is the standard Gibbs energy of formation per mole of Al_2O_3. For mole fractions of Al and O in the liquid, X_{Al} and X_O, lower than 10^{-3}, the activity coefficients γ of these elements are nearly constant (Henry's law) and equal to the activity coefficients at infinite dilution in M, γ_O^∞ and γ_{Al}^∞. Then, equation (6.2) becomes:

$$K_D'(T) = X_{Al}X_O^{3/2} \quad \text{with} \quad K_D'(T) = K_D(T)(\gamma_{Al}^\infty)^{-1}(\gamma_O^\infty)^{-3/2} \tag{6.3}$$

If (i) the vapour phase does not influence reactions at the solid/liquid interface and (ii) the diffusion of components of the solid (Al and O) in bulk Al_2O_3 is slow compared to the dissolution rate at the interface, the values of X_{Al} and X_O must satisfy the condition:

$$3X_{Al} = 2X_O \tag{6.4}$$

The progress of the dissolution reaction (6.1) reached at equilibrium can be characterised by the equilibrium mole fraction of O in the liquid M. Combining equations (6.3) and (6.4):

$$X_O = \left(\frac{3}{2}\right)^{2/5}(K_D'(T))^{2/5} \tag{6.5}$$

Activity coefficients can be related to the partial enthalpies and partial excess entropies of mixing of Al and O at infinite dilution in M as follows:

$$\gamma_{Al}^{\infty} = \exp\left(\frac{\overline{\Delta G}_{Al(M)}^{XS,\infty}}{RT}\right) = \exp\left(\frac{\overline{\Delta H}_{Al(M)}^{\infty} - T\overline{\Delta S}_{Al(M)}^{XS,\infty}}{RT}\right) \tag{6.6.a}$$

$$\gamma_{O}^{\infty} = \exp\left(\frac{\overline{\Delta G}_{O(M)}^{XS,\infty}}{RT}\right) = \exp\left(\frac{\overline{\Delta H}_{O(M)}^{\infty} - T\overline{\Delta S}_{O(M)}^{XS,\infty}}{RT}\right) \tag{6.6.b}$$

Substituting equations (6.6) in equation (6.5) yields:

$$X_O = \left(\frac{3}{2}\right)^{2/5} \exp\left[\frac{\Delta G_R^*}{5RT}\right] \tag{6.7.a}$$

with $\qquad \Delta G_R^* = \Delta G_{f(Al_2O_3)}^0 - 2\overline{\Delta G}_{Al(M)}^{XS,\infty} - 3\overline{\Delta G}_{O(M)}^{XS,\infty}$ $\tag{6.7.b}$

In the general case of an A_nO_m oxide, X_O is given by:

$$X_O = \left(\frac{m}{n}\right)^{n/(n+m)} \exp\left[\frac{\Delta G_R^*}{(n+m)RT}\right] \tag{6.8.a}$$

with $\qquad \Delta G_R^* = \Delta G_{f(A_nO_m)}^0 - n\overline{\Delta G}_{A(M)}^{XS,\infty} - m\overline{\Delta G}_{O(M)}^{XS,\infty}$ $\tag{6.8.b}$

Note that when the mole fraction of dissolved oxygen is higher than the solubility limit of oxygen in M, precipitation of an M_pO_q oxide occurs. However, for the sake of homogeneity, only the dissolution reaction will be considered when establishing the reactivity scale. In other words, the value of X_O used in the reactivity scale is that for equilibrium, stable or metastable, between a M-A-O ternary alloy and the A_nO_m oxide (Figure 6.1).

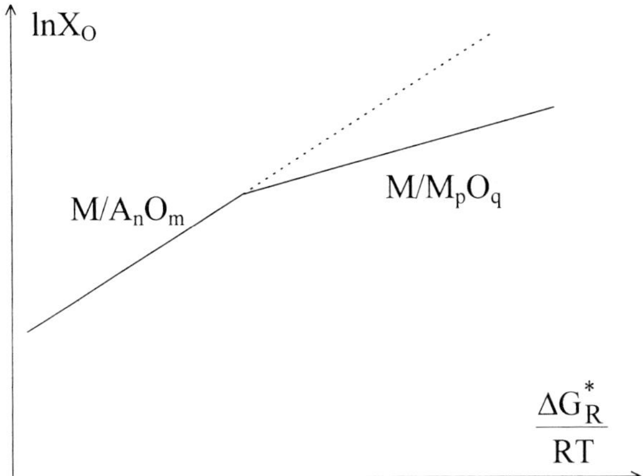

Figure 6.1. Reactivity in liquid metal M/oxide A_nO_m systems: above a certain value of $\Delta G_R^*/(RT)$, where ΔG_R^* given by equation (6.8.b) is a Gibbs energy for dissolution of A_nO_m oxide in liquid metal M, a new oxide (M_pO_q) precipitates at the interface. (Note that the reactivity scale in Figure 6.2 is built using the extrapolated part of the curve (dotted line) corresponding to the equilibrium between the liquid metal and the initial oxide A_nO_m).

Experimental contact angles θ are presented in Figure 6.2 as a function of calculated values of $\log X_O$ for a number of pure metal M/ionocovalent oxide systems. The θ values have been obtained by the sessile drop technique under high vacuum or in neutral or reducing gas atmosphere. Experimental data used in Figure 6.2 come mainly from compilations (Naidich 1981), (Chatain et al. 1986) and (Sangiorgi et al. 1988). Calculations of X_O from equation (6.8), with approximations allowing use of homogeneous thermodynamic data, are detailed by (Eustathopoulos and Drevet 1998).

The main comment that can be made concerning this figure is that for $X_O < 10^{-6}$, contact angles hardly vary with X_O. All θ values (except that for Sb/Al$_2$O$_3$) lie between $110°$ and $140°$ and systems with such θ values henceforth will be considered as *non-reactive* regarding wettability.

For $X_O > 10^{-5}$, θ decreases steeply and tends towards zero as reactivity increases. This reactivity consists first of dissolution reactions (e.g. Cu/NiO, Cu/Fe$_3$O$_4$), then of reactions to form new phases at the interface (e.g., the Ti/MgO and Zr/MgO systems, which form Ti and Zr oxides). In the Sn/NiO and Sn/CoO systems studied in high vacuum, the reduction of NiO and CoO by liquid Sn is possible and as dissolved oxygen is eliminated continuously by pumping the amount of dissolved Ni or Co produced by this reaction can become greater than that needed to precipitate the intermetallics Ni$_3$Sn or Co$_3$Sn$_5$ that are wettable by liquid metals (see Section 5.5).

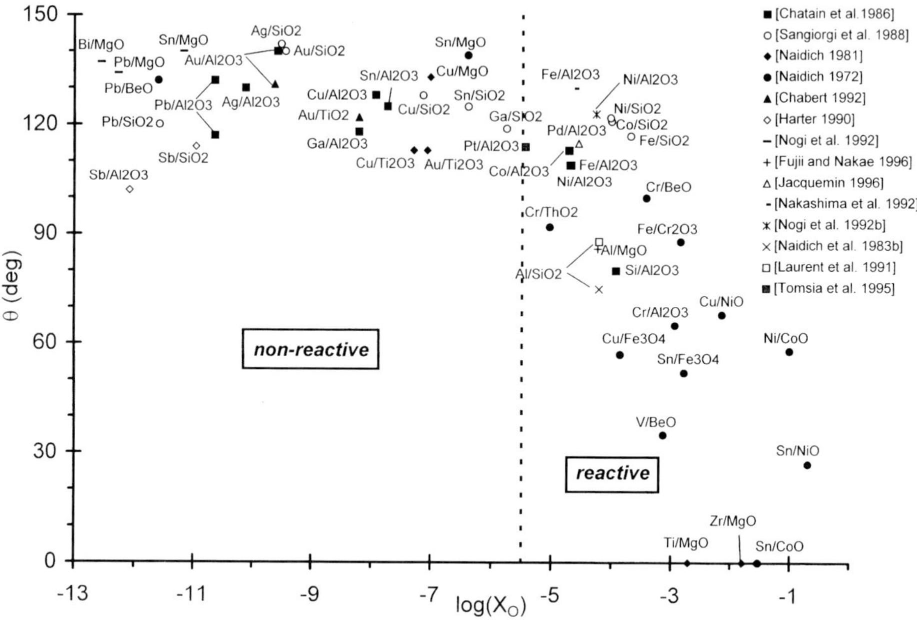

Figure 6.2. Experimental contact angles for pure liquid metal / oxide systems (taken from references given in the figure) versus calculated values of molar fraction of oxygen in the liquid metal caused by dissolution of the oxide. From (Eustathopoulos and Drevet 1998) [17].

The above distinction between non-reactive and reactive systems does not take into account the possible effect on reactivity and wettability of the furnace atmosphere. The concentration of dissolved oxygen in the liquid close to the S/L/V triple line will lie between two limits. The first of these is that for congruent dissolution of the oxide. This can be calculated using equations (6.8) by assuming that local equilibrium is established between the liquid metal and the oxide at the interface. This limit will be identified as X_O^I where the superscript I denotes its relevance to the interface. The second limit, $X_O^S(P_{O2}^f)$ at the liquid surface, is determined by equilibration with the furnace atmosphere in which the oxygen partial pressure is P_{O2}^f (see Figure 6.3). Note that even if X_O^I and X_O^S can be calculated, it is impossible in practice to calculate the actual value of X_O at the triple line, which is the important parameter for the wetting process, because it depends also on the kinetics of the interfacial reaction and oxygen adsorption processes. Despite these difficulties, it is expected that for systems with very small X_O values, such as Ag/Al_2O_3 (Figure 6.2), and for any reasonable value of P_{O2} in the furnace (in high vacuum or in reducing atmosphere P_{O2} is typically lower than 10^{-10} atm), the value of X_O at the triple line will be different from X_O^l but stay still sufficiently small for the system to remain in the non-reactive range of Figure 6.2. For such systems, the influence of P_{O2}^f on wettability is expected to be weak.

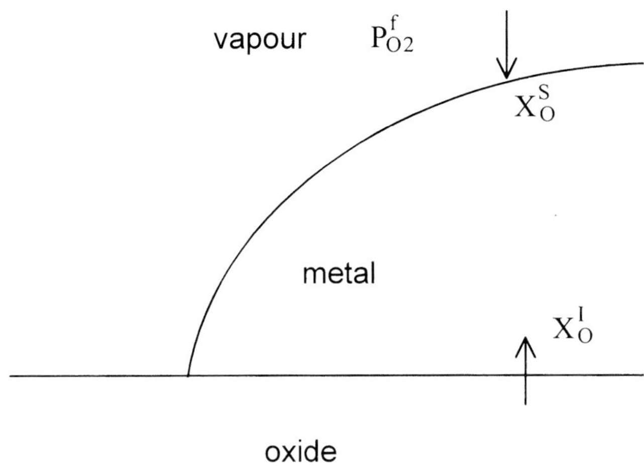

Figure 6.3. Dissolved oxygen in the liquid metal can come from dissolution of the oxide at the interface (molar fraction X_O^l) or from the vapour phase through the drop surface (molar fraction X_O^S).

The situation is different for systems such as Cu/NiO and Cu/Fe_3O_4 that lie in the reactive range or close to the non-reactive/reactive transition as when the high melting point d-metals Ni, Co, Fe, Pd or Pt contact Al_2O_3 or SiO_2. Suppose the dissolved oxygen and/or the dissolved oxide metal (Al in the case of Al_2O_3 or Si in the case of SiO_2) have a significant effect on θ. Any variation in the concentration of these elements due to a shift of equilibrium (reaction (6.1)) caused by varying P_{O2} in the vapour phase would modify θ. An example is Ni/Al_2O_3 (Nogi et al. 1992b) at 1873 K for which the experimental θ-oxygen content curve passes through a maximum (Figure 6.4). The value of θ for the non-reactive Ni/Al_2O_3 system is equal to or higher than the maximum measured value of θ ($\theta_{max} \cong 123°$).

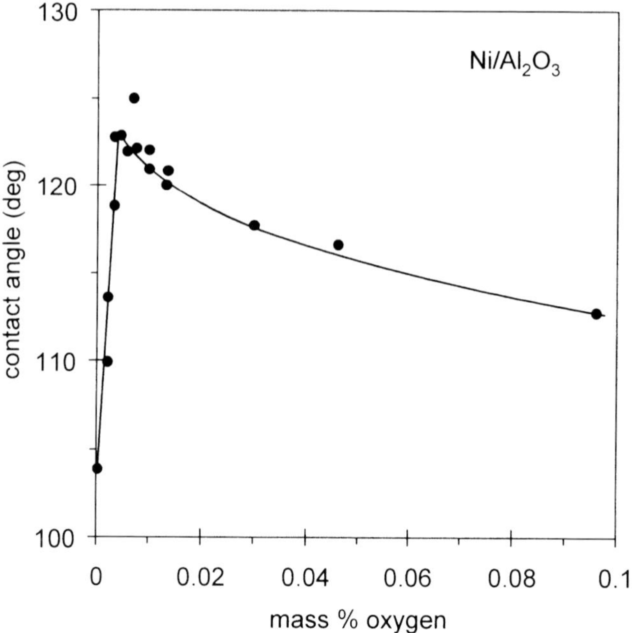

Figure 6.4. Contact angle of Ni on Al$_2$O$_3$ versus oxygen content of Ni at 1873K. 0.01 mass % of O in Ni corresponds to a molar fraction of 3.7×10^{-4}. Data from work reported in (Nogi et al. 1992b) [18].

Lower θ values are due to adsorption of either O and/or Al at the interface (see also Section 6.4.1). Similar behaviour has been observed for the Fe/Al$_2$O$_3$ system (Nakashima et al. 1992) at 1873 K with $\theta_{max} \cong 130°$.

The Al/Al$_2$O$_3$ system, in which the metal is the oxide cation, is not plotted in Figure 6.2. For this system, equation (6.4) is not valid and X$_O$ calculated from equation (6.2) taking a$_{Al}$ = 1 is equal to 3×10^{-10} at 1150 K, at which temperature the contact angle is 82° (Chatain et al. 1986). Similarly, for Si/SiO$_2$ at 1723 K, a value of X$_O$ = 2×10^{-5} is obtained (θ = 87° (Sangiorgi et al. 1988)). These results are discussed in Section 6.3.

A final remark can be made concerning the In/Al$_2$O$_3$, In/SiO$_2$ and In/MgO systems studied by Harding and Rossington (1970) at 429K in ultrahigh vacuum. These systems are not plotted in Figure 6.2 because calculated values of X$_O$ are off the scale: for In/Al$_2$O$_3$, θ is 124° and log X$_O$ is −29, for In/SiO$_2$, θ is 128° and log X$_O$ is −25 and for In/MgO, θ is 133° and log X$_O$ is −25. Such low values of X$_O$ are not realistic because the oxygen content, coming either from the vapour phase or present as an impurity in In will be much higher. However, these results confirm the previously observed tendency of contact angle values to be nearly independent of X_O when X_O is lower than 10^{-6}.

6.2. NON-REACTIVE PURE METAL/IONOCOVALENT OXIDE SYSTEMS

6.2.1 Main experimental features

Ionocovalent oxides are characterized by a high thermodynamic stability with negative values of the Gibbs formation energy of several hundreds of $kJ/g \cdot$ atom O, and gap energies of several eV that make them electrical insulators. Their bonding characteristics lie between that of predominantly ionic compounds like MgO (Pauling ionicity: 73%) and more covalent compounds like SiO_2 (Pauling ionicity: 51%).

For non-reactive metal/ionocovalent oxide systems, contact angles lie between 110 and 140°, as shown by Figure 6.2. Experimental data for the wetting of smooth monocrystalline Al_2O_3 substrates by non-reactive metals in high vacuum or low P_{O_2} neutral gas atmosphere (P_{O_2} between 10^{-14} and 10^{-20} atm typically) are given in Table 6.1. According to the Young-Dupré equation, the work of adhesion W_a is only about 20% of the work of cohesion W_c of the liquid metal (Table 6.1). Because the attraction of the liquid by the solid is very low in these systems, even small barriers to the movement of the triple line, such as surface asperities a few tens of nanometers high, can pin the triple line in a non-equilibrium position (see Figure 1.26). Although no systematic study on spreading kinetics exist for these systems, the stable equilibrium contact angles for millimetre size droplets are expected to be reached in less than 10^{-1} second (Section 2.1.1). Thus significant variations of θ observed in longer times indicate environment effects or interfacial reactions. The variation of θ with temperature is very weak, typically of the order of a few multiples of -0.01 deg/K. For instance, an increase from room temperature to 1000°C produces a decrease of θ for Ga on Al_2O_3 or SiO_2 of only 10 degrees (Figure 6.5). As the liquid metal surface energy σ_{LV} decreases slightly with increasing temperature, the temperature coefficient of the work of adhesion is also weak but always positive (typically 0.1 $mJ.m^{-2}.K^{-1}$) (Figure 6.5). Significantly higher values of $-d\theta/dT$ and dW_a/dT strongly suggest the occurrence of interfacial reactions or environment effects.

No systematic studies have been performed on the *anisotropy* of wetting. However contact angle measurements for pure Cu on very smooth ($R_a = 1–2$ nm) surfaces of α-Al_2O_3 single crystals revealed contact angle differences of less than 10° (Table 6.2) corresponding to changes in the work of adhesion of less than 150 mJ/m^2 (Vikner 1993). These rather weak effects on θ and W_a do not mean that σ_{SV}

Table 6.1. Contact angle data for some pure metals on monocrystalline Al_2O_3 obtained under high vacuum or reducing gas environment. From (Chatain et al. 1986) except for Au at 1365K (Chabert 1992), Fe (Nakashima et al. 1992), Ni (Nogi et al. 1992b) and Sb (Harter 1990).

Metal	T_{exp} (K)	θ (deg)	W_a/W_c
Ag	1373	130	0.18
Au	1373	140	0.12
Au	1365	131	0.17
Cu	1423	128	0.19
Fe	1873	130 [*]	0.18
Ga	1173	118	0.26
In	429	124	0.22
Ni	1873	123 [*]	0.23
Pb	1173	117	0.27
Pb	1173	132	0.16
Sb	998	102	0.40
Sn	1373	125	0.21

[*] maximum experimental values (see Section 6.4.1 and Figure 6.4).

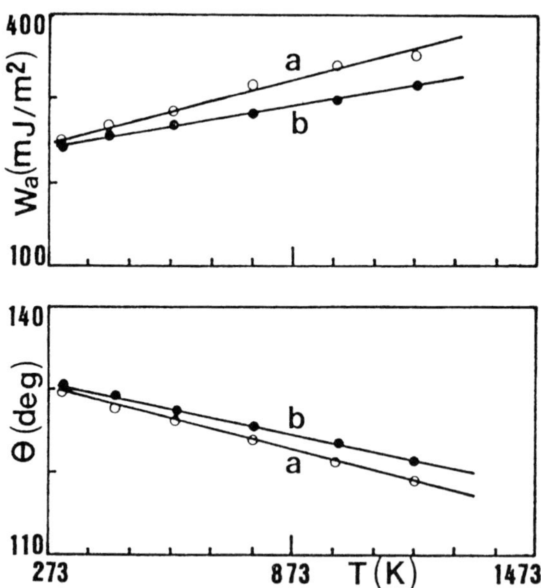

Figure 6.5. Work of adhesion and contact angle of Ga on monocrystalline Al_2O_3 (a) and vitreous SiO_2 (b) in high vacuum as a function of temperature. From data reported in (Naidich and Chuvashov 1983).

and σ_{SL} are nearly isotropic quantities. They may vary in the same sense with surface orientation and, as they appear in the Young and Young-Dupré equations with opposing signs, the effect of anisotropy on σ_{SV} and σ_{SL} will be partially compensated. Differences in contact angle of up to 30° were observed for different liquid metals (Pb, Bi, Sn) on three surfaces of highly ionic MgO single crystals as shown in Table 6.3 for Sn. However, it should be noticed that in some systems with very high contact angles and low-surface-energy metals like Pb, Bi or Sn, small changes in W_a produce relatively large changes in θ.

Table 6.2. Contact angle of pure Cu on smooth as-received α-Al$_2$O$_3$ surfaces at 1373K under high vacuum (Vikner 1993).

Surface	θ (deg)	W_a (mJ/m^2)
0001	120	650
1$\bar{1}$02	123	592
random	128	500

Table 6.3. Contact angle of pure Sn on as-received MgO single crystals at 873K under purified H$_2$ (Nogi et al. 1992).

Surface	θ (deg)	W_a (mJ/m^2)
100	140	121
110	170	8
111	147	84

6.2.2 Nature of metal-oxide bond

To identify the nature of predominant interactions at interfaces between non-reactive metal M and ionocovalent oxide AO, different attempts have been made to correlate the energetic properties of interfaces (work of adhesion, work of immersion) to the energy of formation of M oxide or other quantities characteristic of the contacting phases, such as the surface energy of the metal or the gap energy of the ceramic. Any successful correlation between an energetic quantity of interfaces and the formation energy or enthalpy of MO oxide indicates the occurrence of a chemical *interaction* between M and AO at the interface, even in the absence of a chemical *reaction*. For a series of different metals M on the same oxide AO, the enthalpy of formation of MO oxide, $\Delta H_{f(MO)}$, from pure liquid M and gaseous oxygen roughly quantifies the difference between the bond energy ε_{M-O} in bulk MO and half the bond energy ε_{M-M} in bulk M (the bond energy ε_{O-O} in gaseous oxygen being a constant):

$$\Delta H_{f(MO)} \propto \varepsilon_{M-O} - \frac{\varepsilon_{M-M}}{2} \tag{6.9}$$

According to the definition of the work of adhesion W_a in terms of bond energies (equation (1.12) with A = M and B = AO), W_a is proportional to ε_{M-AO} where ε_{M-AO} is an average interaction energy between M atoms and oxide ions. Clearly, W_a is proportional to an absolute bond energy while $\Delta H_{f(MO)}$ is proportional to a difference of bond energies. As a result, the interfacial quantity to be used in a correlation with $\Delta H_{f(MO)}$ should not be W_a but the work of immersion W_i which, taking into account its definition (equation (1.54)) and expressions (1.9) to (1.11) of σ_{SV} and σ_{SL}, is indeed proportional to:

$$W_i \propto \varepsilon_{M-AO} - \frac{\varepsilon_{M-M}}{2} \tag{6.10}$$

Comparing the expressions (6.9) and (6.10) reveals that for a series of metals M on the same oxide AO, if a correlation is observed between W_i and $\Delta H_{f(MO)}$ it can be concluded that the interfacial energy ε_{M-AO} varies as the chemical bond energy ε_{M-O} in the bulk M oxide (Eustathopoulos and Drevet 1998).

Values of the work of immersion W_i $(= -\sigma_{LV}\cos\theta)$ of non-reactive metals M on Al_2O_3, plotted on Figure 6.6 as a function of the standard enthalpy of formation of M oxides $\Delta H^0_{f(MO)}$, show no correlation between these two quantities. No improvement in correlation is obtained when W_i is plotted versus a parameter taking into account both M-O and M-Al interactions (Eustathopoulos and Drevet 1998). Accordingly, it may be concluded that the contribution to adhesion of chemical interactions between M atoms and oxide ions is negligible or weak. Therefore, the interfacial metal/oxide bond is mainly due to *physical* interactions. This conclusion agrees, in the particular case of the Cu/Al_2O_3 system, with results of a spectrometric study of the interfacial bond (Ohuchi and Kohyama 1991). The Al_2O_3 surface was fabricated *in situ*, under ultrahigh vacuum, by thermal oxidation of Al by microleaks of O_2. Photoemission spectroscopy was used to monitor the evolution of chemical bonding during Cu deposition but no significant reduction of either the Cu or the substrate energy levels was observed, indicating no chemical interaction between Cu and Al_2O_3. Note that in contrast the same technique identified strong chemical interaction in the Ti/Al_2O_3 system, which indeed belongs to the reactive range of Figure 6.2.

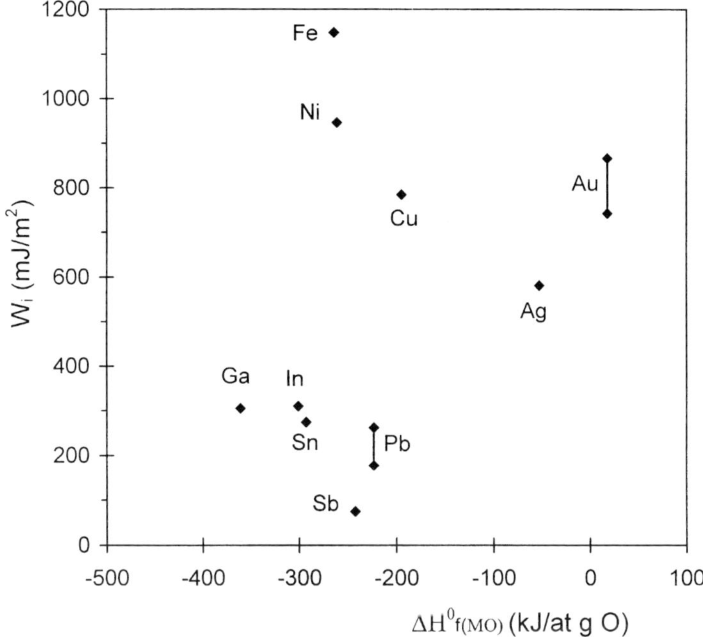

Figure 6.6. Experimental values of the work of immersion determined from the data in Table 6.1 for non-reactive pure metal M/monocrystalline Al_2O_3 systems versus the standard enthalpy of formation of M oxide referred to pure liquid M. From (Eustathopoulos and Drevet 1998) [17].

More information about the physical interactions can be obtained noting that contact angles for non-reactive metals are about 120° whatever the metal (Figure 6.2), so this suggests that W_a varies roughly as the surface energy, σ_{LV}, of the metal in accord with the Young-Dupré equation (1.45). This is verified plotting W_a as a function of σ_{LV} (Figure 6.7) for a series of non-reactive metals on monocrystalline Al_2O_3, which leads to a ratio of the work of adhesion to the work of cohesion of the metal ($W_c = 2\sigma_{LV}$) equal to about 0.20. It may be expected that a similar correlation exists for other ionocovalent oxides.

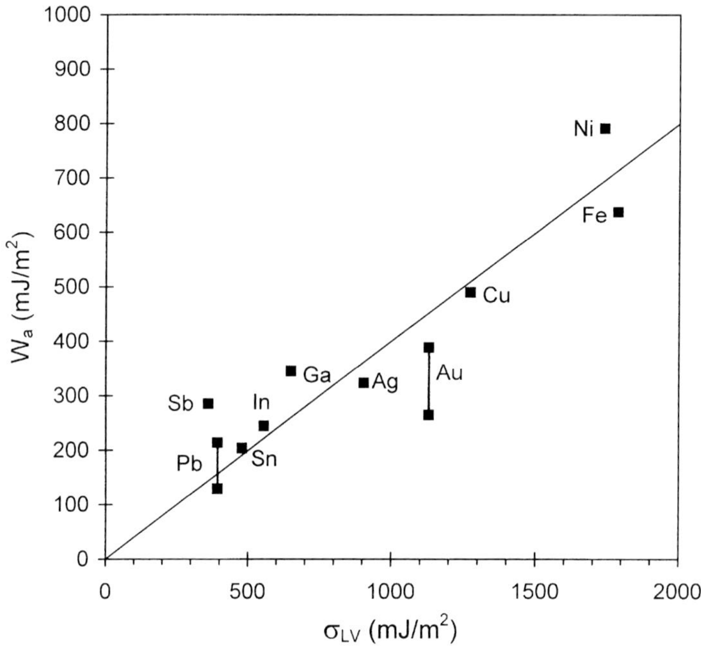

Figure 6.7. Experimental values of the work of adhesion for non-reactive pure metal/monocrystalline Al_2O_3 systems versus the liquid surface energies of metals. From (Eustathopoulos and Drevet 1998) [17].

A possible explanation for this correlation is that σ_{LV} for pure metals increases with the plasmon energy of the metal (Stoneham et al. 1995). The plasmon energy is a parameter explicitly contained in the expression of the adhesion energy due to van der Waals *dispersion forces* (Didier and Jupille 1994). The main features of these forces are first that they can be developed between all species, even between neutral and non-polar molecules (cohesion in liquified or solidified rare gas or in liquid hydrocarbons is ensured by these forces), and, second, that they are long-range forces which can be effective up to a range of several tens of nanometers. To understand their origin, consider a non-polar atom such as argon. Although the time average of its electron dipole moment is zero, at any instant there exists a finite dipole moment determined by the instantaneous position of the electrons around the nuclear protons. This instantaneous dipole generates an electric field that polarizes any nearby neutral atom, inducing a dipole moment in it. The resulted interaction between the two dipoles gives rise to an instantaneous attractive force *between the two atoms* and the time average of this force is finite (Israelachvili 1991). When the two atoms, denoted 1 and 2, at a distance r and separated by vacuum (Figure 6.8.a), are dissimilar, the resultant interaction energy is given by the London expression:

$$E_{VDW} = -\frac{3}{2}\frac{\alpha_1^0\alpha_2^0}{(4\pi\varepsilon_0)^2 r^6}\frac{I_1 I_2}{(I_1 + I_2)} = -\frac{C^*}{r^6} \tag{6.11}$$

where α^0 is the electronic polarizability, I is the first ionization potential and ε_0 is the dielectric constant of vacuum.

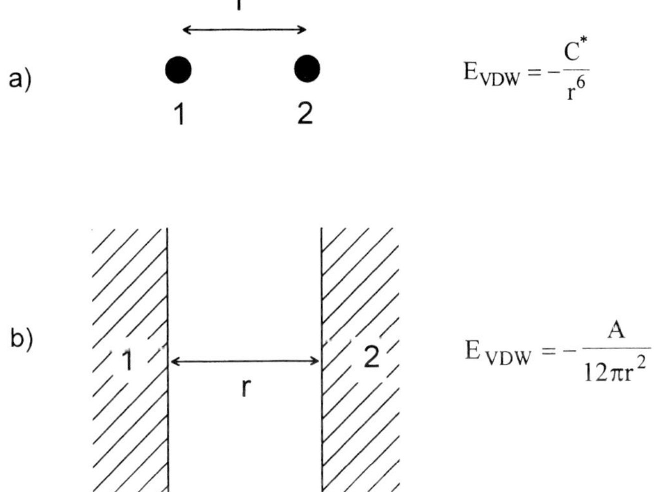

Figure 6.8. Variation with separation distance r of the interaction energy due to van der Waals forces as calculated by the pairwise interaction model, between two different atoms (a) and between two surfaces (b). A, the Hamaker constant, equals $\pi^2 C^* \rho_1 \rho_2$ where C^* is the constant of the atom-atom pair potential in equation (6.11) (case a) and ρ_1 and ρ_2 are the number of atoms per unit volume in the two bodies (Israelachvili 1991).

Equation (6.11) has been used to calculate the contribution of van der Waals interactions to the metal/oxide adhesion energy (McDonald and Eberhart 1965,

Naidich 1981). To this end, only nearest-neighbour interactions between a metal atom M and an oxygen ion O^{2-} of the oxide have been considered, taking r as the sum of the atomic radius of M and the ionic radius of O^{2-}. Note that in equation (6.11) the equilibrium separation distance r is raised to the sixth power, so the calculated values of interaction energy are very sensitive to assumptions on the interfacial atomic structure. Moreover these calculations contain several approximations. First, although the nearest-neighbour interactions are predominant, longer-range interactions are not negligible. While long-range interactions are weak for each pair of atoms, the final contribution can be significant because many pairs can participate. Thus, although atom-atom van der Waals interactions vary as r^{-6}, in order to calculate the interaction energy between two flat surfaces (Figure 6.8.b) one must sum (integrate) the energies of all the atoms in one body with all the atoms in the other, which leads to an interaction potential *between the two surfaces* varying as r^{-2} (Israelachvili 1991). Second, in solids with delocalized electrons like metals, the modes of collective excitation, the *plasmons*, make a supplementary contribution to van der Waals interactions which is not taken into account by pairwise interaction models. Despite these imperfections, equation (6.11) predicts the right order of magnitude for the work of adhesion in metal/oxide systems, namely from 100 to 500 mJ/m^2 (McDonald and Eberhart 1965, Naidich 1981).

Difficulties arising from the use of the pair interaction model are avoided in the Lifshitz theory in which the forces between large bodies, treated as continua, are derived in terms of bulk properties such as dielectric constants and refractive indices. Calculations (Didier and Jupille 1994) of adhesion energy for metal/oxide systems based on the dielectric continuum model of Barrera and Duke (1976) predict that the values for different non-reactive metals on the same oxide (Al_2O_3 or SiO_2) increased with the plasmon energy of the metal, in good agreement with experimental findings. However, the calculated values of W_a exceed experimental measurements by a factor of two. Additionally, these calculations predict that for a given metal on different oxides, the work of adhesion increases as the band gap of the insulator narrows but more accurate wetting experimental results for the same metal on various oxides are needed to test this prediction. Similar calculations using the Lifshitz continuum theory of van der Waals forces were performed by Lipkin et al. (1997) for non-reactive metal/Al_2O_3 systems. These predict the right order of magnitude of W_a, close to 600 mJ/m^2, but did not suggest any significant variation of W_a between low melting point metals like Pb and high melting point metals like Ni.

In these models, the method consists of (i) assuming that a certain type of interaction such as van der Waals dispersion forces (Didier and Jupille 1994, Naidich 1981) or image forces (Stoneham et al. 1995) is predominant at metal/oxide interfaces, (ii) calculation of the adhesion energy resulted from these forces and (iii) comparison with experimental data. However, because of numerous approximations, calculations based on different types of physical interactions all lead to a similar correct order of magnitude for the work of adhesion, so no firm conclusion can be drawn about the nature of interfacial bonding.

Since 1990, a different approach has been used in the so-called first principles or *ab initio* theories which allow calculation of the electronic charge density, the total energy and the relaxed interfacial atomic structure without any need for fit parameters. Because the computer time for these calculations is long, such methods are possible at present only for systems containing a small number of different atoms (typically one hundred) (Magaud and Pasturel 1998). An example is the solid Ag/MgO couple, assuming the (100) face of Ag being in contact with the (100) face of fcc MgO (Schönberger et al. 1992, Hong et al. 1995). In the work of Hong et al. (1995), the model cell contained three layers of MgO in contact with five layers of Ag. The oxide layers parallel to the interface are stoichiometric and therefore charge-neutral. Two translational configurations were studied, both of which led to the right order of magnitude for the work of adhesion (about 1 J/m^2). The translational configuration in which Ag atoms sat on top of the O atoms across the interface was found to have an adhesion energy twice that of the configuration in which the Ag atoms are above the Mg atoms. When, instead of Ag with its filled d-band, a transition metal like Ti with a partially filled d-band is examined (Schönberger et al. 1992), calculations indicate hybridization of the d-band of Ti with the 2p-band of oxygen i.e., a significant covalent/metallic contribution to the adhesion energy. As a result, a higher adhesion energy is suggested for this system than for Ag/MgO (Schönberger et al. 1992). Actually Ti reacts with MgO as indicated by the location of the Ti/MgO couple in Figure 6.2, meaning that the calculations of Schönberger et al. were made for a Ti layer in a metastable equilibrium with MgO. More systematic studies of this kind are needed to draw definitive conclusions about the bonding mechanism in metal/oxide systems but the results obtained so far are encouraging.

To sum up, in non-reactive *liquid* metal/ionocovalent oxide systems, wetting seems to result from physical, van der Waals, interactions which lead to contact angles of typically 120–130° and to a $W_a/(2\sigma_{LV})$ ratio of about 0.20. Any significant difference from the above values indicates the establishment of chemical interactions at the interface.

These conclusions appear to be valid also for non-reactive *solid* metal/oxide systems. Indeed, by studying the equilibrium shape of a solid metallic particle on Al$_2$O$_3$ (see Section 3.1.6), non-wetting contact angles in the range 120–130° have been found for noble metals as Au, Ag and Cu at about 900°C (Pilliar and Nutting 1967, Soper et al. 1996). These values are very close to those obtained for the same metals in the liquid state near their melting point. This closeness can be explained by taking into account that, on melting, the surface energy of the metal and its work of adhesion on the oxide both decrease and, as shown by the Young-Dupré equation, their effects on the contact angle are opposing and apparently nearly compensating. Indeed the metal/vapour surface energy decreases by about 15% on melting, Section 4.1.3, because the cohesion energy of the bulk metal decreases. Moreover, because the van der Waals interaction energy is simply proportional to the atomic density of contacting phases (see Figure 6.8), the work of adhesion would decrease at melting.

The predominance of van der Waals interactions at solid metal/Al_2O_3 interfaces is also shown by the fact that whatever the orientation of monocrystalline Al_2O_3 surface, Cu particles are orientated with (111) faces parallel to the Al_2O_3 surface (Soper et al. 1996). This orientation maximises the number of metal atoms per unit area in nearest-neighbour interactions with Al_2O_3. A similar behaviour was found for non-reactive fcc metals on carbon substrates i.e., for systems with predominant van der Waals interfacial interactions (Section 8.1).

6.3. EFFECT OF ELECTRONIC STRUCTURE OF THE OXIDE

Many oxides of technological importance, such as TiO_2, ZrO_2, UO_2 and Y_2O_3, possess a significant range of non-stoichiometry which introduces charged defects into the crystal. These defects can interact with the metal at the interface, thus affecting adhesion and wetting.

For the Cu/UO_{2+x} system, it was observed that on increasing the value of x from 0.001 to 0.084, the contact angle of Cu at 1446K was reduced from $116°$ to $84°$ (Hodkin and Nicholas 1977). According to thermodynamic data, the partial pressure of oxygen, P_{O2}, in equilibrium with UO_{2+x} for x = 0 at 1446K is 10^{-16} Pa and for x = 0.084 is 10^{-2} Pa. At a P_{O2} value as high as 10^{-2} Pa, molten Cu dissolves a large quantity of oxygen (≈ 0.2 at.% O at 1373K) and as will be seen in Section 6.4.1, oxygen dissolved in liquid metals improves wetting even in concentration as low as a few ppm or tens of ppm. To explain this effect, it was proposed (Naidich 1981) that dissolved oxygen and metal atoms form clusters having a partially ionic character due to charge transfer from the metal to oxygen (Figure 6.9.a). Such clusters can develop coulombic interactions with any ionic substrate and, for this reason, segregate to metal/oxide interfaces, thus improving wetting and thermodynamic adhesion. Accordingly, in the Cu/UO_{2+x} couple, the oxide substrate acts as an oxygen source.

A different mechanism was demonstrated for the U/Y_2O_{3-x} system studied at 1673K (Tournier et al. 1996). In this study, three different experimental configurations were produced leading to decreasing concentrations of dissolved

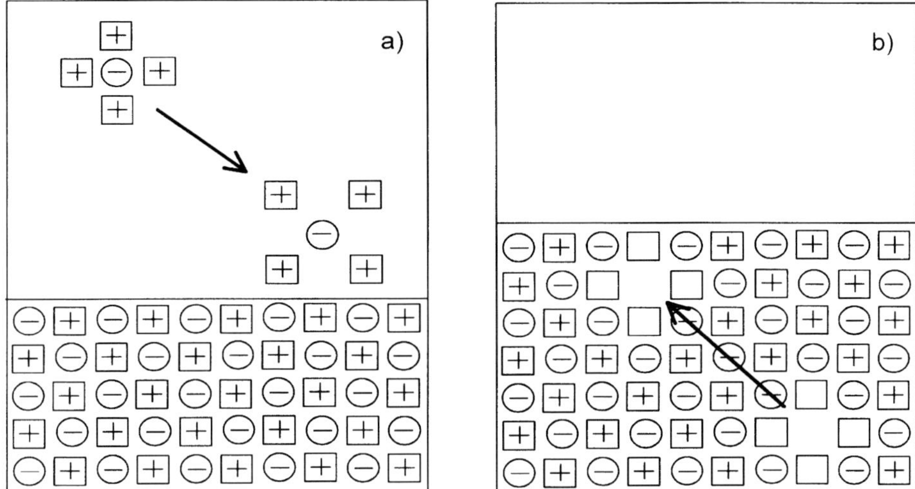

Figure 6.9. (a) Adsorption of oxygen clusters from the bulk metal at the metal/oxide interface can occur when P_{O2} is *high* according to the considerations of Naidich. (b) Adsorption of a cluster centred on an oxygen vacancy from the bulk oxide at the interface can occur when P_{O2} is *low*. For the sake of simplicity, total oxidation (case a) or total reduction (case b) of metal around oxygen atoms (case a) or vacancies (case b) has been assumed.

oxygen in U and to increasing degrees of hypostoichiometry. Table 6.4 shows that these decreases produce a strong improvement in wetting and adhesion. The final value of W_a is four times higher than that predicted to result from physical interactions. Thus, taking $W_a/(2\sigma_{LV}) \approx 0.20$ as typical of van der Waals interactions (see Section 6.2.2) and the surface tension of liquid U $\sigma_{LV} = 1600$ mJ/m², one obtains a W_a of about 640 mJ/m². Therefore, in very low P_{O2} environment, the metal/oxide bond for hypostoichiometric oxide must be chemical in nature.

An explanation of the adsorption capability of oxygen vacancies was proposed (Tournier et al. 1996) by analogy with the mechanism of Naidich for the adsorption of dissolved oxygen from the bulk metal to the metal/oxide interface. What is interesting for our purposes in the diagram of Figure 6.9.a is that the strong metal-oxide interaction results from partial rupture of the metallic bond in the cluster and its replacement by an ionic-type bonding. Formation of oxygen vacancies also can lead to a partial rupture of the ionic bond and it may be expected that some reduction in metal ions around a vacancy would occur. Indeed it has been shown, both experimentally and theoretically, that the presence of oxygen vacancies in

Table 6.4. Experimental contact angles and corresponding W_a values for the U/Y_2O_{3-x} system at 1673K for different partial pressures of oxygen (Tournier et al. 1996).

P_{O2} (Pa)	θ (deg)	W_a (mJ/m^2)
10^{-7}	106	1160
10^{-20}	68	2200
10^{-23}	52	2585

yttria locally leads to a more covalent bond (Jollet et al. 1991). In the diagram of Figure 6.9.b, the interface-active species is a cluster consisting of an oxygen vacancy surrounded by partially reduced metal ions. Segregation of such clusters to the oxide side of the interface would ensure a gradual transition from oxidised metal atoms (yttrium cations in the oxide) to metallic bonded atoms (uranium atoms in the liquid).

α-Al$_2$O$_3$ is an oxide with a negligible non-stoichiometry range but (0001) faces of Al$_2$O$_3$ are known to lose oxygen which leads to a two dimensional reconstruction of the surface in very high vacuum at high temperatures (about 1500K) (French and Somorjai 1970). Similar transformations seem to occur on other faces (French and Somorjai 1970, Gillet and Ealet 1992, Pham Van 1992). Furthermore, Al and Si vapours shift the temperature of this 2D reconstruction towards temperatures of 1200K or less (French and Somorjai 1970). Such a transformation may increase the surface energy of Al$_2$O$_3$ (Brennan and Pask 1968) or the metal/Al$_2$O$_3$ adhesion energy by a mechanism similar to that applicable to U/Y_2O_{3-x}. This could explain why contact angles for the Al/Al$_2$O$_3$ and Si/Al$_2$O$_3$ systems are close to 90° and 80° at the melting point of Al and Si respectively, significantly lower than the values 110–140° observed for non-reactive metal/ionocovalent oxide systems (see Figure 6.2). The *chemical interactions* developed between Al or Si and Al$_2$O$_3$ at the interface explain why the experimental work of adhesion for these two metals (≈ 900 mJ/m^2) deviates strongly from the corresponding van der Waals value, 300 mJ/m^2, taken from the straight line of Figure 6.10.

Some oxides are electric conductors like metals and can be regarded as metallic, for example TiO, VO and ReO$_3$. Thus, TiO has a rocksalt structure (NaCl type) in which each Ti^{2+} is surrounded by six O^{2-} in an octahedral configuration TiO$_6$ and the cohesion energy of the crystal is due both to ionocovalent and metallic interactions. The ionocovalent interactions come from charge transfer (ionic) and orbital overlapping (covalent) between Ti and O ions while the metallic character of TiO is due to both overlapping of d orbitals of Ti^{2+} ions to produce a partly-filled band (Duffy 1990) and to s electrons.

Note that other monoxides like NiO, MnO, FeO and CoO also have a rocksalt structure but they do not conduct. When the above description for TiO is used for NiO, one finds that the overlapping of orbitals leads to a completely-filled band which does not allow conduction. However, the cohesive properties for the NiO compound are known to display a specific behaviour, as discussed in (Anisimov et al. 1991).

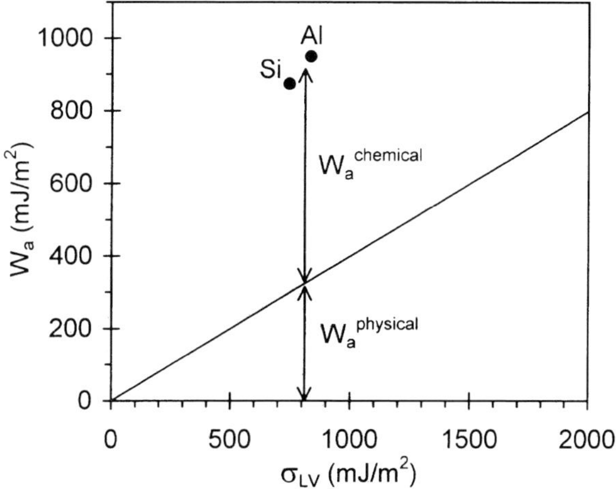

Figure 6.10. Experimental values of the work of adhesion versus the liquid surface energies for molten Al and Si on monocrystalline Al_2O_3. These values exceed significantly the values of the straight line for van der Waals bonded systems shown in Fig. 6.7.

Table 6.5. Contact angle and work of adhesion of Cu on different oxides at 1423K (Naidich 1981). Values of the gap energy E_g of oxides come from Duffy (1990).

Oxide	θ (deg)	W_a (mJ/m^2)	E_g (eV)
SiO_2	128	500	10
MgO	133	410	7.3
$TiO_{1.14}$	82	1480	0
$TiO_{0.86}$	72	1700	0

It may be expected that at the interface between a metal and a metal-like compound, bonding would be at least partially metallic, leading to a high adhesion energy and better wetting than in metal/ionocovalent compound systems. This is confirmed by results in Table 6.5 for the wetting by Cu of oxides with different electrical conductivities, ranging from insulators (including predominantly ionic MgO and the more covalent SiO_2) to metal-like TiO. Similar results have been obtained by Chabert (Chabert 1992) who found $\theta = 122°$ for Au on TiO_2 (which is of semi-conductor type with a gap energy $E_g = 3$ eV (Duffy 1990)) and $\theta = 88°$ on TiO.

6.4. EFFECTS OF OXYGEN

These effects will be described and discussed for non-reactive metal/oxide systems whose behaviour belongs to the plateau of the (θ, X_O^I) curve of Figure 6.2. Conclusions will be drawn and then used to explain results for systems which belong to the reactive range, i.e., those for which the molar fraction of oxygen at the interface, X_O^I, is greater than 10^{-5}.

Consider as an example the Ag/Al_2O_3 couple. Dissolution of Al_2O_3 into liquid Ag is described by the equation:

$$Al_2O_3 \Leftrightarrow 2(Al)_{Ag} + 3(O)_{Ag} \tag{6.12}$$

Ignoring any influence of the vapour phase (Figure 6.3), the equilibrium values of molar fractions of O and Al in Ag, X_O^I and X_{Al}^I, are easily calculated from equations (6.4) and (6.7). Taking $\overline{\Delta G}_{Al(Ag)}^{XS,\infty}$ at 1273K to be -34 kJ/mole (Hultgren et al. 1973), $\Delta G_{f(Al2O3)}^0$ to be -1272 kJ/mole (Chase et al. 1985) and $\overline{\Delta G}_{O(Ag)}^{XS,\infty}$ to be 41 kJ/at.g O (Chang et al. 1988), values of $X_O^I = 1.5 \times 10^{-11}$ and $X_{Al}^I = 10^{-11}$ are obtained. These values are extremely low, justifying consideration of Ag/Al_2O_3 as a non-reactive system (Figure 6.2). The corresponding value of P_{O2}^I for the equilibrium

$$(O)_{Ag} \Leftrightarrow \frac{1}{2}[O_2] \tag{6.13}$$

is given by:

$$(P^I_{O2})^{1/2} = a^I_O = \gamma^\infty_{O(Ag)} \cdot X^I_O \tag{6.14}$$

where a^I_O is the activity of oxygen, defined for a reference state of one atmosphere of pure O_2.

Now, let us take into account the influence of the vapour phase in the furnace, for which we assume the partial pressure of oxygen, P^f_{O2}, to be fixed by an appropriate gas mixture (CO/CO_2, H_2/H_2O, $He+O_2$, etc). Equilibrium between a (Ag,O) solution and Al_2O_3 exists for a range of P^f_{O2} values, the highest value of which corresponds to the formation of the metal oxide, Ag_2O, at a partial pressure of oxygen $P^{ox(Ag)}_{O2}$, (see Figure 6.11 taking M = Ag), in accord with the reaction:

$$2(Ag) + \frac{1}{2}[O_2] \Leftrightarrow Ag_2O \tag{6.15}$$

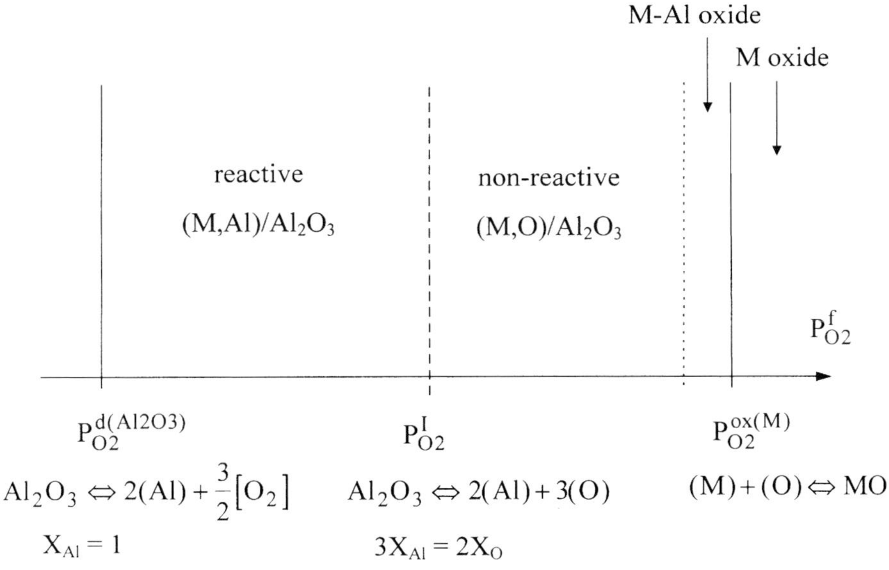

Figure 6.11. Range of partial pressure of oxygen in which a two-phase metal M/Al_2O_3 equilibrium exists. P^I_{O2} corresponds to the congruent dissolution of Al_2O_3.

Formation of oxide films on metal surfaces strongly affects wetting as will be seen in Section 6.4.2. The lowest P_{O2}^f value at which there is equilibrium between (Ag,O) and Al_2O_3 is $P_{O2}^{d(Al2O3)}$ below which there is decomposition of Al_2O_3, indicated by the superscript d (Figure 6.11).

If P_{O2}^f is higher than P_{O2}^I, the concentration of dissolved oxygen in the bulk liquid will increase. According to equation (6.3), this will produce a decrease of the molar fraction of Al towards values even lower than $X_{Al}^I = 10^{-11}$. Clearly, for $P_{O2}^I < P_{O2}^f < P_{O2}^{ox(Ag)}$, the only species dissolved in Ag is oxygen. In this range, the Ag/Al_2O_3 system is still non-reactive and the only phenomenon which can modify wetting is adsorption of oxygen at the different surfaces of the system. For some metals, at P_{O2}^f lower than but close to $P_{O2}^{ox(M)}$, mixed oxides M-Al-O can form, modifying the nature of the interface (Figure 6.11). Typical examples are $CuAlO_2$ (Gonzalez and Trumble 1996) and $NiAl_2O_4$ (Mehrotra and Chaklader 1985).

If $P_{O2}^{d(Al2O3)} < P_{O2}^f < P_{O2}^I$, dissolution of the substrate into the liquid metal will occur until X_{Al} satisfies equation (6.2) with $a_O = (P_{O2}^f)^{1/2}$. The equilibrium contact angle will be that of a Ag-Al alloy on Al_2O_3 (see Section 6.5 for the effect of alloying elements on wetting).

6.4.1 Dissolved Oxygen

Now, we will describe in some detail experimental results and their interpretation for the $Ag-O/Al_2O_3$ system in the range of P_{O2} from 10^{-12} to 10^5 Pa, in which no complications arise from interfacial reactions or from oxidation of Ag.

The contact angles of Ag on monocrystalline Al_2O_3 are shown in Figure 6.12.a as a function of P_{O2}^f. The change of σ_{LV} with P_{O2}^f is presented in the same figure, as well as the change of work of immersion W_i, which is equal to $(\sigma_{SL} - \sigma_{SV})$ and hence to $-\sigma_{LV}\cos\theta$. As P_{O2}^f increases, the contact angle remains nearly constant in a wide range of P_{O2}^f, then decreases sharply from 130° to about 100° and thereafter more slowly. A more regular decrease is observed for σ_{LV} values that can be analysed using the Gibbs adsorption equation (Defay et al. 1966):

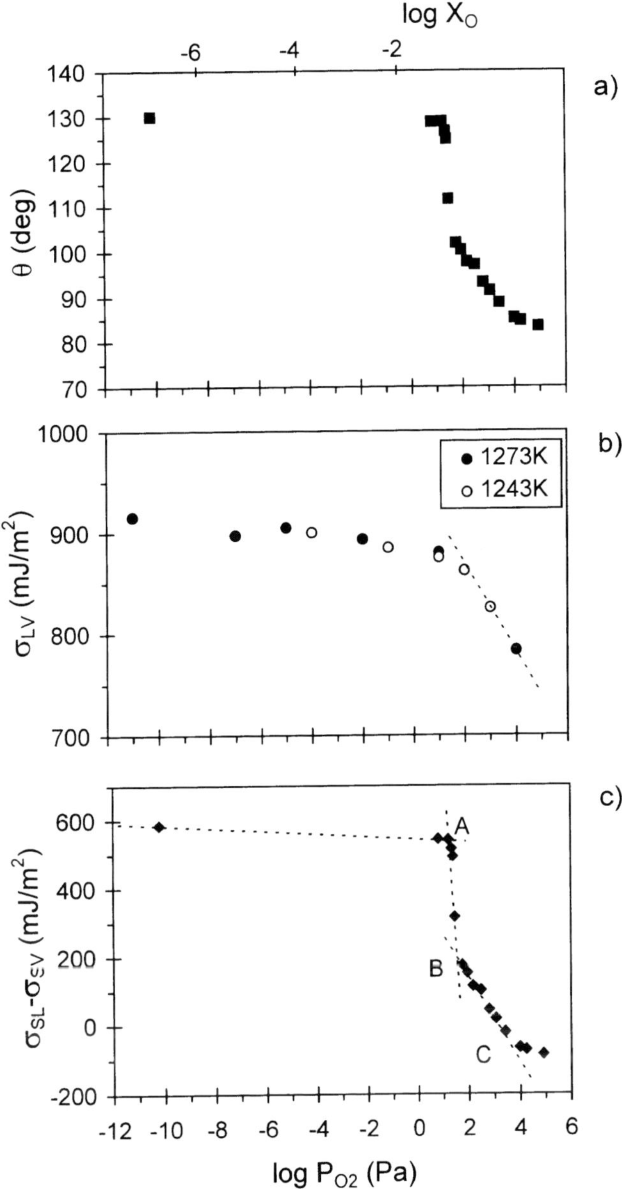

Figure 6.12. The variation with partial pressure of oxygen or the corresponding molar fraction of oxygen in Ag of the a) contact angle at 1253K (Gallois 1980), b) surface energy of liquid Ag (Mehrotra and Chaklader 1985) and c) ($\sigma_{SL} - \sigma_{SV}$) in the Ag/$Al_2O_3$ system.

$$\frac{1}{RT}d(\sigma_{LV}) = -\Gamma_O^{Ag}d(\ln a_O) = -\frac{1}{2}\Gamma_O^{Ag}d(\ln P_{O2}) \qquad (6.16)$$

where Γ_O^{Ag} is the *relative adsorption* of oxygen with respect to Ag at the liquid surface. A positive value of Γ_O^{Ag} means that the surface is enriched in oxygen with respect to the bulk liquid, while a negative value of this quantity indicates the opposite situation. Γ_O^{Ag} is related to the *absolute adsorptions* Γ_O and Γ_{Ag} by the equation:

$$\Gamma_O^{Ag} = \Gamma_O - \Gamma_{Ag}\frac{X_O}{X_{Ag}} \qquad (6.17)$$

For dilute solutions, X_O/X_{Ag} is much lower than unity, so that $\Gamma_O^{Ag} \cong \Gamma_O$. As it is usually assumed that only the first atomic layer of the liquid has a different composition from that of the bulk liquid, Γ_O is given by:

$$\Gamma_O^{Ag} \cong \Gamma_O = \frac{Y_O - X_O}{\Omega_m^O} \qquad (6.18)$$

Y_O is the mole fraction of oxygen within the surface monolayer and Ω_m^O the area occupied by a mole of oxygen at the surface. Figure 6.12.b shows that the absolute value of the slope of σ_{LV}-log P_{O2}^f curve is negligible for a large range of P_{O2} values and then begins to increase and attains a nearly constant value. In this region, the surface is saturated in oxygen with a maximum value Γ_O of 4×10^{-6} mole/m^2 obtained from equations (6.16) and (6.18). Taking $Y_O = 1$ at saturation and neglecting X_O, the atomic surface area of oxygen ω_O ($= \Omega_m^O/N_a$) is equal to about 40 Å^2/atom. This is close to the area occupied by an oxygen ion on the (111) face of bulk Ag$_2$O oxide (Gallois and Lupis 1981). In other words, although this oxide is not stable as a 3D compound at this temperature and value of P_{O2} (much higher P_{O2} values are needed for Ag$_2$O to be stable at 1273K), a 2D "compound" would form *gradually* as a result of Gibbsian adsorption. Similar situations have been found for systems such as Cu-O and Fe-O (Gallois and Lupis 1981) (see also Section 4.1.1).

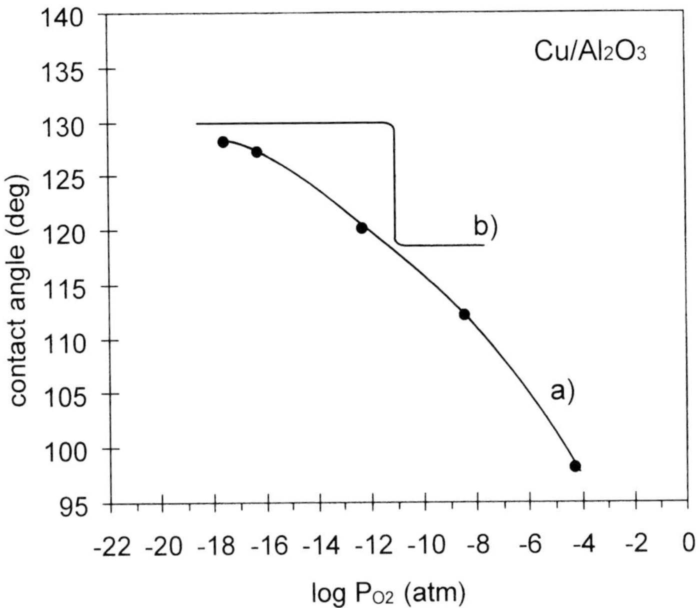

Figure 6.13. The variation of contact angle with oxygen partial pressure for Cu/Al$_2$O$_3$ at 1100°C. a) (Ownby and Liu 1988), b) (Ghetta et al. 1996).

In Figure 6.12.c, the decrease of ($\sigma_{SL} - \sigma_{SV}$) occurs in two stages, a first step A→B takes place at a nearly constant activity of oxygen, followed by a progressive decline B→C, similar to that for σ_{LV}. The B→C part of the curve may be explained by Gibbsian adsorption. Thus because it has been found that the surface energy of Al$_2$O$_3$ does not depend on P$_{O2}$ in the range 10^{-15}–10^{5} Pa (Ghetta et al. 1996), the B→C decrease must result from a reduction of σ_{SL}. Gibbsian adsorption cannot explain the A→B part of the curve since such a rapid decrease causes equations (6.16) to (6.18) to predict an extremely low value of the atomic surface area of oxygen, that is without physical meaning. Because this decrease takes place at a nearly constant P$_{O2}$ value, it may be considered instead as equivalent to a *first-order* 2D phase transition. For instance, it has been suggested that the complex oxide AgAlO$_2$, which is unstable as a bulk compound at low P$_{O2}$, forms as a 2D structure at the Ag/Al$_2$O$_3$ interface (Chaklader et al. 1981).

At present, available experimental results do not allow precise definitions of (θ-log P$_{O2}^{f}$) isotherms. For instance, one study of the couple Cu/Al$_2$O$_3$ (Figure 6.13) (Ownby and Liu 1988) found behaviour consistent with Gibbsian adsorption (curve a in Figure 6.14) and another one (Ghetta et al. 1996) behaviour indicative of a 2D phase transition (curve b in Figure 6.14). These divergences underline the difficulties of performing wetting experiments, using stable and controlled P_{O2} values in the gas mixture. Moreover, because this type of experiment usually lasts several hours or tens of hours, segregation at the interface of impurities from the substrate (Ca, Mg, Si, etc) can occur and affect results (Chabert 1992).

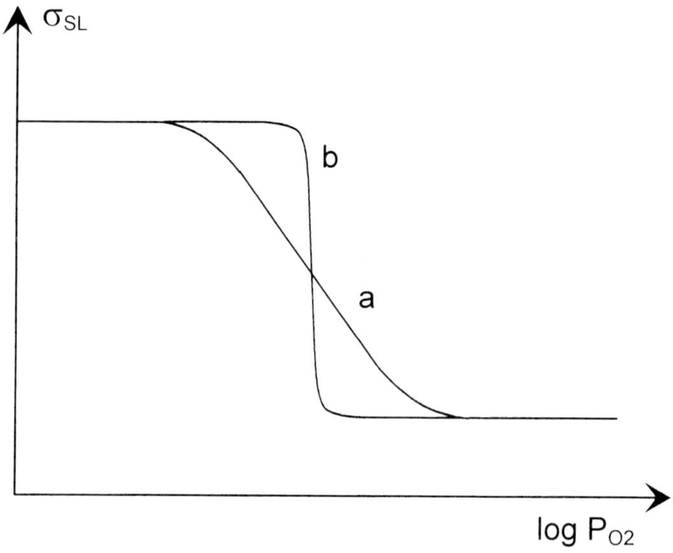

Figure 6.14. Possible isotherms of liquid metal/oxide interfacial energy in the non-reactive range when enrichment of the interface in oxygen occurs by a) Gibbsian adsorption or b) 2D phase transition.

Despite these difficulties, except for the noble metals Au, Pt and Ir on oxides, for which P_{O2} has little or no effect on contact angles, published data (Tables 6.6 and 6.7) show that oxygen decreases both σ_{LV} and σ_{SL}. This effect is higher for σ_{SL} than for σ_{LV}, causing a decrease in the contact angle of several degrees or tens of degrees (Figure 6.15). An attempt to explain these effects was made by Naidich (1981) using the oxygen-metal cluster model already cited in Section 6.3. There is no direct evidence on the formation and lifetime of such clusters. However, analysis of thermodynamic data concerning interaction of dissolved oxygen with metallic atoms in bulk liquid provides strong arguments in favour of their existence. These interactions are usually quantified by the *activity coefficient of oxygen* at infinite dilution in the melt, γ_O^∞. Because it is usually easier to calculate variations of a thermodynamic quantity than its absolute value models have been proposed to predict the variation of γ_O^∞ in a A-B binary alloy when the alloy composition changes from pure A to pure B (Belton and Tankins 1965, Jacob and Alcock 1972, Chang et al. 1988). Although differences exist among these models, they all assume that strong metal-oxygen bonds distort the distribution of electrons around the metal atoms. In the framework of a simple nearest-neighbour interaction model (see Section 1.1), such a distortion modifies the metal-metal bonds around the oxygen, the energy of which is only a fraction $(1 - \alpha)$ of that of normal metal-metal bonds. In other words, a partial rupture of the metallic bond occurs in the vicinity of oxygen.

Table 6.6. Metal/oxide systems for which determinations of contact angle as a function of P_{O2} exist.

System	Reference
Cu/Al$_2$O$_3$	(Chaklader et al. 1981)
	(Ownby and Liu 1988)
	(Ghetta et al. 1996)
Ag/Al$_2$O$_3$	(Gallois 1980)
	(Chatain et al. 1994)
	(Chaklader et al. 1981)
Au/Al$_2$O$_3$	(Ghetta et al. 1996)
Ni/Al$_2$O$_3$	(Nogi et al. 1992b)
Fe/Al$_2$O$_3$	(Nakashima et al. 1992)
Pb/SiO$_2$	(Sangiorgi et al. 1995)
Ag/SiO$_2$	(Sangiorgi et al. 1982)

Table 6.7. Wetting of oxides by Pt-group metals in air showing that the contact angle values for Ag and Rh are much lower than the values of 120-130° observed in inert gas (Grigorenko et al. 1995).

Metal	T (°C)	θ (deg) on MgO	θ (deg) on ZrO$_2$	θ (deg) on Al$_2$O$_3$
Ag	1100	83 ± 4	73 ± 8	84 ± 3
Rh	2100	97 ± 4	98 ± 4	–
Pt	1900	103 ± 5	118 ± 5	139 ± 6
Au	1200	119 ± 2	125 ± 4	136 ± 6
Ir	2500	138 ± 4	–	–

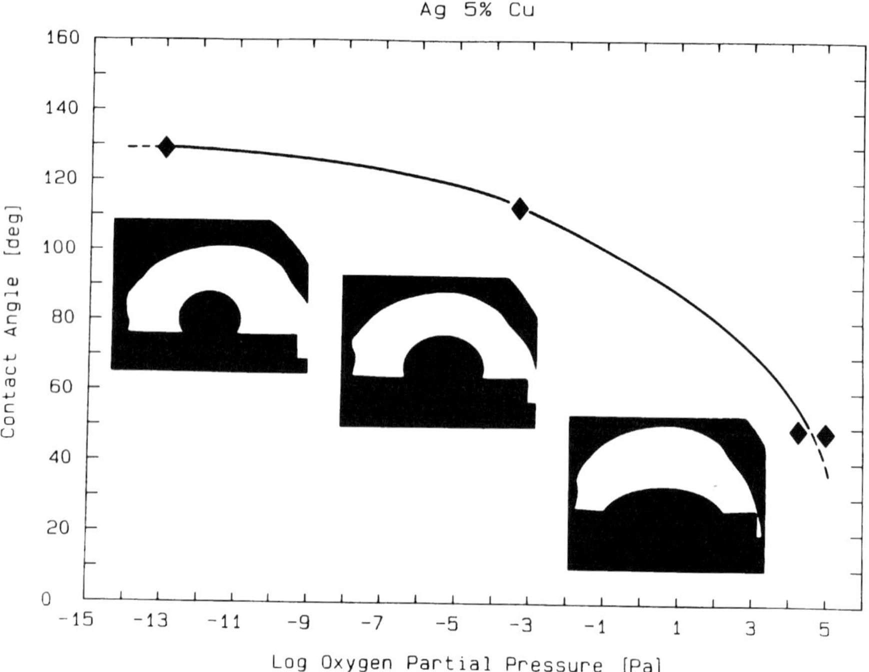

Figure 6.15. Effect of the oxygen partial pressure on the wetting of a sapphire substrate by a Ag-5 at.% Cu alloy at 1373K. Results from (Chatain et al. 1993). Photographs kindly provided by Muolo of ICFAM-CNR (Genova).

Consider an A-B lattice with a coordination number Z and, inside this lattice, an oxygen atom forming a total of n bonds with metal atoms A or B. The atomic fraction of B atoms around O, Y_B, is given as a function of atomic fraction of B in the bulk alloy, X_B, by an equation obtained from the quasichemical treatment of solutions (Jacob and Alcock 1972):

$$\frac{Y_B}{1 - Y_B} = \frac{X_B}{1 - X_B} \exp\left(\frac{-\Delta G_{ex}^{xs}}{RT}\right) \tag{6.19}$$

where ΔG_{ex}^{xs} is the excess Gibbs energy change that occurs when a A-O pair is changed to a B-O pair by *exchanging* a A atom in the coordination shell of the oxygen atom with a B atom in the melt bulk. Calculations lead to the equations:

$$\frac{Y_B}{1 - Y_B} = \frac{X_B}{1 - X_B} \left(\frac{\gamma_{O(A)}^\infty}{\gamma_{O(B)}^\infty}\right)^{1/n} \left(\frac{\gamma_B}{\gamma_A}\right)^\alpha \qquad (6.20)$$

$$(\gamma_O^\infty)^{-1/n} = X_B(\gamma_B)^\alpha(\gamma_{O(B)}^\infty)^{-1/n} + (1 - X_B)(\gamma_A)^\alpha(\gamma_{O(A)}^\infty)^{-1/n} \qquad (6.21)$$

This model allows calculation of the activity coefficient of oxygen at infinite dilution in a binary A-B alloy, γ_O^∞, from known values of the activity coefficient of oxygen at infinite dilution in pure A, $\gamma_{O(A)}^\infty$, and pure B, $\gamma_{O(B)}^\infty$, and the activity coefficients of A and B, γ_A and γ_B, in the A-B alloy. This model provides a satisfactory prediction of γ_O^∞ in several binary A-B alloys taking $n = 4$ and $\alpha = 1/2$ (Jacob and Alcock 1972). An example is given in Figure 6.16 for the Ag-Sn couple in which γ_O^∞ differs by five orders of magnitude between pure Ag and pure Sn. It can be seen that the first additions of Sn to Ag produce a strong decrease of γ_O^∞. This behaviour results from the fact that even for small X_{Sn} values, Y_{Sn} is high (the calculated value of $(Y_{Sn}/X_{Sn})_{X_{Sn} \to 0}$ is 20), so oxygen clusters are enriched in Sn atoms. Conversely, the first additions of Ag to Sn have relatively weak effects, so that Ag atoms have a negligible tendency to segregate around O atoms.

Henry's law (γ_B nearly constant and equal to γ_B^∞) holds for A-B alloys dilute in B. For this range of composition, another quantity can be introduced which takes into account interactions between dissolved O and dissolved B in an A matrix. This is the Wagner's first-order interaction parameter, ε_O^B, defined by an equation in which high-order terms are neglected:

$$\ln \gamma_O = \ln \gamma_{O(A)}^\infty + \varepsilon_O^B X_B + \dots \qquad (6.22)$$

In the quasichemical model of Jacob and Alcock, ε_O^B is given by:

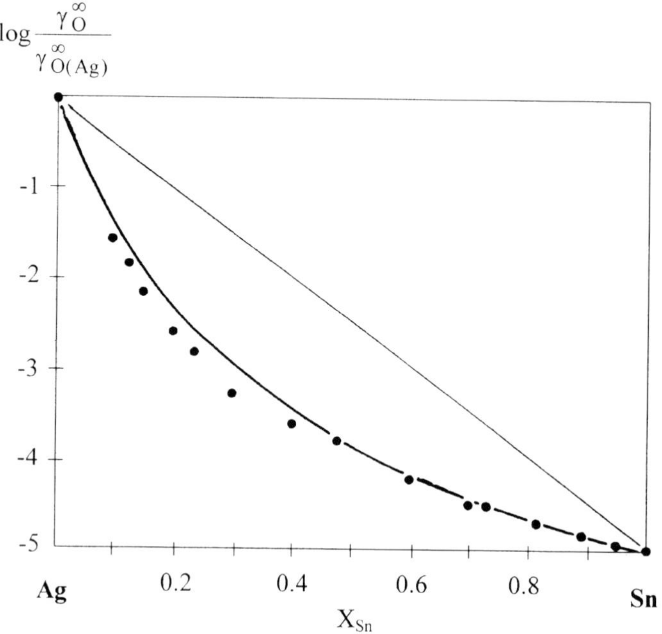

Figure 6.16. Effect of alloy composition on the activity coefficient of oxygen in Ag-Sn alloys at 1473K. Circles: experimental results, full line: values calculated according to the quasichemical model of Jacob and Alcock (1972) [19] (equation (6.21) with n = 4 and α = 1/2).

$$\varepsilon_O^B = -n\left[\left(\frac{\gamma_{O(A)}^\infty}{\gamma_{O(B)}^\infty}\right)^{1/n}(\gamma_B^\infty)^\alpha - 1\right] \tag{6.23}$$

A strong solute-solute interaction between O and B atoms in an A matrix results in a very negative value of ε_O^B and in a strong decrease of the activity coefficient of oxygen in the alloy. For instance, for Ag-Sn alloys dilute in Sn (Figure 6.16), $\varepsilon_O^{Sn} = -50$. It must be emphasized, however, that values of ε_O^B calculated by equation (6.23) are very sensitive to values of the model parameters, α and n. Errors in calculated values of ε_O^B can be such that they must be considered at best as semi-quantitative estimates. However, equation (6.23) is interesting because it identifies the parameters determining the sign and the absolute value of ε_O^B. Thus, the favourable conditions for $\varepsilon_O^B \ll 0$ are (i) a strong O-B interaction in pure B that yields a low value of $\gamma^\infty_{O(B)}$, (ii) a weak O-A interaction in pure A that results in a high value of $\gamma^\infty_{O(A)}$ and (iii) a weak A-B interaction so the partial enthalpy of mixing of B in A, $\Delta H_{B(A)}$, is positive or close to zero and γ_B has a high value. All these conditions are satisfied for instance for binary Cu-B alloys with B = Fe, Cr, Ti resulting in very negative values of ε^B_O ($\varepsilon^{Fe}_O = -540$ (Chang et al. 1988), $\varepsilon^{Cr}_O = -2800$ (Kritsalis et al. 1990), $\varepsilon^{Ti}_O = -10^4$ (Kritsalis et al. 1991)). For such values of ε^B_O, the calculated values of $(Y_B/X_B)_{XB\to0}$ $(= 1 - \varepsilon^B_O/n)$ can attain very high values, from several 10^2 to several 10^3, meaning that for Cu alloys containing a few atomic percent of Fe, Cr or Ti, dissolved oxygen is in the form of O-B clusters rather than O-Cu clusters.

This quasichemical model is compatible with the Naidich hypothesis of oxygen clusters and allows a rough description of their structure and energy to be given: oxygen atom is surrounded by four metal atoms (n = 4) and the charge transfer from these atoms to oxygen halves the strength of the metal-metal bonds around oxygen ($\alpha = 1/2$), Figure 6.17.

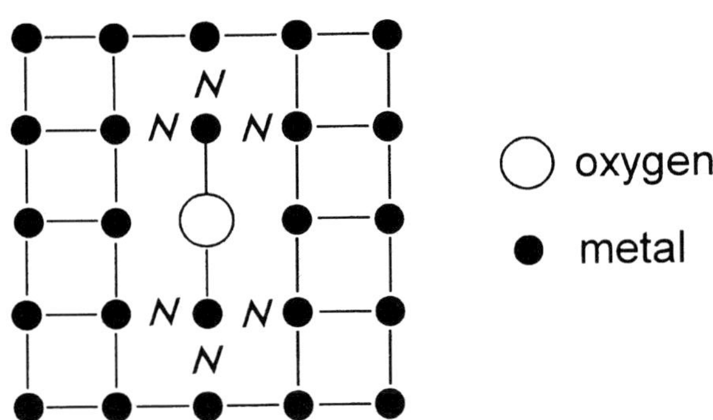

Figure 6.17. Schematic illustration of an oxygen cluster according to Jacob and Alcock (1972). For 2D representation the number of metal atoms bonded to oxygen, n, is 2 and the coordination in bulk metal is 4. The symbol *N* means partial rupture of the metal-metal bond.

Consider now the influence of oxygen clusters on metal/vapour and metal/oxide interfacial energies. Because oxygen clusters cause a major perturbation of the structure and energy of the bulk metal, they are expected to segregate at 2D defects such as surfaces and interfaces, where the perturbation can be partially relaxed. This may explain the strong tendency for oxygen adsorption at both solid metal/vapour and liquid metal/vapour surfaces, to decrease the corresponding surface energies. The adsorption energy at the metal/oxide interface i.e., the change in energy of the system when an oxygen atom segregates at the interface from the bulk, will be even higher in absolute value because of the ionic interactions between the clusters and the oxide surface (Figure 6.9.a). Thus, the reduction of σ_{LV} due to oxygen adsorption will be lower than the reduction of σ_{SL}, resulting in an increase of the work of adhesion, $W_a = \sigma_{SV} + \sigma_{LV} - \sigma_{SL}$, and a decrease of θ, since $\cos\theta = W_a/\sigma_{LV} - 1$.

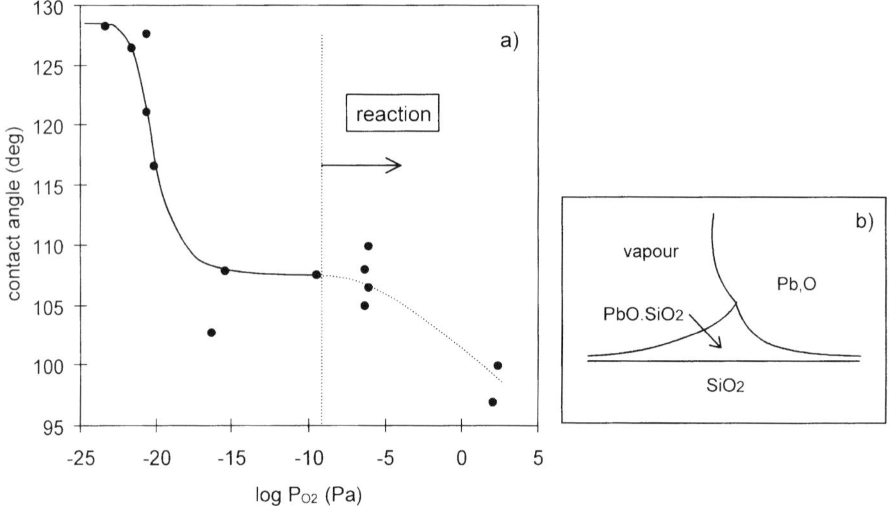

Figure 6.18. a) Contact angle of the Pb/SiO$_2$ system plotted as a function of log P$_{O2}$ at 1050K. b) At high P$_{O2}$, formation of a liquid interfacial reaction product (PbO.SiO$_2$) occurs so the apparent contact angle is not a Young contact angle. From work reported in (Sangiorgi et al. 1995).

A complex oxide is formed at the interface in certain systems at high P$_{O2}$ values by reaction between the liquid metal and the oxide substrate (see Figure 6.11). An example is the Pb/SiO$_2$ system (Figure 6.18). With increasing P$_{O2}$, the contact angle decreases owing to adsorption of oxygen at the Pb/SiO$_2$ interface. At P$_{O2}$ > 10^{-10} Pa, a reaction occurs to form *liquid* PbO-SiO$_2$ oxide at the interface that wets the SiO$_2$ surface such that the wetting of Pb occurs on a liquid PbO-SiO$_2$ layer rather than on SiO$_2$. The configuration at the PbO-SiO$_2$/Pb/vapour triple line deviates strongly from that assumed by Young (Figure 6.18.b) and obeys Smith's equation (1.30). As a result, the macroscopic contact angles in this type of system are not Young contact angles.

In the studies cited in Table 6.6, the oxygen affecting surface and interface properties of metal/oxide systems comes from the vapour phase. "Pure" metals and alloys contain oxygen as an impurity, but its concentration in many cases is sufficient to affect wettability. An example is given in Figure 6.19.a, in which the contact angle of Cu-Cr alloys is represented as a function of the molar fraction of Cr for two series of experiments performed using Cr with different oxygen contents. Significant differences were observed in θ that are mainly due to oxygen. Thus Figure 6.19.b shows that all θ values belong to the same curve when plotted against $\log X_O$ whatever the type of Cr used. In this example, the improved wettability of the Cu/Al_2O_3 system by Cr additions in Cu is not an intrinsic effect of Cr, but is due to adsorption at the $Cu-Cr/Al_2O_3$ interface of oxygen fixed by Cr in the melt. Similar effects may be observed in other alloys consisting of a noble metal matrix and an alloying element with a greater affinity for oxygen, for example the $Au-Ni/Al_2O_3$ (Rivollet et al. 1987) and $Au-Si/Al_2O_3$ (Drevet et al. 1990) systems.

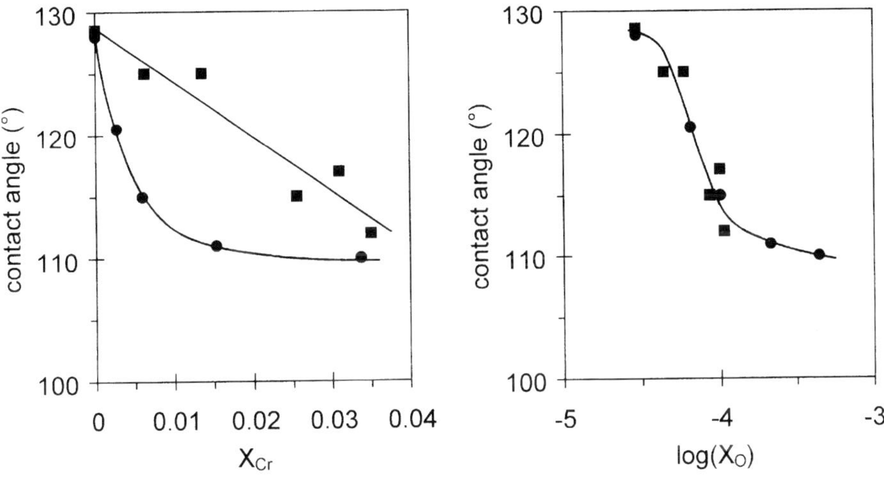

Figure 6.19. Contact angles of Cu-Cr alloys on monocrystalline Al_2O_3 at 1423K plotted as a function of molar fraction of Cr (left) and molar fraction of oxygen (right) in the alloy. Circles : Cr containing (3700 ± 500) ppm of oxygen. Squares: Cr containing (700 ± 150) ppm of oxygen. From (Kritsalis et al. 1990) [20].

contact angle towards a value close to 90° occurs in a few hundreds of seconds. In an alumina chamber furnace with a relatively high P_{O2}, no decrease of contact angle is observed even after several hours at temperature. To obtain a curve similar to that shown as (a) in Figure 6.20, a further increase of temperature is needed, for example up to 1000°C.

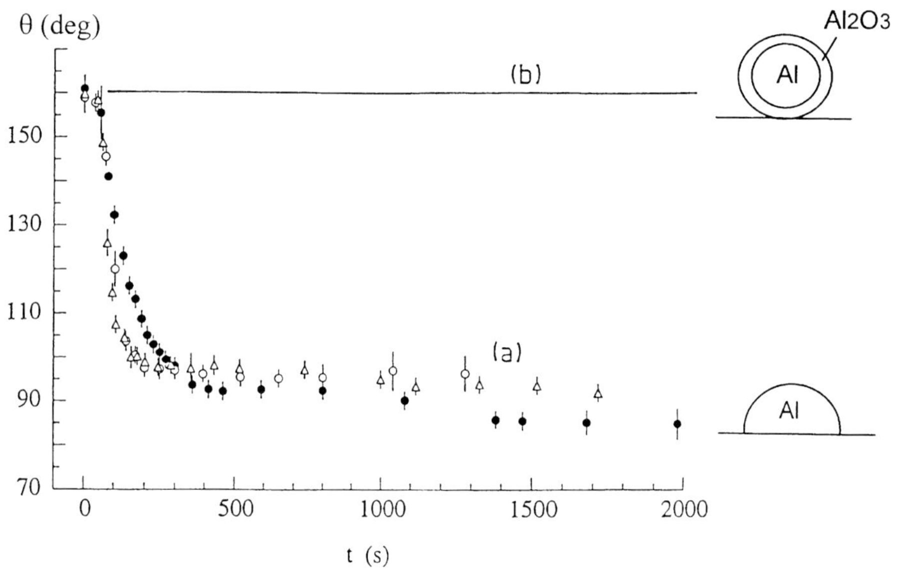

Figure 6.20. Experimental contact angles in high vacuum versus time of Al on Al_2O_3 substrates at 827°C in (a) a metal chamber furnace (low P_{O2}) and in (b) an alumina chamber furnace (high P_{O2}). From (Landry et al. 1998) [17].

Deoxidation of Al by *dissociation* of Al_2O_3 according to reaction (6.24) needs extremely low values of P_{O2} in the furnace (for instance 10^{-43} Pa at the melting temperature of Al and 10^{-35} Pa at 900°C). Such low P_{O2} values are not easily achieved in a furnace in which P_{O2} is usually in the range 10^{-15}–10^{-5} Pa. Therefore, deoxidation of Al observed under high vacuum does not occur by dissociation of Al_2O_3 and has been attributed to the reduction of the oxide film by liquid Al to form volatile Al_2O by the reaction (Laurent et al. 1988, Castello et al. 1994):

$$4(Al) + \langle Al_2O_3 \rangle \Leftrightarrow 3[Al_2O] \tag{6.25}$$

The condition for such deoxidation is that the total flow of oxygen, Φ_O, leaving the drop surface is higher than the flow of oxygen impinging on the drop. This egress of O is proportional to P_{Al2O} but the ingress is proportional to the partial pressure of O_2 in the furnace, P_{O2}^f. The maximum P_{Al2O} value is the value corresponding to the equilibrium of three phases: liquid (Al), solid (Al_2O_3) and vapour (equation (6.25)). This equilibrium can be established if the oxide film on the Al drop surface is disrupted as shown in Figure 6.21.a. The value of P_{Al2O}^{eq} can be easily calculated from thermodynamic data (Chase et al. 1985) and increases strongly with temperature (Figure 6.22). Accordingly, for a fixed value of P_{O2}^f, there is a threshold value of temperature, T^*, below which the oxide film thickens and above which it is eroded to deoxidise the surface (Figure 6.22).

If the operating conditions (T, P_{O2}^f) permit deoxidation, this will occur at a time that depends mainly on the temperature (because $P_{Al2O}^{eq} = f(T)$) and the thickness of the oxide skin. This thickness increases from its initial value e_0 during heating between room temperature and T^* by an amount that is strongly dependent on $P_{O_2}^f$ (Figure 6.23).

When, at a temperature T, the condition $P_{Al2O}^{eq} \gg P_{O2}^f$ is fulfilled, the impinging flow of oxygen can be neglected. Then, the deoxidation time t_d for a *monolayer* of oxide is simply given by (Laurent et al. 1988) (Figure 6.24):

$$ t_d = \frac{1}{\Omega_m \Phi_O} \tag{6.26} $$

Ω_m is the molar surface area of the oxide and the flow of oxygen leaving the drop surface, Φ_O, is given by:

$$ \Phi_O = \frac{\alpha_{Al_2O} P_{Al_2O}^{eq}}{(2\pi m_{Al_2O} kT)^{1/2}} \tag{6.27} $$

where k is the Boltzmann constant, m_{Al2O} is the mass of a molecule of Al_2O and α_{Al2O} is the "evaporation coefficient" which takes into account a possible departure from equilibrium (6.25) at the surface of the oxide film on the drop ($\alpha_{Al2O} < 1$). This mechanism has also been used with success to explain deoxidation of other pure molten metals as Si (Drevet et al. 1990) and U (Tournier et al. 1996).

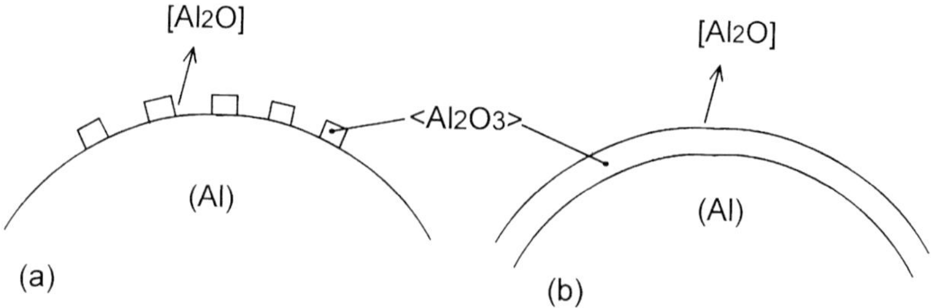

Figure 6.21. The model of Laurent et al. (1988) for deoxidation of an Al drop under high vacuum is based on the assumption that deoxidation is controlled by the rate of evaporation of a volatile oxide formed according to reaction (6.25). This is justified for disrupted oxide films (a) or for continuous but very thin films (b), through which diffusion is very fast, but not for thick oxide films.

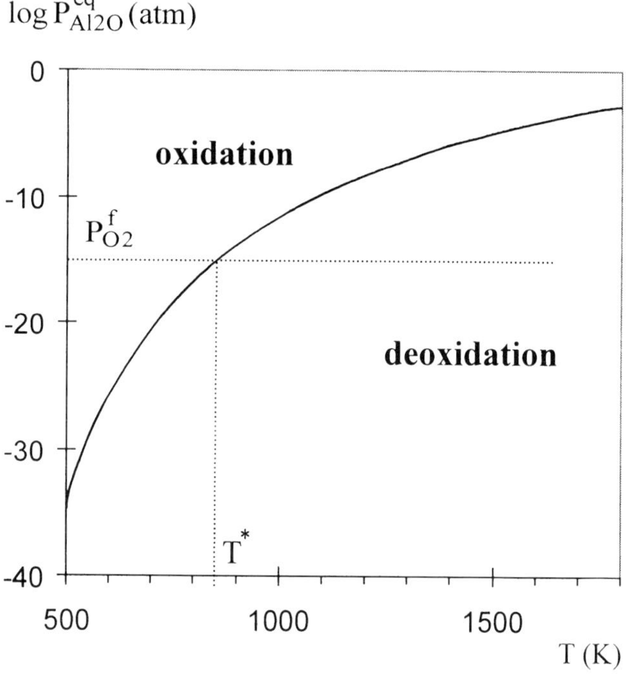

Figure 6.22. Equilibrium partial pressure of Al_2O, derived from equation (6.25), plotted as a function of temperature. For a given value of the partial pressure of oxygen in the furnace, $P^f_{O_2}$, there is a threshold temperature, T^*, below which the oxide film thickens and above which it is eroded.

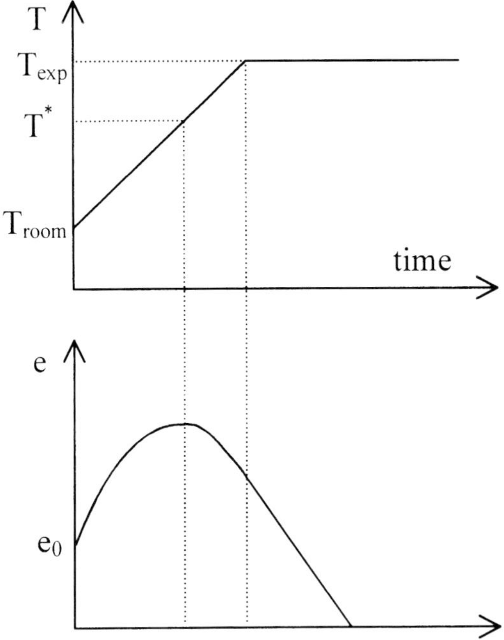

Figure 6.23. Schematic variation of thickness, e, of an oxide film on a metallic drop during a thermal cycle from room temperature to the experimental temperature.

In such calculations, the diffusion through a continuous oxide layer (Figure 6.21.b) has been neglected. This may be a good approximation for films a few monolayers thick but for thicker films, diffusion may control the overall deoxidation rate such that the deoxidation time is much higher than the product of t_d by the number of monolayers of the film (Chatillon et al. 1996).

For a binary alloy containing an oxidable metal such as Al or Si, the partial pressure of volatile suboxide (Al_2O or SiO) depends not only on temperature but also on the activity of the oxidable element in the alloy. Thus, for a Cu-Al alloy at constant temperature, P^{eq}_{Al2O} (i.e. roughly $1/t_d$) is proportional to $a_{Al}^{4/3}$ according to reaction (6.25). As a result, under high vacuum, it is much easier to deoxidise Cu-Al (Li et al. 1987–88) or Au-Si (Drevet et al. 1990) alloys that are rich in Al or Si. This straightforwardly explains the apparent contact angle maxima observed by many authors for alloys diluted in the oxidable element (Figure 6.25): for these alloys, the oxidable element is present in a concentration sufficient to produce oxidation at $T < T^*$, but not enough to produce deoxidation in a reasonable time at $T > T^*$. When experiments are performed in a metallic furnace with a very low P_{O2}

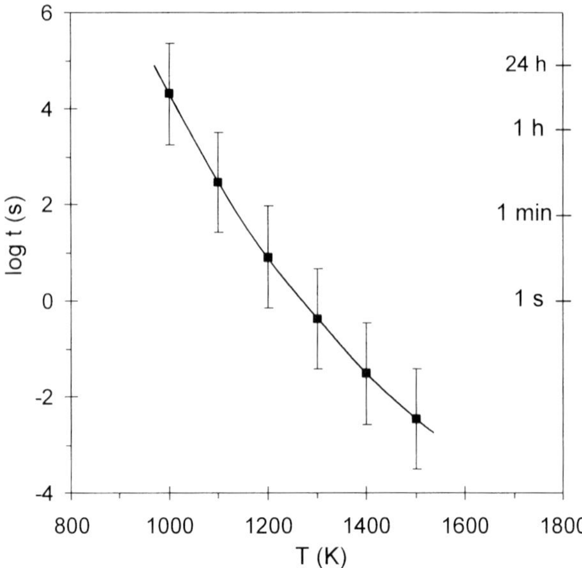

Figure 6.24. Times required for the disappearance of an oxide monolayer from the surface of liquid Al at different temperatures. The error bars are due to the uncertainty of the Gibbs energy of reaction (6.25). From (Laurent et al. 1988) [21].

using the dispensed drop technique (see Figure 3.7.c), no oxidation occurs and the maximum in the θ-composition curve is suppressed as shown in Figure 6.25.

When experiments are performed in a static neutral gas atmosphere, deoxidation depends on diffusion of volatile species into the gas. For the same temperature and P_{O2} values, deoxidation takes longer in such atmospheres than in a high vacuum but some acceleration can be achieved by using dynamic conditions i.e., gas flow (Ricci et al. 1994).

The disappearance of an oxide film may be achieved not only by reduction by the liquid metal but also by mechanical disruption or dissolution in the liquid. For instance, mechanical disruption can occur when the liquid drop is dispensed onto the substrate surface through a small hole in a tube end (Figure 3.7.c). When this technique is used for Al on Al_2O_3, the deoxidation stage is suppressed and a contact angle close to 80° is rapidly achieved (Landry et al. 1996).

In the case of liquid Al, the solubility of oxygen is too low for a dissolution process to be effective. However, for other metals like Sn, it may be the principal mechanism. The variation in contact angle of Sn on Al_2O_3 shown in Figure 6.26 as a function of temperature displays three ranges. At low temperatures, high apparent

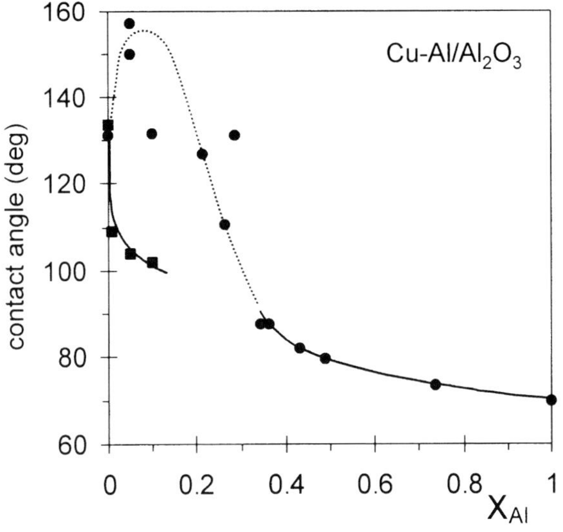

Figure 6.25. Variation of contact angle with the Al molar fraction for the Cu-Al/Al$_2$O$_3$ system at 1423 K in high vacuum. Circles: in an alumina furnace using the sessile drop technique (Li et al. 1987–88). Squares: in a metal chamber furnace using the dispensed drop technique (Labrousse et al., to be published). All points represented by circles for $0 < X_{Al} < 0.3$ are for oxidised drops.

θ values are caused by the SnO$_2$ film around melted Sn; when the temperature increases, the oxide film is dissolved in Sn since the concentration of oxygen in Sn in *equilibrium with tin oxide* increases. When the oxide film disappears, enough oxygen is present in the melt to affect the metal/oxide interface and the contact angle decreases. When the temperature is further increased, the concentration of oxygen in Sn in *equilibrium with the vapour phase* decreases and oxygen desorbs from the interface leading to dewetting. At high temperatures, the normal wetting behaviour of non-reactive pure metal/oxide systems is observed.

6.5. ALLOYING ELEMENTS

6.5.1 Non-reactive solutes

Consider a non-reactive system consisting of a binary liquid alloy A-B and an oxide substrate such as Al$_2$O$_3$ at constant temperature. A simple statistical thermodynamic model has been developed (Li et al. 1989) to predict the contact angle and the work of adhesion isotherms, $\theta(X_B)$ and $W_a(X_B)$, from the known values of contact angles

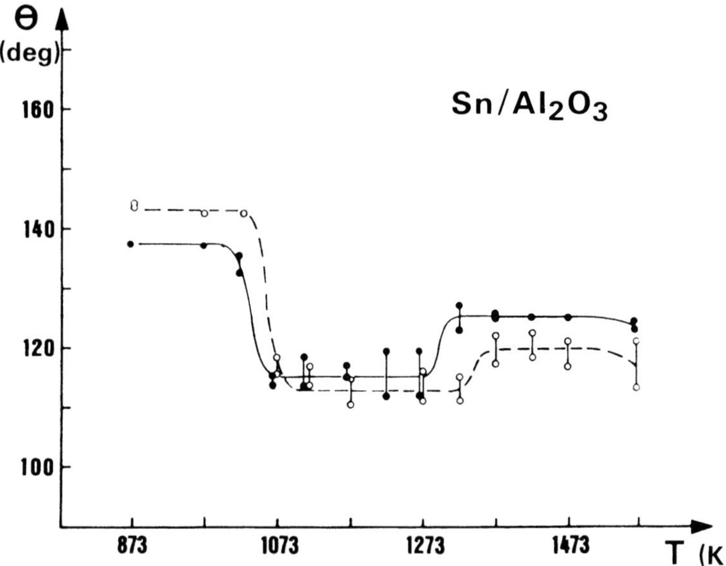

Figure 6.26. Contact angle variation with temperature for Sn on Al_2O_3 under one atmosphere of He with $P_{O2} = 10^{-16}$ Pa (Rivollet et al. 1987) [10].

θ_A and θ_B, works of adhesion W_a^A and W_a^B, and liquid surface energies σ_{LV}^A and σ_{LV}^B of the pure metals. Adsorption of metal atoms on free Al_2O_3 surfaces is negligible for non-wetting metals (see Section 1.4.2), so σ_{SV} is constant and calculation of the contact angle and work of adhesion isotherms is reduced to calculating the σ_{LV} and σ_{SL} isotherms.

These calculations were performed by Li et al. (1989) using a regular solution model (Section 4.1.2) and assuming that the interface between the liquid and the external phase, solid or vapour, is a liquid monolayer (Figure 6.27). Moreover, when the external phase is a solid, interpenetration of the liquid and solid phases at the atomic level was neglected. Indeed, for metal-ceramic interfaces any roughening on an atomic level would be more expensive in energy than the gains due to configurational entropy.

Although the model was developed for a A-B alloy over the whole composition range (Li et al. 1989), application to the case of a solution that is infinitely dilute in B leads to simple expressions for the slope of σ-X_B curve and for the "enrichment factor" at the interface Y_B/X_B where Y_B is the molar fraction of B in the interfacial monolayer:

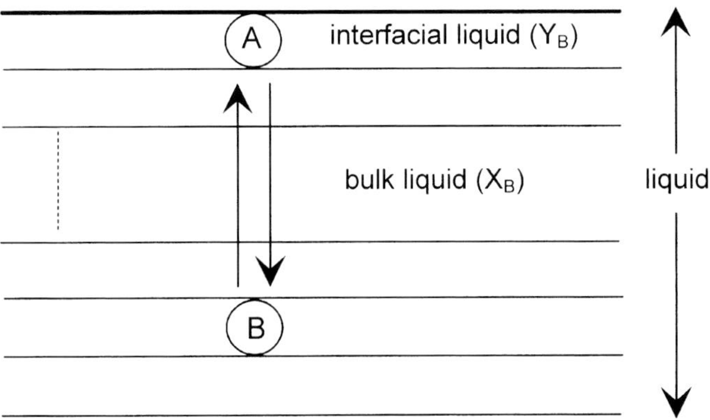

Figure 6.27. Schematic representation of a monolayer interface in a binary liquid A-B/external phase system according to the model of Li et al. (1989).

$$\left(\frac{d\sigma}{dX_B}\right)_{X_B \to 0} = \frac{RT}{\Omega_m}\left[1 - \exp\left(-\frac{E^\infty_{B(A)}}{RT}\right)\right] \tag{6.28}$$

$$\left(\frac{Y_B}{X_B}\right)_{X_B \to 0} = \exp\left(-\frac{E^\infty_{B(A)}}{RT}\right) \tag{6.29}$$

Ω_m is equal to ωN_a, where ω is the surface area per atom and is readily derived from the molar volume of the alloy ($\Omega_m \approx v_m^{2/3}$).

The pertinent parameter in equations (6.28) and (6.29) is the quantity $E^\infty_{B(A)}$, which is the adsorption energy of solute B from the bulk liquid A to the surface or interface (Figure 6.27): a negative value of $E^\infty_{B(A)}$ indicates attraction of B by the interface, while a positive value indicates attraction by the bulk liquid. The adsorption energy is equal to the difference in energy of the system after and before segregation of B. If m_2 and m_1 are, for an atom, the fractions of nearest neighbours located respectively in the same layer and in an adjacent one ($m_2 + 2 m_1 = 1$, see Figure 1.3) and $\varepsilon_{i\text{-ex}}$ is the energy of a bond formed between the atom i and the external phase, by calculating the variation of the number of pairs A-A and A-B during segregation, it holds:

$$E_{B(A)}^{\infty} = [Z(m_2 + 2m_1)\varepsilon_{AA} + Z(m_2 + m_1)\varepsilon_{AB} + Zm_1\varepsilon_{B-ex}] \tag{6.30}$$

$$- [Z(m_2 + 2m_1)\varepsilon_{AB} + Z(m_2 + m_1)\varepsilon_{AA} + Zm_1\varepsilon_{A-ex}]$$

$$= -Zm_1\varepsilon_{AB} + Zm_1\varepsilon_{AA} + Zm_1\varepsilon_{B-ex} - Zm_1\varepsilon_{A-ex}$$

$$= -Zm_1\left[\varepsilon_{AB} - \frac{\varepsilon_{AA} + \varepsilon_{BB}}{2}\right] + Zm_1\left[\frac{\varepsilon_{AA}}{2} - \frac{\varepsilon_{BB}}{2}\right] + Zm_1[\varepsilon_{B-ex} - \varepsilon_{A-ex}]$$

The first term results from the heat of mixing of A-B alloy (see equations (4.3) and (4.4)). The second term comes from the cohesion energies of pure A and pure B and is related to the liquid surface energies σ_{LV}^A and σ_{LV}^B (see equations (1.9) and (1.10)). The third term comes from the adhesion energy of pure A and B on the external phase. When the external phase is the vapour, $\varepsilon_{B-ex} = \varepsilon_{A-ex} = 0$ and the molar adsorption energy is reduced to equation (4.8):

$$E_{B(A)}^{\infty,LV} = (\sigma_{LV}^B - \sigma_{LV}^A)\Omega_m - m_1\lambda \tag{6.31}$$

When the external phase is a solid substrate S, ε_{B-ex} and ε_{A-ex} are related to the work of adhesion of pure B and pure A on the solid (see equation (1.12)):

$$W_a^i = -\frac{Zm_1 N_a \varepsilon_{i-S}}{\Omega_m} \qquad i = A, B \tag{6.32}$$

and the adsorption energy is given by:

$$E_{B(A)}^{\infty,SL} = (\sigma_{LV}^B - \sigma_{LV}^A)\Omega_m - (W_a^B - W_a^A)\Omega_m - m_1\lambda \tag{6.33}$$

Following equation (6.28), if $E_{B(A)}^{\infty}/RT$ is very negative, the slope $(d\sigma/dX_B)_{X_B\to 0}$ will also be very negative. Conversely, if $E_{B(A)}^{\infty}/RT$ is very positive, the slope will be positive but small (Figure 6.28.a). Indeed, the maximum value of $(d\sigma/dX_B)_{X_B\to 0}$ will be equal to RT/Ω_m and substitution of typical values of temperature (1000K) and molar surface area ($\Omega_m \approx 5 \times 10^4$ m^2.mol^{-1}), gives the negligible value of 2 mJ/m^2 per 1% of solute. This behaviour predicted by the

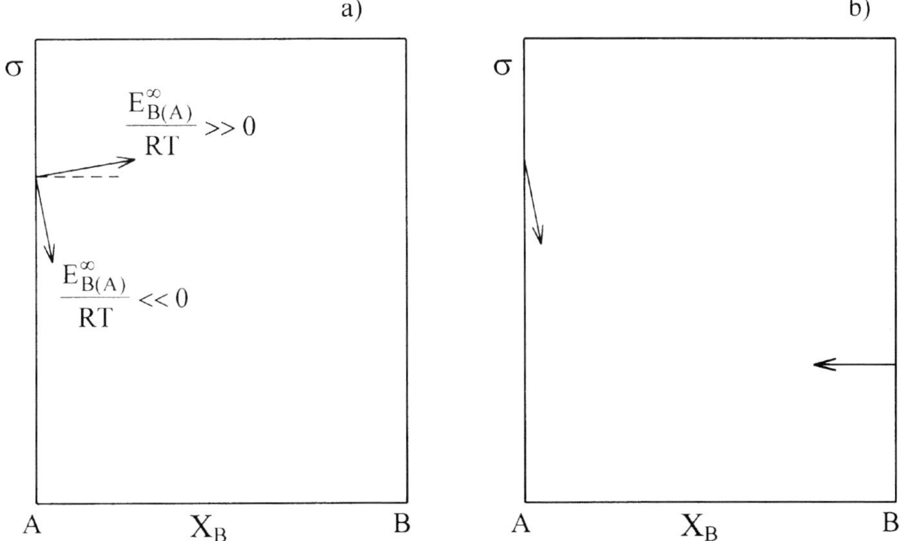

Figure 6.28. a) Effect of small additions of an alloying element on the interfacial or liquid surface energies of a non-reactive binary alloy/ceramic system for very positive and very negative values of adsorption energy. b) A very negative value of the slope of σ when $X_B \rightarrow 0$ implies a negligible slope when $X_B \rightarrow 1$.

monolayer model is confirmed by the experimental surface energy values of binary alloys (Brunet et al. 1977).

Moreover, since the absolute value of the interaction term in equations (6.31) and (6.33), $|m_1\lambda|$, is usually much less than that of the capillary terms, $|\sigma^B_{LV} - \sigma^A_{LV}|\Omega_m$ and $|W^B_a - W^A_a|\Omega_m$, the following relation applies for both the E^{LV} and E^{SL} energies of adsorption for a binary alloy:

$$E^{\infty}_{B(A)} \ll 0 \Leftrightarrow E^{\infty}_{A(B)} \gg 0 \tag{6.34}$$

Thus, if the slope of the σ-X_B curve at $X_B \rightarrow 0$ is negative, that at $X_B \rightarrow 1$ will nearly equal zero (see Figure 6.28.b and experimental results in Figure 4.5.a).

For A-B/oxide systems, and more generally non-reactive liquid A-B/solid systems, the model predicts three main types of $W_a(X_B)$ and $\theta(X_B)$ isotherms, depending on the absolute and relative values of the adsorption energies $E^{\infty,LV}_{B(A)}$ and $E^{\infty,SL}_{B(A)}$ (Figure 6.29). In the particular case of isotherms (b), two different

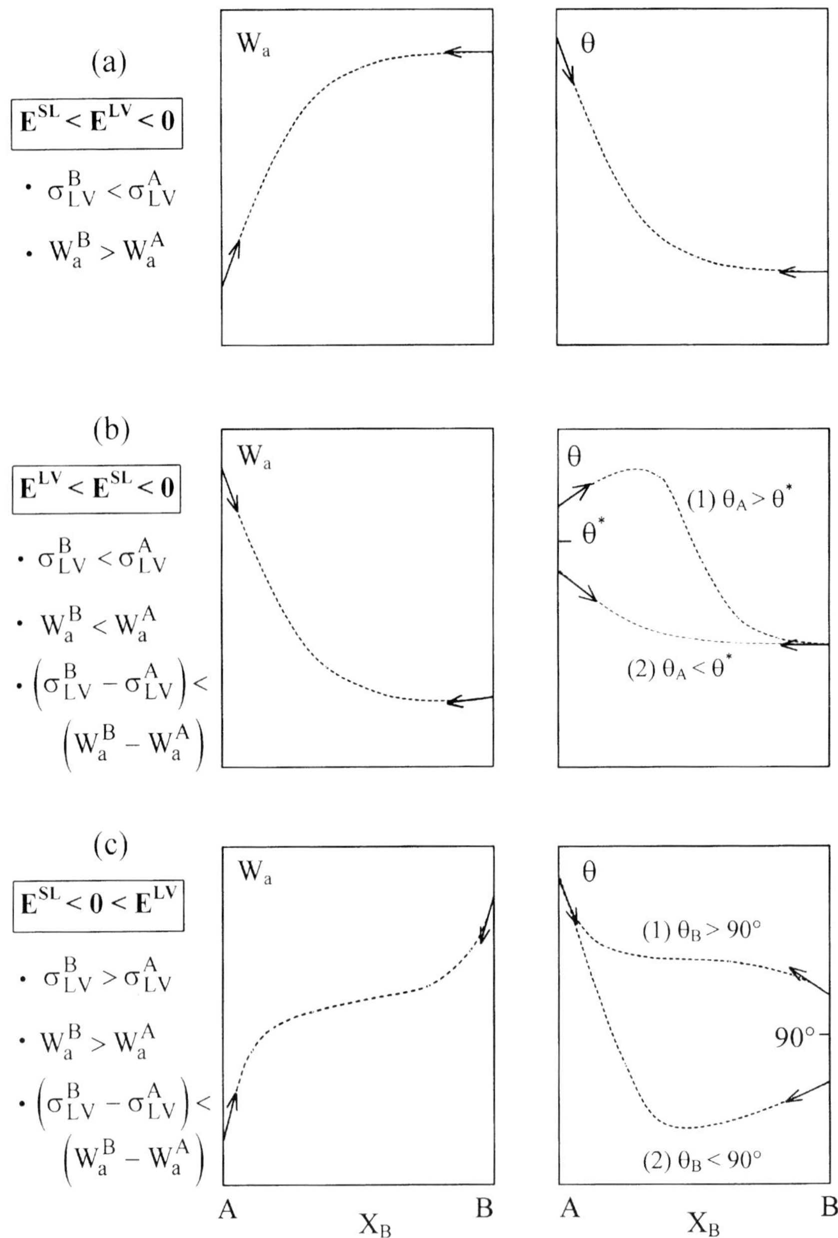

Figure 6.29. Main forms of work of adhesion and contact angle isotherms for non-reactive A-B liquid alloy/ceramic systems (Li et al. 1989) [2].

curves can be obtained for $\theta(X_B)$ depending on whether θ_A is greater or smaller than a value θ^* given by (Li et al. 1989):

$$\cos\theta^* = -\frac{1 - \exp[-E_{B(A)}^{\infty,SL}/RT]}{1 - \exp[E_{B(A)}^{\infty,LV}/RT]} \tag{6.35}$$

For alloys consisting of two non-wetting, non-reactive metals developing van der Waals interactions with the oxide, we have seen in Figure 6.7 that W_a is nearly proportional to σ_{LV} ($W_a \cong 0.4\sigma_{LV}$). It can be readily shown that the behaviour of such binary alloys on Al_2O_3 belongs to the type (b) isotherms of Figure 6.29. Thus, addition of B to A (with $\sigma_{LV}^B \ll \sigma_{LV}^A$) hardly decreases θ_A but in all cases decreases the work of adhesion of A on Al_2O_3 and any other ceramic for which adhesion is mainly due to van der Waals interactions. Examples are Cu-Sn on Al_2O_3 (Figure 6.30), Ga-Sb on Al_2O_3 and SiO_2 (Harter et al. 1993) and Cu-Au on Al_2O_3 (Ghetta et al. 1996).

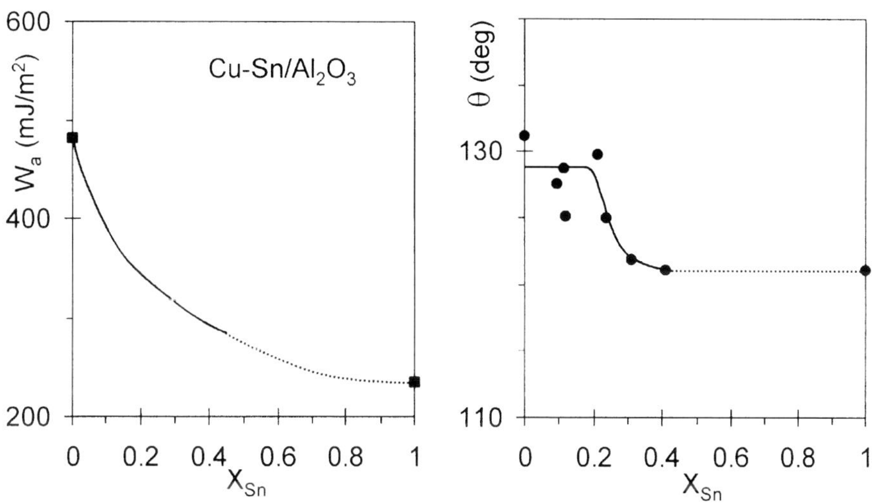

Figure 6.30. Experimental $W_a(X_{Sn})$ and $\theta(X_{Sn})$ isotherms for the Cu-Sn/Al_2O_3 system at 1423K (Li et al. 1989) illustrating case (b) in Figure 6.29.

In order to obtain the maximum improvement in wetting and adhesion for a A/oxide system, the alloying element B must satisfy the conditions that generated

a ternary Cu-Mg-Ag alloy. At $500°C$ and for a molar fraction of Ag in Cu of 5×10^{-3}, a strong adsorption of Ag was found that corresponded to a Ag monolayer at the Cu/MgO interface. Such an adsorption cannot be explained by the model used for liquid alloys, that is, taking into account only differences in cohesion energy and adhesion energy of pure components and the regular solution parameter. It is possible that a term of elastic energy due to atomic size differences becomes predominant in some cases, as for the segregation of alloying elements at grain boundaries or free surfaces of solid metals (see also Section 4.1.3).

6.5.2 Reactive solutes

As a general rule, the addition of non-reactive solutes to a liquid metal does not cause contact angle decreases below $60°$. In principle, lower contact angles can be achieved using reactive solutes.

Consider a reactive solute B dissolved in a non-reactive matrix M in contact with an oxide substrate such as Al_2O_3. The chemical interaction in this system can be described by the dissolution of Al_2O_3 in the alloy:

$$\frac{1}{2}Al_2O_3 \Leftrightarrow (Al) + \frac{3}{2}(O) \tag{6.36}$$

possibly followed by the precipitation of a B oxide at the interface, for example:

$$(B) + \frac{3}{2}(O) \Leftrightarrow \frac{1}{2}B_2O_3 \tag{6.37}$$

For small molar fractions of B and Al, and taking into account equation (6.4), the equilibrium mole fraction of dissolved oxygen X_O^D for reaction (6.36) (the superscript D stands for dissolution) will be given by the equation:

$$X_O^D = \left(\frac{3}{2}\right)^{2/5} (K_D'(T))^{2/5} \exp\left(-\frac{3}{5}\varepsilon_O^B X_B\right) \tag{6.38}$$

where the constant $K_D'(T)$ is given by equation (6.3). ε_O^B is Wagner's first order interaction parameter between the O and B solutes defined by equation (6.22). If $\varepsilon_O^B < 0$, then X_O^D increases rapidly at high values of X_B: strong O-B interactions will promote Al_2O_3 dissolution in the melt (curves D in Figure 6.33).

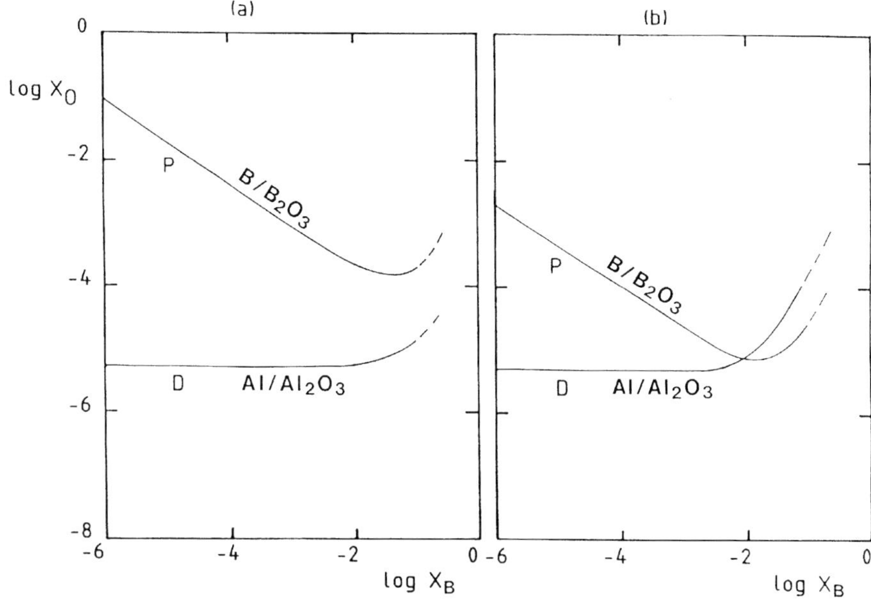

Figure 6.33. Thermodynamics of (matrix M-solute B)/Al$_2$O$_3$ systems with $\varepsilon_O^B < 0$. Curve D presents the logarithm of the molar fraction of oxygen coming from stoichiometric dissolution of Al$_2$O$_3$ in M-B alloys as a function of log X$_B$. Curve P illustrates the variation of X$_O$ in equilibrium with B$_2$O$_3$ as a function of X$_B$. In a) only dissolution of Al$_2$O$_3$ in the melt occurs, as for Ni-Cr alloys at 1773K (Kritsalis et al. 1992). In b) the solute B reduces Al$_2$O$_3$ to the right of the intersection point of the two curves, as for Ni-Ti at 1773K (Naidich et al. 1974, Merlin et al. 1992).

Moreover, a solute B satisfying the condition $\varepsilon_O^B \ll 0$ can cause precipitation of an oxide B$_2$O$_3$ by reaction with the excess oxygen in the alloy. The mole fraction of dissolved oxygen in equilibrium with the B$_2$O$_3$ precipitate (reaction (6.37)), denoted X$_O^P$, is given by:

$$X_O^P = K_P'(T)X_B^{-2/3}\exp(-\varepsilon_O^B X_B) \tag{6.39}$$

In this equation, K$'_P$(T) is a constant given by (Kritsalis et al. 1992):

$$K_P'(T) = (K_P(T))^{-2/3}(\gamma_B)^{-2/3}(\gamma_{O(M)}^\infty)^{-1} \tag{6.40}$$

where $K_P(T)$ is the equilibrium constant of reaction (6.37), γ_B is the activity coefficient of B in the alloy and $\gamma^\infty_{O(M)}$ is the activity coefficient of oxygen at infinite dilution in pure M. When $X_B \to 0$, X_O^P varies as $X_B^{-2/3}$, i.e. it is a decreasing function of X_B. However, for $\varepsilon_O^B \ll 0$, the exponential term in equation (6.39) predominates at higher X_B values and X_O^P increases rapidly with X_B (curves P in Figure 6.33). Thus, the concentration of dissolved oxygen increases above a certain value of X_B for both dissolution and precipitation.

If for a particular value of X_B the inequality $X_O^D > X_O^P$ is verified, the solute B will reduce Al_2O_3 forming B_2O_3:

$$Al_2O_3 + 2(B) \Leftrightarrow B_2O_3 + 2(Al) \tag{6.41}$$

This condition is satisfied for the Ni-Ti alloy because interactions between O and Ti solutes are very strong ($\varepsilon_O^{Ti} = -100$), but not for Ni-Cr alloys ($\varepsilon_O^{Cr} = -25$). Thus, additions of Cr up to 20 at.% in Ni do not cause formation of a chromium oxide, the only effect is to increase the dissolution of Al_2O_3 in the alloy. As shown in Figure 6.34, Cr additions in Ni significantly decrease the contact angle of Ni on Al_2O_3 and this could result from two effects, both of which require that $\varepsilon_O^B < 0$:
i) an increased concentration of dissolved oxygen due to interfacial reactions i.e., an increased concentration of O-M clusters which can be adsorbed at the metal/oxide interface;
ii) an increased adsorption capability of the O-Cr clusters as compared with the O-Ni clusters. Indeed, as Cr is more electropositive than Ni, the charge transfer from Cr to O (or, in other terms, the degree of ionicity of O-Cr clusters) would be greater than in the case of Ni.

Other examples of alloying elements leading to strong improvements in wetting without forming 3D reaction products at the interface are Ti in Cu in contact with Y_2O_3 (Figure 6.36) and Ti in Sn in contact with Al_2O_3 (Naidich 1981, Tomsia et al. 1998).

In the Ni-Ti/Al_2O_3 system, as $\varepsilon_O^{Ti} \ll 0$, O-Ti clusters in Ni are very active at metal/oxide interfaces but wettability is affected also by the formation of a continuous Ti_2O_3 layer at the Ni/Al_2O_3 interface. The replacement of ionocovalent Al_2O_3 by the semi-metallic oxide Ti_2O_3 increases W_a and decreases θ, as explained in Section 6.3. The combination of these two effects, i.e., adsorption of O-Ti clusters at the *liquid-side* of the interface and formation of a semi-metallic oxide

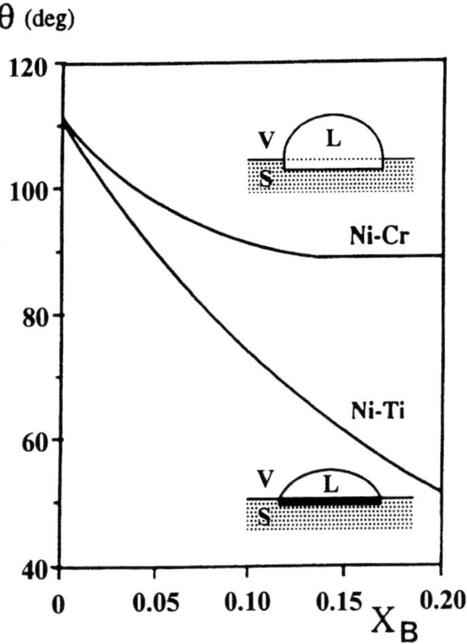

Figure 6.34. Contact angles at 1773 K of Ni-based alloys on Al_2O_3 plotted as a function of the Cr (Ti) molar fraction for Ni-Cr (Kritsalis et al. 1992) and Ni-Ti (Naidich et al. 1974). The Ni-Cr alloys cause dissolution of Al_2O_3 but the Ni-Ti alloys cause dissolution of Al_2O_3 followed by precipitation of a Ti oxide.

such as Ti_2O_3 at the *solid-side* of the interface, explains the strong decrease in contact angle caused by Ti additions in Ni (Figure 6.34).

This influence of Ti is even greater in the case of Cu-Ti alloys on Al_2O_3 (Naidich et al. 1990, Kritsalis et al. 1991) and NiFeCr-Ti alloys on Al_2O_3 (Wan et al. 1996) where the metal-like oxide TiO is formed at the interface owing to the higher thermodynamic activity of Ti (Figure 6.35). Wetting is nearly perfect as opposed to the Cu-Ti/Y_2O_3 system (Figure 6.36) where only dissolution of Y_2O_3 into the alloys occurs, to produce a significant but limited decrease of θ from 140° to 80°. However, adding Ca to Al on Al_2O_3 or SiO_2 substrates has no effect on wetting in spite of the formation of CaO at the interface (Mori et al. 1983) because the ionocovalent substrate is replaced by an oxide of the same type. A similar result was obtained by Kritsalis et al. (1992) for NiPd-Cr alloys on Al_2O_3 at 1523K: Cr additions do not affect wetting in spite of the precipitation of Cr oxide at the interface.

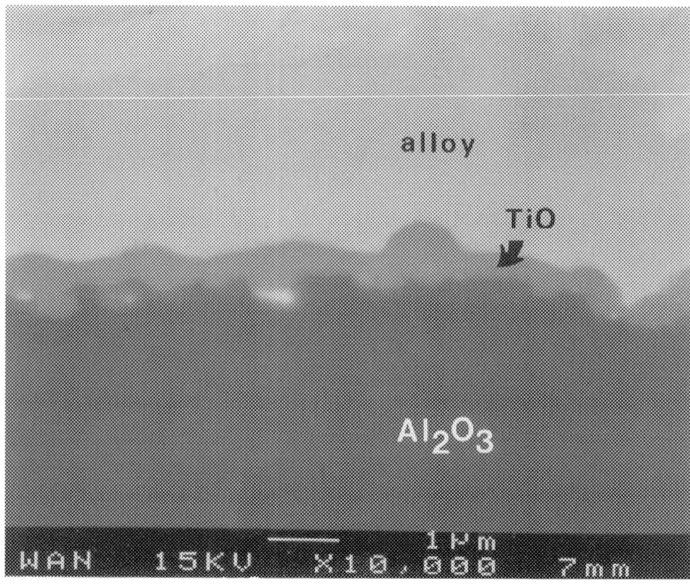

Figure 6.35. SEM micrograph of the cross-section of a Ni alloy containing 16 at.% Ti with Al_2O_3 showing the TiO reaction product at the interface (holding time of 30 min at 1463K). The final contact angle in this sytem is about 20°. From (Wan et al. 1996) [22].

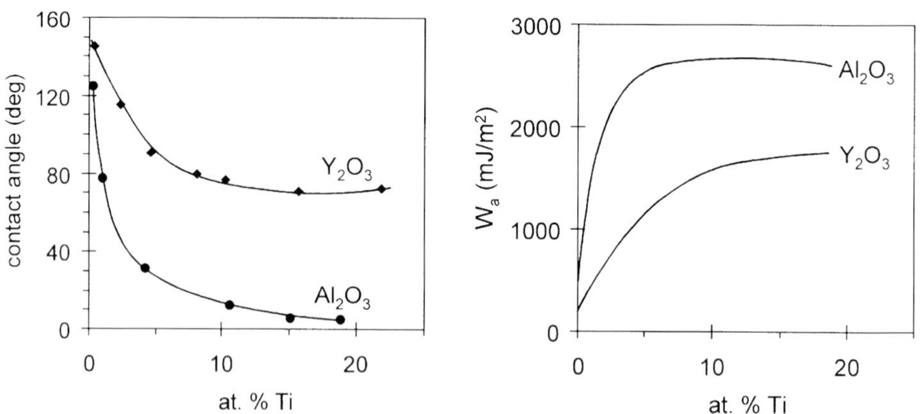

Figure 6.36. Contact angle and work of adhesion values for Cu-Ti melts on Y_2O_3 and Al_2O_3 at 1423K. Reactions cause dissolution of the oxide substrate in the first case, and dissolution + precipitation of a Ti oxide at the interface in the second. Data from work reported in (Naidich et al. 1990) [2].

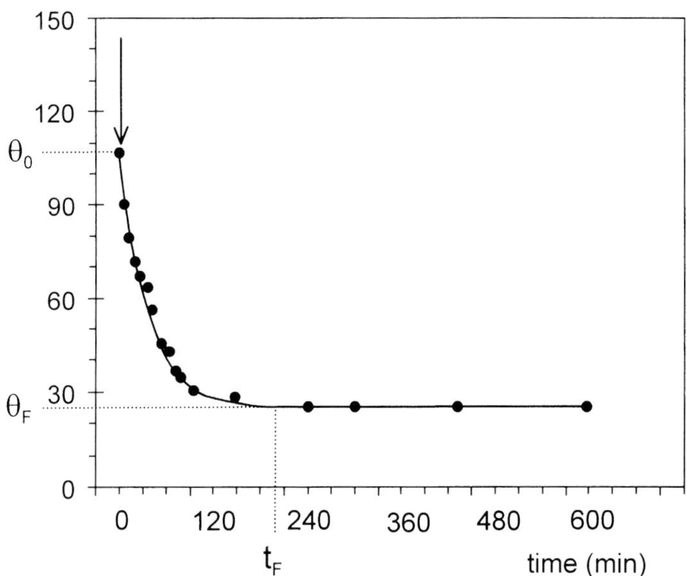

Figure 6.37. Contact angle of a Cu-9.5 at.% Ti alloy on Al_2O_3 as a function of time at 1373K (Kritsalis et al. 1991) [2].

Figure 6.37 shows the change with time of the contact angle of a reactive Cu-Ti alloy on Al_2O_3. This result is also discussed in Section 2.2.2.2 (sub-section "diffusion-limited spreading"). The addition of 9.5 at.% Ti decreases the contact angle from the 130° of pure Cu to about 30°. However, the θ versus time curve exhibits two stages. The first corresponds to a change of θ from 130° to 100° that is achieved in less than 1 second and attributed to adsorption of Ti at the interface. The second is much slower and controlled by the lateral extension of the interfacial reaction product (TiO) at the triple line. Similar two stage kinetics have been observed in (Espie et al. 1994) for other Ti containing alloys on various oxide substrates.

These considerations can explain the results of Figure 6.2 for reactive pure metal/oxide couples in which a new phase is formed at the interface. Wettability then depends on the bonding character of this interfacial layer and two cases can be distinguished:

1) First, replacement at the interface of an ionocovalent oxide by another of the same type. No significant improvement of wetting is expected for this type of

system. Al on SiO_2 provides an example of intense reactivity, causing reduction of SiO_2 to form Al_2O_3 and a slight liquid enrichment in Si. Because the ionocovalent oxide SiO_2 is replaced by an oxide of the same type, the contact angle of Al on SiO_2 ($\theta \approx 80° \pm 5°$ at 1073K) is similar to that of Al on Al_2O_3. The same explanation is valid for the Al/MgO system in which the reaction leads to formation of Al_2O_3 or $MgAl_2O_4$ at the interface.

2) Second, replacement at the interface of an ionocovalent oxide by a metallic one. A typical example of this is provided by the Ti/MgO system. Liquid Ti can react to dissolve several at.% O and form metal-like oxides such as TiO or even solid solutions of Ti with high oxygen contents. The perfect wetting observed for this system (Figure 6.2) can be explained by the double *in-situ* modification of the interface: the adsorption of oxygen at the liquid-side and the formation of a metallic phase at the solid-side.

6.6. WETTING OF FLUORIDES

Like many oxides, alkaline-earth metal fluorides such as CaF_2 and BaF_2 are very stable thermodynamically and have a high degree of ionocity, for instance that of CaF_2 on the Pauling's scale is 89%. Only van der Waals dispersion force interactions are expected to develop at interfaces with non-reactive metals, resulting in large, obtuse, contact angles. For instance, sessile drop experiments carried out for pure Cu and Sn on smooth, monocrystalline CaF_2 surfaces in high vacuum at 1100°C yield contact angles of 140° and 118° (Naidich and Krasovsky 1998). For molten Al on CaF_2, a contact angle of 130° was found at 700°C, decreasing to 100° at 1000°C (Naidich et al. 1983b).

Using alloying elements with high affinities for F, such as Ti, Zr and Hf, improves wetting as shown by the examples in Figure 6.38. The addition of 1.3 at.% Ti to a Cu-Sn alloy at 900°C produces only a limited improvement in wetting, but addition of 35 at.% Ti causes the contact angle to decrease strongly to about 30° owing to the formation of TiF_3 at the interface. However, as TiF_3 is volatile with a boiling temperature of 1037°C, it is eliminated from the interface by evaporation in high vacuum and the contact angle increases with time to a value greater than 90°. When the wetting experiment is repeated at 1150°C, i.e., at a temperature higher than the boiling temperature of TiF_3, no minimum in the θ-t curve appears and the final contact angle is close to 120°.

Alkaline-earth metal fluorides have been proposed for use as refractories when melting and casting Ti, Zr and Hf alloys which react with and wet most current materials (Naidich and Krasovsky 1998).

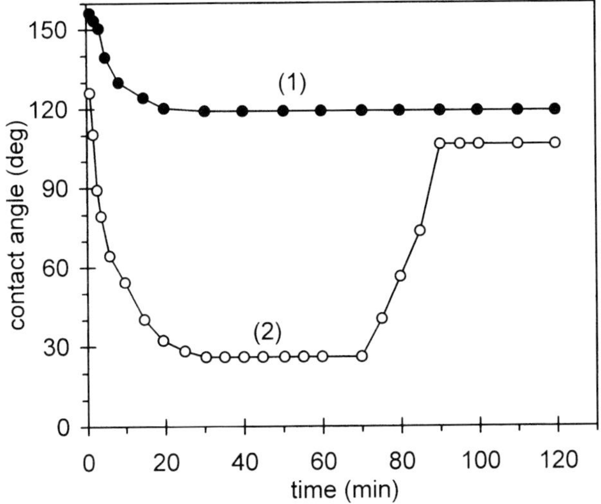

Figure 6.38. Contact angle vs time at 900°C in high vacuum for a Cu-11.9 at.% Sn alloy containing 1.3 at.% Ti (curve 1) and 35 at.% Ti (curve 2) on CaF$_2$. Data from work reported in (Naidich and Krasovsky 1998) [9].

6.7. CONCLUDING REMARKS

Pure metal/oxide systems can be divided into three categories. Systems for which the molar fraction of oxygen dissolved in the liquid metal coming from the oxide, X_O, is lower than 10^{-6} have contact angles of 110–140°. Weak physical metal/oxide interactions are localized at a sharp interface and result in work of adhesion values that are small compared to the work of cohesion of the metal. For systems with $X_O > 10^{-5}$ that do not react to form a new phase, dissolved oxygen and metal atoms from the oxide can produce significant improvements in wetting by forming 2D adsorption layers (i.e., with a typical thickness of 1 nm) at the metal/oxide interface. These adsorptions can lead to contact angles lower than 90° but usually θ remains above 60°. Even lower contact angles can be achieved by metals that react with the oxide forming 3D compounds at the interface which are, at least partially, metallic in their bonding characteristics.

For a non-reactive and non wetting pure metal on an oxide, improved wetting can be achieved either by adsorption at metal/oxide interface of oxygen supplied by a gas with a controlled P_{O2} or by introducing certain specific alloying elements modifying *in situ* the interface. In practice, it is difficult to operate in a gas atmosphere with a fixed P_{O2} using a buffer gas and hence wetting behaviour is more often modified by introducing alloying elements capable of producing a favourable interfacial chemistry.

REFERENCES FOR CHAPTER 6

Anisimov, V. I., Zaanen, J. and Andersen, O. K. (1991) *Phys. Rev. B*, **44**, 943

Barrera, R. G. and Duke, C. B. (1976) *Phys. Rev. B*, **13**, 4477

Belton, G. R. and Tankins, E. S. (1965) *Trans. Am. Inst. Min. (metall.) Engrs*, **233**, 1892

Brennan, J. J. and Pask, J. A. (1968) *J. Amer. Ceram. Soc.*, **51**, 569

Brunet, M., Joud, J. C., Eustathopoulos, N. and Desré, P. (1977) *J. Less-Common Metals*, **51**, 69

Castello, P., Ricci, E., Passerone, A. and Costa, P. (1994) *J. Mater. Sci.*, **29**, 6104

Chabert, F. (1992) Ph.D. Thesis, INP Grenoble, France

Chaklader, A. C. D., Gill, W. W. and Mehrotra, S. P. (1981) in *Surfaces and Interfaces in Ceramic and Ceramic-Metal Systems*, ed. J. Pask and A. Evans, Materials Science Research, Vol. 14, Plenum Press, New York, p. 421

Chang, Y. A., Fitzner, K. and Zhang, M. X. (1988) *Progr. Mater. Sci.*, **32**, 97

Chase, M. W., Davies, C. A., Downey, J. R., Frurip, D. J., McDonald, R. A. and Syverud, A. N. (1985) *JANAF Thermochemical Tables*, ed. D. R. Lide, 3rd ed., J. of Physical and Chemical Reference Data, vol. 14, American Chemical Society and the American Institute of Physics for the National Bureau of Standards

Chatain, D., Rivollet, I. and Eustathopoulos, N. (1986) *J. Chim. Phys.*, **83**, 561

Chatain, D., Muolo, M. L. and Sangiorgi, R. (1993) in *Proc. 2ⁿᵈ Eur. Coll. "Designing Ceramic Interfaces II"*, Petten, Nov. 1991, ed. S. D. Peteves, p. 359

Chatain, D., Chabert, F., Ghetta, V. and Fouletier, J. (1994) *J. Am. Ceram. Soc.*, **77**, 197

Chatillon, C., Coudurier, L. and Eustathopoulos, N. (1996) *Materials Science Forum*, Transtec Publications, Switzerland, **251-254**, 704

Defay, R., Prigogine, I., Bellemans, A. and Everett, D. H. (1966) in *Surface Tension and Adsorption*, Longmans, London, p. 161

Didier, F. and Jupille, J. (1994) *Surface Science*, **314**, 378

Drevet, B., Chatain, D. and Eustathopoulos, N. (1990) *J. Chim. Phys. (Fr)*, **87**, 117

Duffy, J. A. (1990) *Bonding, Energy Levels and Bands in Inorganic Solids*, Longman, London

Espie, L., Drevet, B. and Eustathopoulos, N. (1994) *Metall. Mater. Trans. A*, **25**, 599

Eustathopoulos, N., Joud, J. C., Desré, P. and Hicter, M. (1974) *J. Mat. Sci.*, **9**, 1233

Eustathopoulos, N. and Drevet, B. (1994) *J. Phys. III France*, **4**, 1865

Eustathopoulos, N. and Drevet, B. (1998) *Mater. Sci. Eng. A*, **249**, 176

French, T. M. and Somorjai, G. A. (1970) *J. Phys. Chem.*, **74**, 2489

Fujii, H. and Nakae, H. (1996) *Acta Mater.*, **44**, 3567

Gallois, B. (1980) Ph.D. Thesis, Carnegie Mellon Univ., Pittsburgh, USA

Gallois, B. and Lupis, C. H. P. (1981) *Metall. Trans.*, **12B**, 549

Ghetta, V., Fouletier, J. and Chatain, D. (1996) *Acta Mater.*, **44**, 1927

Gillet, E. and Ealet, B. (1992) *Surface Science*, **273**, 427

Gonzalez, E. J. and Trumble, K. P. (1996) *J. Am. Ceram. Soc.*, **79**, 114

Grigorenko, N. F., Stegny, A. I., Kasich-Pilipenko, I. E., Naidich, Y. V. and Pasichny, V. V. (1995) in *Proc. Int. Conf. High Temperature Capillarity*, Smolenice Castle, May 1994, ed. N. Eustathopoulos (Reproprint, Bratislava) p. 123

Grigorenko, N., Zhuravlev, V., Poluyanskaya, V., Naidich, Y. V., Eustathopoulos, N., Silvain, J. F. and Bihr, J. C. (1998) in *Proc. 2nd Int. Conf. on High Temperature Capillarity*, Cracow (Poland) 29 June-2 July 1997, ed. N. Eustathopoulos and N. Sobczak, published by Foundry Research Institute (Cracow), p. 133

Harding, F. L. and Rossington, D. R. (1970) *J. Amer. Ceram. Soc.*, **53**, 87

Harter, I. (1990) Ph.D. Thesis, INP Grenoble, France

Harter, I., Dusserre, P., Duffar, T., Nabot, J. P. and Eustathopoulos, N. (1993) *J. Crystal Growth*, **131**, 157

Heinz, M., Koch, K. and Janke, D. (1989) *Steel Res.*, **60**, 246

Hodkin, E. N. and Nicholas, M. G. (1977) *J. Nucl. Mater.*, **67**, 171

Hondros, E. D. (1978) in *Precipitation Processes in Solids*, The Metallurgical Society of AIME, Warrendale, PA, p. 1

Hong, T., Smith, J. R. and Srolovitz, D. J. (1995) *Acta Metall. Mater.*, **43**, 2721

Hultgren, R., Desai, P. D., Hawkins, D. T., Gleiser, M. and Kelley, K. K. (1973) *Selected Values of the Thermodynamic Properties of Binary Alloys*, American Society for Metals, Metals Park, Ohio 44073

Israelachvili, J. N. (1991) *Intermolecular and Surface Forces*, Academic Press, Harcourt Brace Jovanovich Publishers, London, 2nd Edition

Jacob, K. T. and Alcock, C. B. (1972) *Acta Met.*, **20**, 221

Jacquemin, J. P. (1996) DEA report, INP Grenoble, France

Jollet, F., Noguera, C., Gautier, M., Thormat, N. and Duraud, J. P. (1991) *J. Amer. Ceram. Soc.*, **74**, 358

Joud, J. C., Eustathopoulos, N., Bricard, A. and Desré, P. (1973) *J. Chimie Physique*, **9**, 1290

Kritsalis, P., Li, J. G., Coudurier, L. and Eustathopoulos, N. (1990) *J. Mater. Sci. Lett.*, **9**, 1332

Kritsalis, P., Coudurier, L. and Eustathopoulos, N. (1991) *J. Mater. Sci.*, **26**, 3400

Kritsalis, P., Merlin, V., Coudurier, L. and Eustathopoulos, N. (1992) *Acta Metall. Mater.*, **40**, 1167

Labrousse, L., Gasse, A. and Eustathopoulos, N., to be published

Landry, K., Kalogeropoulou, S., Eustathopoulos, N., Naidich, Y. and Krasovsky, V. (1996) *Scripta Mater.*, **34**, 841

Landry, K., Kalogeropoulou, S. and Eustathopoulos, N. (1998) *Mat. Sci. Eng. A*, **254**, 99

Laurent, V., Chatain, D., Chatillon, C. and Eustathopoulos, N. (1988) *Acta Metall.*, **36**, 1797

Laurent, V., Chatain, D. and Eustathopoulos, N. (1991) *Mater. Sci. Engineering*, **A135**, 89

Li, J. G., Coudurier, L. and Eustathopoulos, N. (1987-88) *Rev. Int. Hautes Temper. Refract. Fr.*, **24**, 85

Li, J. G., Coudurier, L., Ansara, I. and Eustathopoulos, N. (1988) *Ann. Chim. (Fr)*, **13**, 145

Li, J. G., Coudurier, L. and Eustathopoulos, N. (1989) *J. Mater. Sci.*, **24**, 1109

Lipkin, D. M., Israelachvili, J. N. and Clarke, D. R. (1997) *Phil. Mag.*, **76**, 715

Magaud, L. and Pasturel, A. (1998) in *Proc. 2nd Int. Conf. on High Temperature Capillarity*, Cracow (Poland) 29 June-2 July 1997, ed. N. Eustathopoulos and N. Sobczak, published by Foundry Research Institute (Cracow), p. 3

McDonald, J. E. and Eberhart, J. G. (1965) *Trans. Metall. Soc. AIME*, **233**, 512

Mehrotra, S. P. and Chaklader, A. C. D. (1985) *Metall. Trans.*, **16B**, 567

Merlin, V., Kritsalis, P., Coudurier, L. and Eustathopoulos, N. (1992) *Mat. Res. Soc. Symp. Proc.*, **238**, 511

Merlin, V. and Eustathopoulos, N. (1995) *J. Mater. Sci.*, **30**, 3619

Mori, N., Sorano, H., Kitahara, A., Ogi, K. and Matsuda, K. (1983) *J. Jpn. Inst. Metals*, **47**, 1132

Naidich, Y. V. (1972) *Kontaktnie Javlenia v Metallicheskih Rasplavakh*, Naukova Dumka, Kiev

Naidich, Y. V., Zhuravlev, V. S. and Chuprina, V. G. (1974) *Sov. Powd. Metall. Met. Ceramics*, **13**, 236

Naidich, Y. V. (1981) in *Progress in Surface and Membrane Science*, vol. 14, ed. by D. A. Cadenhead and J. F. Danielli, Academic Press, New York, p. 353

Naidich, Y. V. and Chuvashov, Y. N. (1983) *J. Mater. Sci.*, **18**, 2071

Naidich, Y., Chubashov, Y., Ishchuk, N. and Krasovskii, V. (1983b) *Poroshkovaya Metallurgiya*, **246**, 67 (English translation : Plenum Publishing Corporation p. 481)

Naidich, Y. V., Zhuravljov, V. S. and Frumina, N. I. (1990) *J. Mater. Sci.*, **25**, 1895

Naidich, Y. V. and Krasovsky, V. P. (1998) in *Proc. 2nd Int. Conf. on High Temperature Capillarity*, Cracow (Poland) 29 June-2 July 1997, ed. N. Eustathopoulos and N. Sobczak, published by Foundry Research Institute (Cracow), p. 87

Naidich, Y. V. and Taranets, N. Y. (1998) *J. Mater. Sci.*, **33**, 3993

Nakashima, K., Takihira, K., Mori, K. and Shinozaki, N. (1992) *Materials Transactions, JIM*, **33**, 918

Nogi, K., Tsujimoto, M., Ogino, K. and Iwamoto, N. (1992) *Acta Metall. Mater.*, **40**, 1045

Nogi, K., Iwamoto, N. and Ogino, K. (1992b) *Transactions of JWRI*, **21**, 141

Ogino, K., Nogi, K. and Yamase, O. (1983) *Transactions ISIJ*, **23**, 234

Ohuchi, F. S. and Kohyama, M. (1991) *J. Am. Ceram. Soc.*, **74**, 1163

Ownby, P. D. and Liu, J. (1988) *J. Adhesion Sci. Technol.*, **2**, 255

Pham Van, L. (1992) Ph.D. Thesis, Université Paris VI, France

Pilliar, R. M. and Nutting, J. (1967) *Phil. Mag.*, **16**, 181

Ricci, E., Castello, P. and Passerone, A. (1994) *Mater. Sci. Eng. A*, **178**, 99

Rivollet, I., Chatain, D. and Eustathopoulos, N. (1987) *Acta Metall.*, **35**, 835

Sangiorgi, R., Passerone, A. and Muolo, M. L. (1982) *Acta Met.*, **30**, 1597 and Sangiorgi, R., Passerone, A. and Minisini, R. (1981) in *Surfaces and Interfaces in Ceramic and Ceramic-Metal Systems*, Materials Science Research, vol. 14, ed. J. Pask and A. Evans, Plenum Press, New-York, p. 445

Sangiorgi, R., Muolo, M. L., Chatain, D. and Eustathopoulos, N. (1988) *J. Am. Ceram. Soc.*, **71**, 742

Sangiorgi, R., Muolo, M. L. and Eustathopoulos, N. (1995) in *Proc. Int. Conf. High Temperature Capillarity*, Smolenice Castle, May 1994, ed. N. Eustathopoulos (Reproprint, Bratislava) p. 148

Schönberger, U., Andersen, O. K. and Methfessel, M. (1992) *Acta Metall. Mater.*, **40**, S1

Shashkov, D. A. and Seidman, D. N. (1996) *Materials Science Forum*, Transtec Publications, Switzerland, **207-209**, 429

Soper, A., Gilles, B. and Eustathopoulos, N. (1996) *Materials Science Forum*, Transtec Publications, Switzerland, **207-209**, 433

Stoneham, A. M., Duffy, D. M. and Harding, J. H. (1995) in *Proc. Int. Conf. High Temperature Capillarity*, Smolenice Castle, May 1994, ed. N. Eustathopoulos (Reprint, Bratislava) p. 1

Tomsia, A. P., Saiz, E., Dalgleish, B. J. and Cannon, R. M. (1995) in *Proc. 4th Japan International SAMPE Symposium*, Sept. 25-28, p. 347

Tomsia, A. P., Saiz, E., Foppiano, S. and Cannon, R. M. (1998) in *Proc. 2nd Int. Conf. on High Temperature Capillarity*, Cracow (Poland) 29 June-2 July 1997, ed. N. Eustathopoulos and N. Sobczak, published by Foundry Research Institute (Cracow), p. 59

Tournier, C., Lorrain, B., Le Guyadec, F. and Eustathopoulos, N. (1996) *J. Mater. Sci. Lett.*, **15**, 485

Vikner, P. (1993) DEA report, LTPCM, INP Grenoble, France

Wan, C., Kritsalis, P., Drevet, B. and Eustathopoulos, N. (1996) *Mater. Sci. Eng. A*, **207**, 181

Weirauch, D. A. (1988) in *Ceramic Microstructures '86 : The Role of Interfaces*, ed. J. A. Pask and A. G. Evans, Plenum Press, New York, p. 329

Chapter 7

Wetting properties of metal/non-oxide ceramic systems

Oxides are not the only technically important ceramic materials. Carbides (SiC, B_4C, TiC) and nitrides (BN, AlN, TiN, Si_3N_4) are of prime importance as bonded components in advanced developments for the machine tool, microelectronics, aircraft, automobile and heat exchanger industries.

Although some of the concepts established for metal/oxide systems are also valid for non-oxide ceramics, there are other concepts which are specific to these kinds of ceramics, owing to their predominantly covalent (SiC, BN, AlN) or metallic (TiC, TiN, WC) character. These materials seldom can be obtained as the high-purity monocrystalline specimens desirable for fundamental wetting studies. Usually, they are sintered materials with impurity contents higher than 0.1% and they often contain open porosity. Further difficulties arise from the high oxidization tendency of many of them, the presence of an oxide layer dramatically changing their wettability by liquid metals.

The first Section of this Chapter deals with predominantly covalent ceramics (Table 7.1), with a particular emphasis on SiC for which a recent detailed study of wetting by liquid metals combined with a study of surface chemistry has been reported. In the second Section, wetting by liquid metals of metal-like ceramics is presented.

7.1. METALS ON PREDOMINANTLY COVALENT CERAMICS

7.1.1 Wetting of SiC

SiC is a covalent ceramic of technological interest because of its semi-conductor character, good mechanical properties, high thermal conductivity and good thermal shock behaviour. SiC is easily oxidizable but the oxide layer is protective so the ceramic can be used as a structural material in high temperature applications or as reinforcement in metal matrix composites. Also, it is an interesting material for basic studies of wetting as it can be obtained in the form of high purity monocrystals.

There are two main allotropic forms of SiC: hexagonal (α-SiC) and cubic (β-SiC). The lattice planes of SiC consist of tetrahedrons of SiC_4 (four C atoms around a Si atom) or CSi_4 (Figure 7.1). If we consider a SiC_4 tetrahedron, the distance

Table 7.1. Degree of covalence of different ceramics according to the scale of electronegativity of Pauling (1960).

Compound	B_4C	SiC	Al_4C_3	BN	Si_3N_4	AlN	SiO_2	Al_2O_3	MgO
Degree of covalence (%)	94	89	78	78	70	57	49	37	27

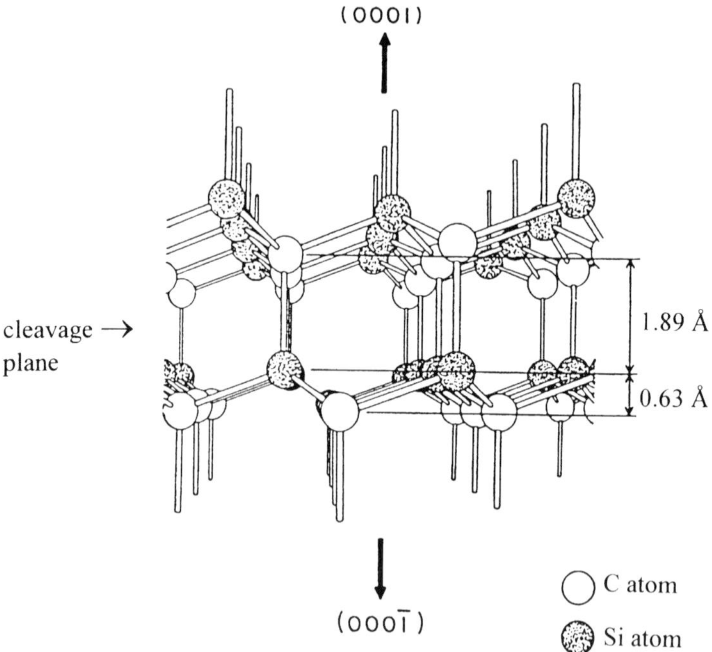

Figure 7.1. Si face (0001) and C face (0001̄) of the hexagonal structure of α-SiC.

between Si and the plane containing the three C atoms at the base of the tetrahedron (0.63 Å for α-SiC) is much smaller than the distance between this Si atom and the fourth C atom located at the top of the tetrahedron (1.89 Å for α-SiC). As a consequence, the cleavage of SiC occurs only beween the more widely separated planes of Si and C. Thus, one gets two parallel faces with different surface chemistry, the Si face terminated by Si atoms ((0001) for α-SiC and (111) for β-SiC) and the C face terminated by C atoms ((0001̄) for α-SiC and ($\bar{1}\,\bar{1}\,\bar{1}$) for β-

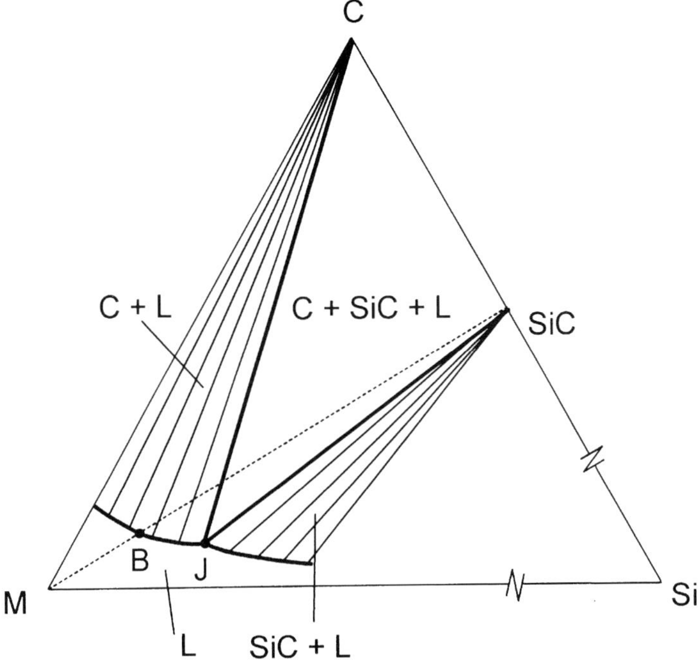

Figure 7.2. Schematic representation of isothermal section of M-Si-C phase diagram when M does not form a stable carbide (otherwise the liquid of composition J is not in equilibrium with SiC and C but with SiC and M carbide).

SiC). In α-SiC, whatever their chemical nature, basal planes are faces of lowest surface energy which, even at high temperatures, present a negligible atomic roughness that is reflected at the macroscopic scale by the formation of very stable facets.

When a drop of metal M *which does not form a stable carbide* is placed on a SiC substrate, SiC dissolves in the liquid metal:

$$\langle SiC \rangle \Leftrightarrow (Si)_M + (C)_M \tag{7.1}$$

Thus, M is continuously enriched with Si and C and the composition of the metallic phase varies following the straight line M-SiC of Figure 7.2 up to point B where precipitation of graphite starts:

$$(C)_M \Leftrightarrow \langle C \rangle \tag{7.2}$$

Dissolution of SiC continues, leading to an increase of the Si content in the liquid M while the excess C precipitates. The overall reaction is:

$$\langle SiC \rangle \Leftrightarrow (Si)_M + \langle C \rangle \tag{7.3}$$

This reaction stops when the liquid composition reaches point J, where SiC becomes stable in contact with the liquid and precipitated C. At this point, the equilibrium molar fraction of Si dissolved in M, X_{Si}^J (Figure 7.2), is related to the partial enthalpy of mixing of Si in M, $\overline{\Delta H}_{Si(M)}$ (neglecting the partial excess entropy of mixing), and to the molar Gibbs energy of formation of SiC, $\Delta G_{f(SiC)}^0$, by the equation:

$$X_{Si}^J = \exp\left(\frac{\Delta G_{f(SiC)}^0 - \overline{\Delta H}_{Si(M)}}{RT}\right) \tag{7.4}$$

Semi-quantitative calculations of X_{Si}^J can be made for all metals by equating $\overline{\Delta H}_{Si(M)}$ to the value at infinite dilution, $\overline{\Delta H}_{Si(M)}^\infty$ (data for this last quantity are given in Appendix G). Figure 7.3 suggests that, except for some low melting point elements like Pb and Sn (and also Bi, Sb), metals have a significant reactivity with SiC even at a temperature as low as 1000K. For high-melting-point transition metals like Ni and Pd, the reactivity is very high and results in substantial dissolution of SiC that occurs rapidly and is accompanied by the formation of large graphite precipitates (Figure 7.4).

In wetting experiments, such graphite precipitation makes the spreading process an extremely complex phenomenon owing to the fact that graphite is less wetted by M-Si alloys than SiC: graphite precipitates act as barriers to wetting (see Section 2.2.2.1, Figures 2.28 and 2.29). Because of these complications, results of wetting for non-reactive, Si saturated liquids i.e., M-Si alloys with $X_{Si} > X_{Si}^J$, will be presented first. However, even for such M-Si/SiC systems, spreading on SiC surfaces is a complex process because the chemistry of SiC surfaces is very different from that of the bulk.

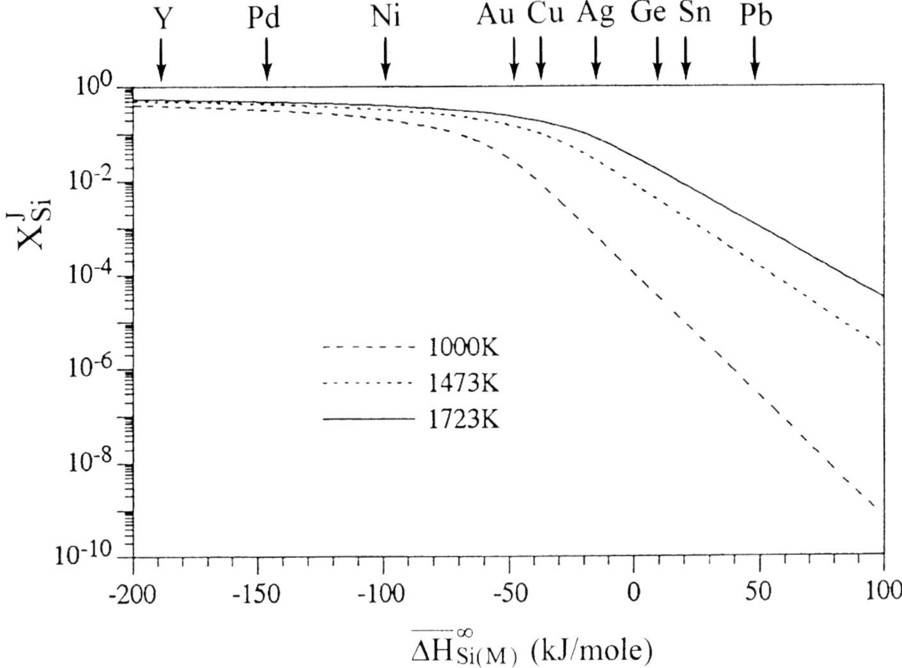

Figure 7.3. Calculated values of molar fraction of Si in liquid metal M at the three phase $M/SiC/C_g$ equilibrium (equation 7.4) as a function of the partial enthalpy of mixing of Si at infinite dilution in M. Arrows indicate values of $\overline{\Delta H}^{\infty}_{Si(M)}$ for some metals M given in Appendix G.

7.1.1.1 Non-reactive systems: silicon alloys on SiC

Equilibrium contact angles. At the Si melting point, the Si/SiC couple is a nearly non-reactive system, with only a slight dissolution of SiC into Si. Pure Si wets SiC well ($\theta = 30$–$50°$) regardless of whether the furnace atmosphere is a high vacuum or neutral gas and whether the ceramic is monocrystalline, polycrystalline, CVD or sintered α-SiC or β-SiC (Naidich and Nevodnik 1969, Yupko and Gnesin 1973, Nikolopoulos et al. 1992, Drevet et al. 1993, Rado 1997). For monocrystalline SiC, the degree of wetting depends on the polarity of the (111) face for β-SiC and of the (0001) face for α-SiC. However, the contact angle on Si-terminated faces is lower than on C-terminated faces by only 5–10° for both α-SiC (Rado 1997) and β-SiC (Bakovets 1977). The good wetting observed for Si/SiC corresponds to a value of the work of adhesion of about 90% of the work of cohesion W_c of liquid Si ($W_c = 2\sigma_{LV}$). Such a value is too high to be attributed to van der Waals interactions so chemical interactions must be responsible. In solid Si, bonds are covalent (with a

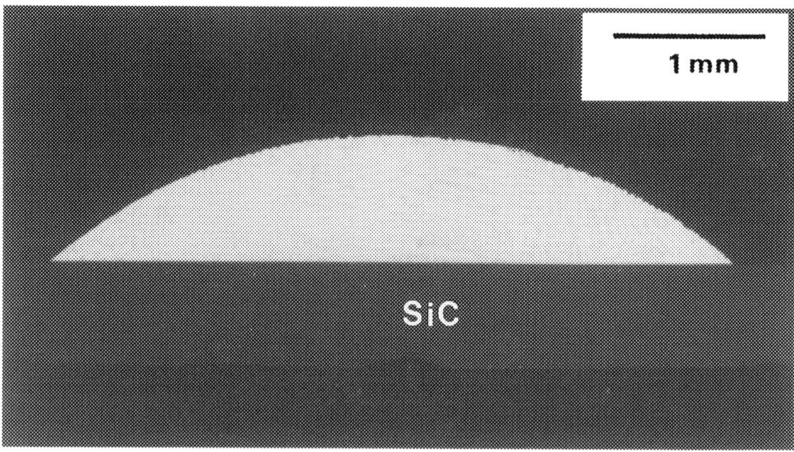

Figure 7.8. Solidified drop of eutectic Ag-Si alloy on α-SiC after heating at 1200°C in a purified He atmosphere. In this non-reactive system, wetting is good and the interface is strong after cooling (Rado 1997).

contact angles were observed that could be explained by strong chemical interactions at the interface between atoms of the transition metal M and those of SiC. High values of W_a^M and low contact angles were calculated by extrapolating data for binary M-Si alloys to pure M. These findings have been confirmed also by the experiment described in Section 2.2.2.1 showing that the contact angle of pure Ag on unreacted SiC is much lower than 90°.

To determine if these interactions are mainly between the M and Si or C atoms of SiC, Rado attempted to correlate the work of immersion of pure M, W_i^M, with different thermodynamic quantities characterising the energy of M-Si and M-C interactions. He found that W_i^M scaled rather well with the parameter $\lambda_{\text{M-Si}} / \Omega_m$, Ω_m being the molar surface area (Figure 7.10). The regular solution parameter $\lambda_{\text{M-Si}}$ is proportional to $\varepsilon_{M-Si} - (\varepsilon_{M-M} + \varepsilon_{Si-Si})/2$ (equation (4.3)) while $W_i^M.\Omega_m$ varies as $\varepsilon_{M-S} - (\varepsilon_{M-M}/2)$ where ε_{M-S} is an average interaction energy between M and Si or C atoms at the interface (see the similar equation (6.10)). Therefore, in a series of different metals M on SiC, the correlation observed between W_i^M and $\lambda_{\text{M-Si}}/\Omega_m$ shows that the bond energy ε_{M-S} at the interface scales with ε_{M-Si}, suggesting that the interactions predominant at the M/SiC interface are M-Si rather than M-C.

The extrapolation of the straight line of Figure 7.10 at $W_i^M = 0$ defines a threshold value $(\lambda_{\text{M-Si}}/\Omega_m)^* = 1.72$ J/m^2 corresponding to a transition from non-

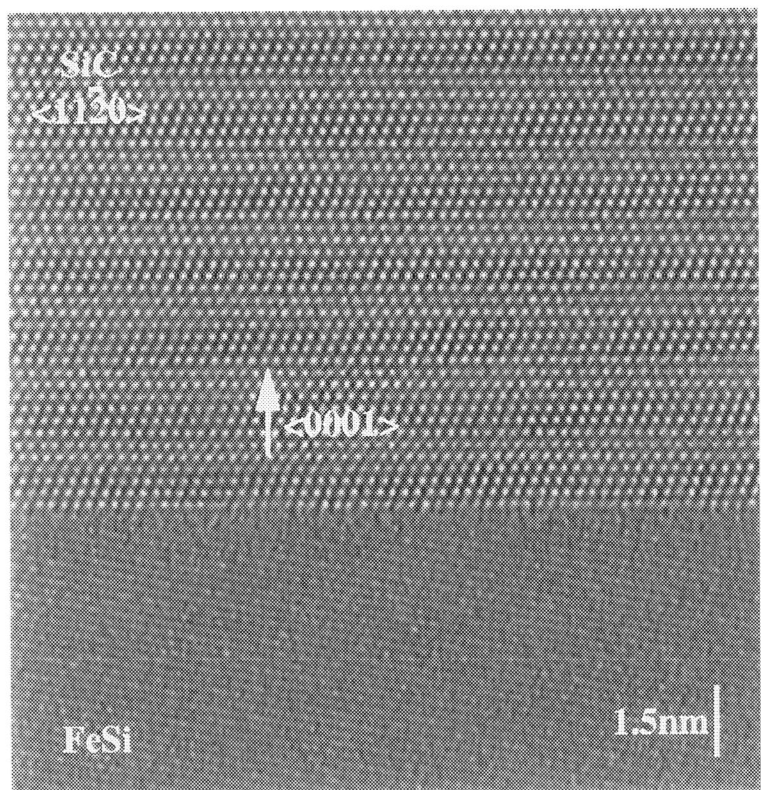

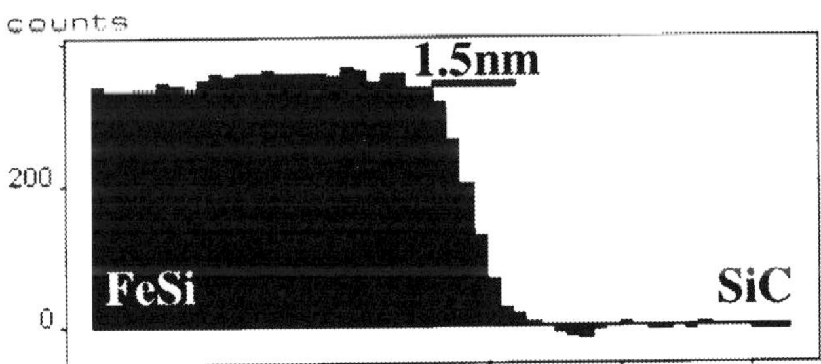

Figure 7.9. Interface between a solidified non-reactive Fe-Si alloy and monocrystalline α-SiC (for this alloy $\theta \cong 40°$ (Kalogeropoulou et al. 1995)). Top: high resolution transmission electron Micrograph showing that the interface is sharp at the atomic scale. Bottom : Fe concentration profile across the interface. The thickness of the "chemical" interface is the 1.5 nm resolution of the technique (electron energy-loss spectroscopy). Reprinted from (Lamy, private communication) with kind permission.

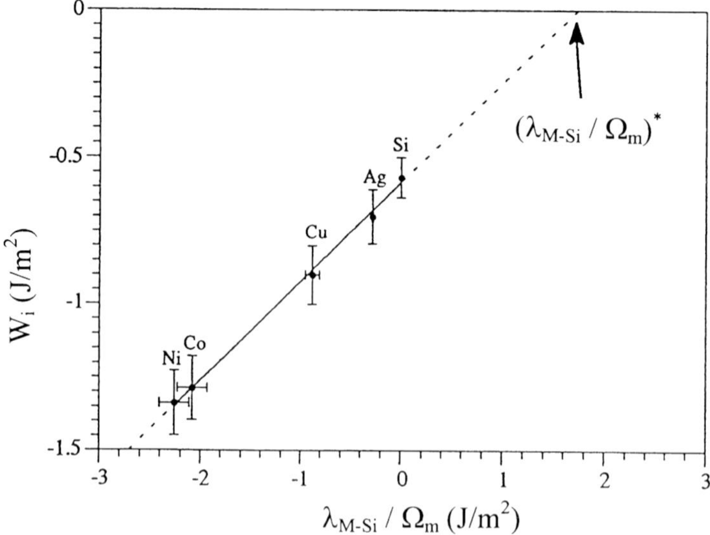

Figure 7.10. Work of immersion for pure metals M on SiC, evaluated from data for binary M-Si alloys, plotted as a function of ($\lambda_{\text{M-Si}}/\Omega_m$). The value $W_i = 0$ corresponds to $\theta = 90°$. From (Rado 1997).

wetting ($W_i > 0$) to wetting ($W_i < 0$). Almost all pure metals, even those with high positive values of $\lambda_{\text{M-Si}}$, such as Pb, Sn and In, have $\lambda_{\text{M-Si}}/\Omega_m$ values less than 1.72 J/m^2 and *should* wet SiC surfaces. This prediction is not confirmed by experimental results obtained in the systematic studies performed by Naidich and Nevodnik (1969) and by Nogi and Ogino (1988), which show large non-wetting contact angles for all metals that have a negligible or a weak reactivity with SiC (Table 7.3). The reasons for this disagreement are discussed in the next Section.

Spreading kinetics. The wetting behaviour cited in Table 7.3 is that for metals experiencing weak van der Waals interactions with SiC, while the analysis of Rado predicts chemical interactions at their interfaces. To understand this discrepancy, Rado used Auger Electron Spectroscopy (AES) and X-ray Photoelectron Spectroscopy (XPS) in a study of the surface chemistry of the basal planes of α-SiC as a function of temperature in a high vacuum environment typical of sessile drop experiments (Rado 1997, Rado et al. 1998). Figure 7.11 shows the evolution of the Auger peaks of Si, C and O for SiC substrates heated at various temperatures for 5 minutes. The AES spectra of substrates heated at 900°C and 1000°C are nearly identical to that of the reference substrate (i.e., non-heated) and indicate only some pollution by oxygen. XPS analysis of this substrate showed that the thickness of the

Table 7.3. Wettability of SiC by different liquid metals M with a negligible or weak reactivity with SiC.

Metal M	α-SiC [*]	T (°C)	Atmosphere	θ (deg)	$(\lambda_{\text{M-Si}} / \Omega_{\text{m}})$ (J/m^2)
Ga	M	800	high vacuum	118 [1]	0.03
In	M	800	high vacuum	130 [1]	0.46
Sn	M	1024	Ar	137 [2]	0.46
	M	1050	high vacuum	135 [1]	
	S	1100	Ar	150 [2]	
Ge	M	1050	high vacuum	113 [1]	0.19
	M	1230	Ar	135 [2]	
Ag	S	1010	Ar	140 [2]	−0.28
	M	1100	high vacuum	128 [1]	
	M	1100	He	120 [3]	
Au	M	1100	Ar	118 [2]	−1.12
	M	1150	high vacuum	138 [1]	
	M	1100	high vacuum	138 [4]	

[*] S = sintered, M = monocrystal
[1] (Naidich and Nevodnik 1969), [2] (Nogi and Ogino 1988), [3] (Rado 1997), [4] (Drevet et al. 1993)

contamination layer was less than 1 nm and resembled a very thin SiO$_x$ (with x $\leq$ 2) or SiO$_x$C$_y$ layer more than one of pure SiO$_2$. However, heating to 1150°C produces two significant changes. First, the Auger O peak intensity decreases and second, the C peak increases while that of Si decreases. At the same time, a small shoulder appears at 250 eV, characteristic of C graphite. Clearly, some carbidic C at the SiC surface was transformed into graphite. These two phenomena i.e., enrichment of the surface in C and transformation of surface carbidic C to graphite, are enhanced when the heating temperature is raised to 1250°C. At this temperature and using ion beam erosion analysis, the thickness of the C rich layer was found to be about 3 nm.

From Rado's study, it can be concluded that the nature of SiC surface changes dramatically at a temperature T* which depends on the oxygen partial pressure in the furnace (in Rado's experiments T* $\approx$ 1050°C). At lower temperatures, the SiC surface is *oxidised* while at higher temperatures it is *graphitized*. The transition is continuous around T* i.e., without an intermediate state of a non-contaminated,

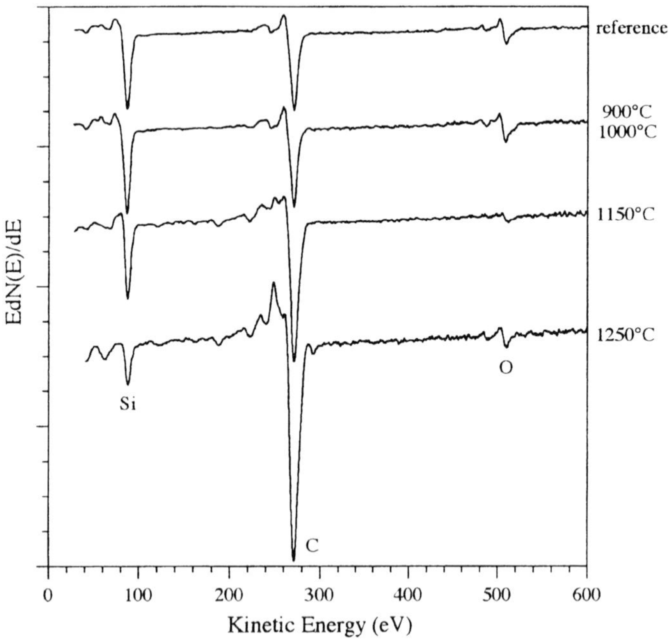

Figure 7.11. Auger spectra of the α-SiC Si face after 5 min holding in a high vacuum metallic furnace at different temperatures (except for the reference specimen) (Rado et al. 1998) [9].

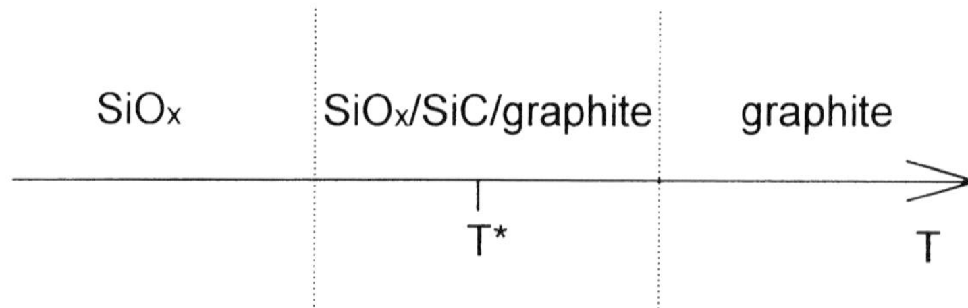

Figure 7.12. Surface chemistry of SiC in high vacuum as a function of temperature. Wettability of SiC by non-reactive metals is mainly dictated by the oxidation layer on SiC at $T \ll T^*$ and by the graphite-rich layer formed by Si evaporation at $T \gg T^*$.

stoichiometric, SiC surface (Figure 7.12). Deoxidation of SiC can occur by reaction between SiO_2 and SiC leading to formation of volatile SiO and CO (Rado et al. 1995):

$$2\langle SiO_2 \rangle + \langle SiC \rangle \Leftrightarrow 3[SiO] + [CO] \qquad (7.6)$$

Graphitization of SiC in a high vacuum is known to occur above 1100°C by preferential evaporation of Si (Muehlhoff et al. 1986):

$$\langle SiC \rangle \Leftrightarrow [Si] + \langle C_{graphite} \rangle \qquad (7.7)$$

At temperatures much lower than T*, deoxidation by the reaction (7.6) is too slow to be effective during wetting experiments typically lasting 10^3 seconds, so contact angles for low melting-point metals such as In and Ga in a high vacuum (and *a fortiori* in a commercially pure neutral gas) are expected to be close to those on SiO_2 i.e., in the range 120°–130°. However, in high vacuum, some metals such as Al and Si can reduce SiO_2 layers to form the volatile Al_2O and SiO. For alloys containing such elements, the contact angle is expected to change from an initial value typical of that on SiO_2 to a final value characteristic of SiC. An example is the non-reactive Al-19.5 at.% Si/SiC system under high vacuum at 800°C, for which spreading times as long as 150 minutes were observed (Figure 7.13). In this case, the spreading kinetics are controlled by the kinetics of SiC deoxidation occurring at the triple line.

At temperatures higher than T*, the SiC surface is covered by a thin graphite layer that should strongly affect wetting and reaction kinetics, graphite acting as a barrier in both cases. An example is Au on monocrystalline SiC. Several authors have found that at temperatures close to its melting point, the contact angle for pure Au is in the range 120–140° (Table 7.3). However, it is clear that these contact angles are for Au on graphite and not on SiC. Moreover, no significant reactivity was observed after 150 minutes of contact between Au and monocrystalline SiC at 1200°C (Rado et al. 1994) despite the fact that Au dissolves SiC up to 15 at.% of Si according to the experimental Au-Si-C phase diagram. Therefore, for metals with very weak affinity for C, the contact angles on SiC at T > 1100°C are expected to be close to those on carbon i.e., close to or higher than 120° (Figure 7.14.a) (see Tables 8.3 and 8.4).

For Au-Si alloys with molar fractions of Si higher than $X_{Si}^J \approx 0.15$ (equation (7.4)), the graphite layer is not stable in contact with Au-Si (Figure 7.14.b). The spreading kinetics are thus controlled by the degraphitization process at the triple line, either simple dissolution of C (the solubility of C in Au is enhanced by Si) or dissolution of C followed by precipitation of SiC. In both cases the final contact angle is that on SiC and, as for other binary alloys of Si (Table 7.2), this angle is very low (Figure 7.15). Note that the "cleaning" of SiC surface can also be achieved by alloying Au (or another metal which does not react with C) with Ni

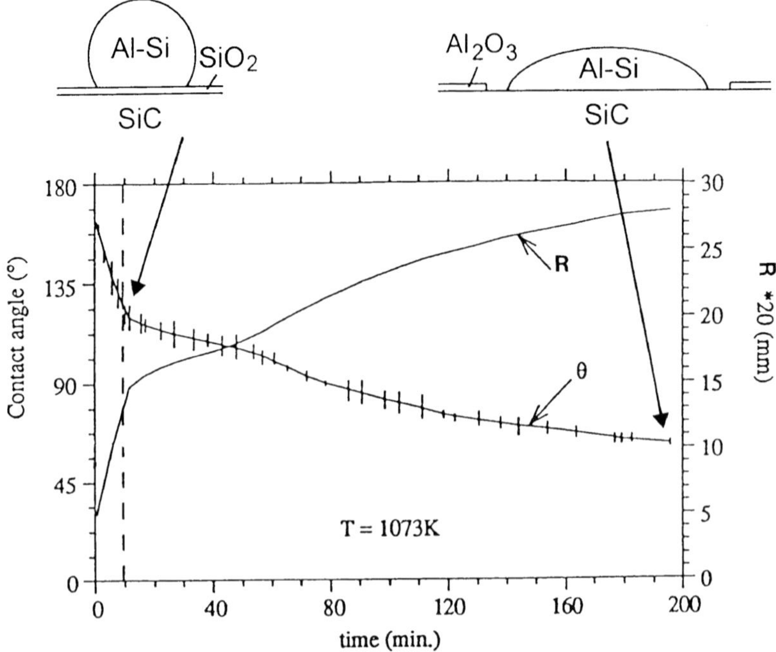

Figure 7.13. Contact angle and radius of the sessile drop base versus time for Al-19.5 at.% Si on monocrystalline α-SiC at $800°C$ ($P = 1$ to 3×10^{-7} mbar). After a comparatively rapid spreading due to deoxidation of the drop (between $t = 0$ and $t = 10$ min), spreading is controlled by deoxidation of SiC surface close to the triple line. In front of the triple line, the SiO_2 layer was transformed into Al_2O_3 by reaction with Al coming from the liquid drop. From (Laurent et al. 1996) [22].

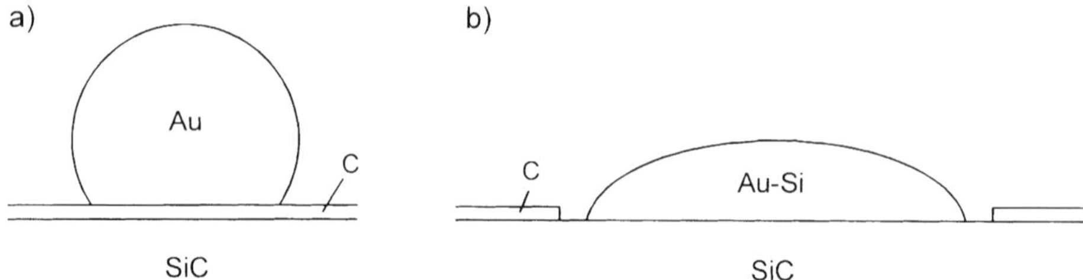

Figure 7.14. a) In contact with metals with a negligible affinity for C (such as Au, Ag, Sn), graphitization of SiC surface at high temperature inhibits both wetting and chemical reaction with the metal. b) Additions of Si to Au causes C dissolution followed eventually by SiC precipitation. This configuration corresponds to a true contact angle on SiC.

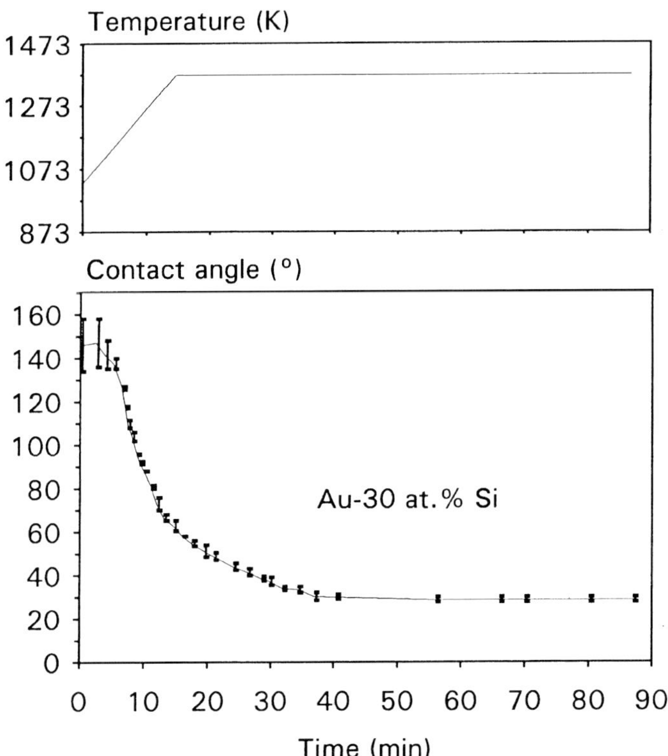

Figure 7.15. Contact angle changes with temperature and time during temperature rise for Au-30 at.% Si on monocrystalline α-SiC. The experiment was performed in an alumina chamber furnace under high vacuum (Drevet et al. 1993) [24].

and other metals that dissolve large quantities of C. However, Ni aggressively attacks the SiC itself after dissolving the graphite layer (Drevet 1987).

The results shown in Figure 7.16 for Sn on α-SiC under Ar were obtained for a wide range of temperatures. Large non-wetting contact angles were observed at both low and high temperatures, whereas a contact angle minimum was found at intermediate temperatures close to 1500K. These results can be explained easily by the oxidation of SiC at low temperatures and graphitization of SiC at high temperatures. Similarly Table 7.4 presents contact angles measured by Allen and Kingery (1959) for Sn on Al_2O_3 and SiC in two different atmospheres. In the case of Al_2O_3, very large contact angles were observed at all temperatures. These values are much higher than those measured on smooth, monocrystalline Al_2O_3 surfaces (120–130°, see Table 6.1) and could result from the roughness of the sintered

quantified by the enthalpy of formation of Al_4C_3, $\Delta H_f = -72$ kJ per mole of C) is much higher than that for Si (the enthalpy of mixing of Si in Al being $\overline{\Delta H}^{\infty}_{Si(Al)} = -9$ kJ/mole).

At temperatures higher than 1000°C, the formation of Al_4C_3 at the interface is so greatly accelerated that the stationary contact angles observed during the first few minutes or tens of minutes (depending on the temperature, the furnace atmosphere and whether monocrystalline or sintered SiC is the substrate), lying between 35° and 55° (Naidich 1981, Nogi and Ogino 1988, Shimbo et al. 1989, Ferro and Derby 1995), correspond to those for Al on Al_4C_3. These values are close also to the contact angles observed at low temperatures for Al on an nearly unreacted SiC, indicating that the types of interactions developed at Al/Al_4C_3 and Al/SiC interfaces are similar.

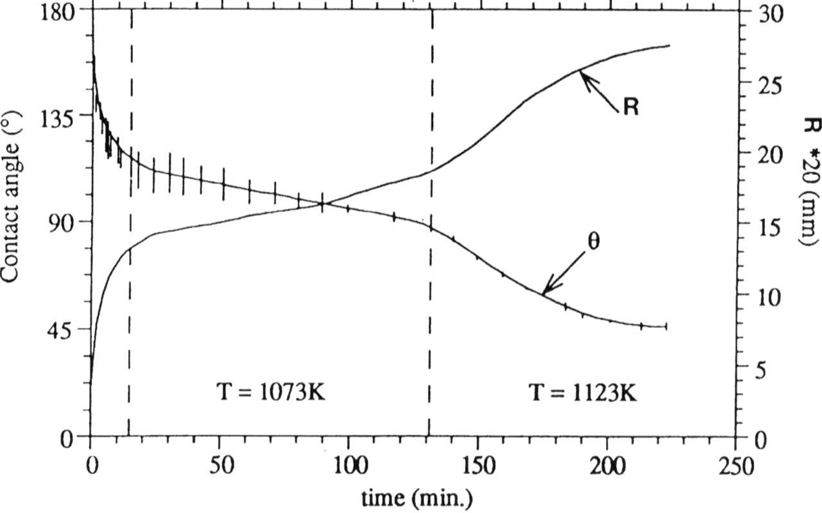

Figure 7.17. Contact angle and radius of a sessile drop base versus time for Al on monocrystalline α-SiC. The increase in temperature by 50K leads to a significant acceleration of spreading rate (Laurent et al. 1996) [22].

Alloying elements. As mentioned above, Cu reacts strongly with SiC and the graphite formed during the reaction inhibits wetting. With polycrystalline SiC substrates, this phenomenon results in large non-wetting apparent contact angles, which do not change with time. By alloying Cu with Cr, the C liberated by SiC decomposition is combined to form a continuous layer of Cr_3C_2 a few microns thick according to the reaction:

$$2\langle SiC \rangle + 3(Cr)_{Cu} \Leftrightarrow \langle Cr_3C_2 \rangle + 2(Si)_{Cu} \qquad (7.9)$$

The Cr_3C_2 is wettable (see Table 7.10) and as a result, low wetting contact angles can be achieved by Cu-Cr alloys (Figure 7.18). However, this reaction layer is not protective and Cu penetrates the grain boundaries of the SiC substrate far from the interface to react and form graphite (Landry and Eustathopoulos unpublished, Xiao and Derby 1998).

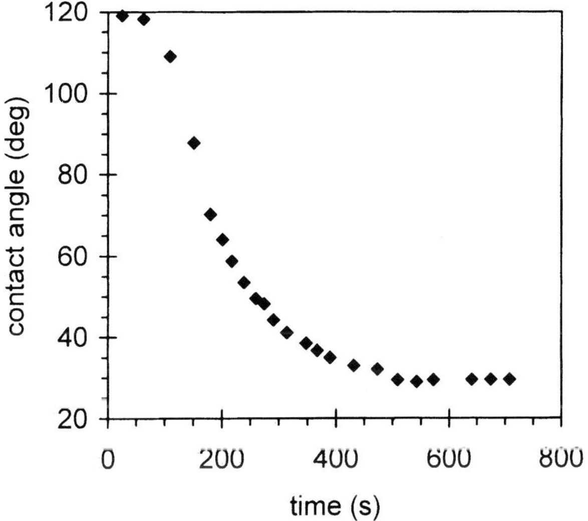

Figure 7.18. Contact angle versus time for Cu-1 wt.% Cr on pressureless-sintered α-SiC at 1100°C in a high vacuum (Landry and Eustathopoulos unpublished).

Additions of 3.1 at.% Ti to an Ag-Cu alloy with a composition close to the eutectic allow nearly perfect wetting of SiC to be achieved at 900°C in a few minutes. In this case, the interfacial reaction leads to the formation of two layers: a layer of TiC about 1 μm thick on the SiC side and a layer a few microns thick on the metal side with a composition close to Ti_2Si (Ljungberg 1992). The reaction mechanisms at a (Ag-Cu-Ti/monocrystalline α-SiC) interface were studied by *in situ* high resolution transmission electron microscopy by Iwamoto and Tanaka (1998) who found the reaction started by dissolution of SiC basal planes and continued by epitaxial nucleation and growth of TiC particles on the SiC. No silicide was observed, suggesting that Ti_2Si found by Ljungberg (1992) probably formed during cooling.

Additions of Ti to Al suppress the formation of Al_4C_3 at the Al/SiC interface and lead to contact angles close to $20°$ at $1150°C$. *Post mortem* analysis reveals the formation of a reaction product layer consisting mainly of TiC and $TiSi_2$ phases (Sobczak et al. 1998).

7.1.2 Wetting of nitrides

The nitrides of Al, Si and B are nearly stoichiometric compounds with a predominantly covalent character. Their ionicity, Table 7.5, and other characteristics are intermediate between those of oxides and carbides. However, individual members possess specific characteristics that have led to the emergence of several nitrides as important engineering materials during the last two decades. Thus AlN can be used as a heat-dissipating electronic substrate and cubic BN finds applications as an ultra-hard cutting material. Si_3N_4 and sialon (a complex Si_3N_4-Al_2O_3 ceramic) have found applications in the aerospace and automobile industries where lightness and high temperature strength are required.

At high temperature, nitrides can dissociate to form N_2. In the case of AlN, the dissociation reaction at a temperature higher than the melting point of Al is:

$$\langle AlN \rangle \Leftrightarrow (Al) + \frac{1}{2}[N_2] \tag{7.10}$$

Taking $a_{Al} = 1$, the dissociation pressure of AlN, P^d_{N2}, can be readily calculated from the Gibbs energy of formation of nitride (Kubaschewski and Alcock 1979, Chase et al. 1985). The values given in Table 7.5 indicate a rather weak thermal stability for the three nitrides. In the absence of kinetic barriers, nitrides would decompose rapidly at 1673K in a high vacuum and decomposition of Si_3N_4 would occur at a significant rate even at 1273K. Actually, little or no dissociation is observed because of kinetic barriers resulted from a low atomic mobility, on surface and in the bulk, for these compounds (Chatillon and Massies 1990). These barriers lead to a "retarded vaporisation" in which the effective pressure in the vicinity of the nitride surface is α. P^d_{N2}, where the *evaporation coefficient* α ($\alpha < 1$) can attain very low values (Table 7.5).

Table 7.5. Nature of bonding and thermochemical properties of nitrides.

Nitride MeN	Degree of ionicity (%) [1]	P_{N2}^d (atm)		Evaporation coefficient α	Bond energy [2]	
		1273K	1673K		Me-N (kJ/at. g N)	Me-O (kJ/at. g O)
AlN	43	2×10^{-15}	5×10^{-9}	$<2\times10^{-3}$ (1873K) [3]	-322	-558
BN	22	5×10^{-12}	4×10^{-7}	6×10^{-3} (1200–1800K) [4] $<6\times10^{-3}$ (2000K) [3]	-252	-424
Si_3N_4	30	3×10^{-7}	10^{-3}	10^{-4} (1673K) [5]	-185	-354

[1] Pauling's scale

[2] (Kubaschewski and Alcock 1979), [3] (Hildenbrand and Hall 1964), [4] (Chatillon and Massies 1990)

[5] (Rocabois et al. 1996)

The three Me nitrides (Me = Al, Si, B) are oxidizable materials, the Me-O bond being stronger than the Me-N bond (Table 7.5). As a result, the surface of these nitrides is complex and consists of oxides and oxynitrides even in a high vacuum.

7.1.2.1 Wetting of AlN. AlN crystallizes with a hexagonal structure of the wurtzite ZnS type. The Al-N bond is covalent of the sp^3 type but with a substantial degree of ionicity (Table 7.5). Nearly all the wettability data for AlN by liquid metals have been obtained using polycrystalline substrates fabricated by sintering that often contain several % of additives, mainly Y_2O_3. Even when these substrates are prepared without additives, they can contain several thousands ppm of various impurities, mainly O and C, which can enter the AlN lattice substituting for N atoms. Because of the partially ionic character of AlN, N substitution by O is associated with Al vacancies to maintain electrical neutrality. Dissolved O can migrate to the surface and modify wetting and bonding at high temperatures.

In contact with a liquid metal M, AlN dissociates according to the reaction:

$$\langle AlN\rangle \Leftrightarrow (Al)_M + \frac{1}{2}[N_2] \tag{7.11}$$

Therefore, the stability of AlN depends both on the strength of the Al-M interactions in the liquid and on the partial pressure of nitrogen in the furnace. For values of the molar fraction of Al in M at equilibrium $X_{Al}^{eq} \ll 1$, this quantity is given by:

$$X_{Al}^{eq} = P_{N2}^{-1/2} \exp\left[\frac{\Delta G_{f(AlN)}^{0} - \overline{\Delta H}_{Al(M)}^{\infty}}{RT}\right]$$ (7.12)

The results shown in Figure 7.19.a have been obtained taking P_{N2} to be 10^{-9} atm which is typical for furnaces evacuated using an oil diffusion pump (Nicholas et al. 1990). These results show high reactivity for all metals at temperatures above 1273K in vacuum. When using a neutral gas with a total pressure of one atmosphere, P_{N2} is approximately 10^{-5} atm. In this case, the reactivity, as expressed by X_{Al}^{eq}, is decreased by a factor 10^{2} (Figure 7.19.b).

Table 7.6. Contact angle of Al on sintered AlN at 1000°C.

AlN characteristics	Atmosphere [*]	Final contact angle θ_F (deg)	Reference
not specified	HV $(2\times10^{-4}\,Pa)$ $P_{O2} = 10^{-11}\,Pa$	53	(Tomsia et al. 1989)
2 wt.% O 0.2 wt.% C	HV $(2\times10^{-3}\,Pa)$ $P_{O2} = 3\times10^{-6}\,Pa$	60	(Nicholas et al. 1990)
99% AlN porosity <5%	HV $(2\times10^{-3}\,Pa)$	53	(Naidich and Taranets 1995)

[*] HV = high vacuum

At temperatures below 1500°C, the solubility of nitrogen in molten Al in equilibrium with AlN is negligible, so Al/AlN can be considered as a non-reactive couple. Table 7.6 summarises the experimental conditions and results obtained in a high vacuum by three different investigations using sintered AlN. In all cases, good wetting is ultimately observed, with small contact angles indicating a chemical bond across the interface. Taking into account the differences in AlN substrates and furnace atmospheres, the agreement between different investigations on the contact angle at 1000°C can be regarded as remarkable.

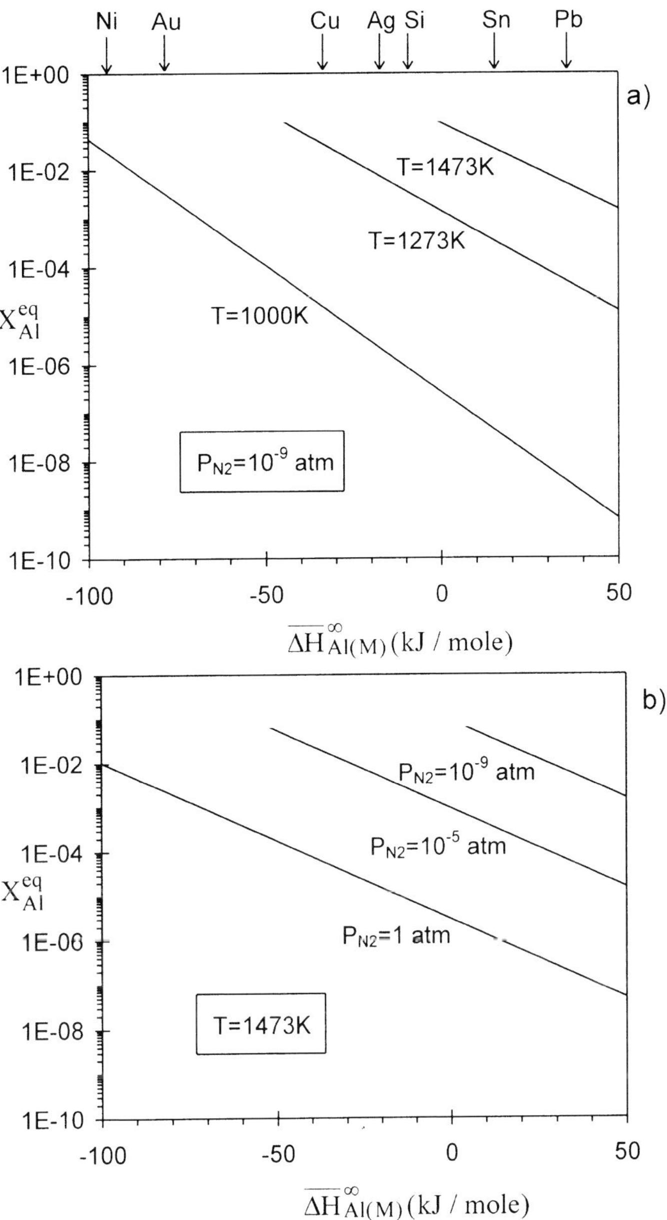

Figure 7.19. Calculated values of molar fraction of Al in liquid metal M in equilibrium with AlN at a fixed P_{N2} and temperature plotted as a function of the partial enthalpy of mixing of Al at infinite dilution in M. Arrows indicate values of $\overline{\Delta H}_{Al(M)}^{\infty}$ for some metals M that do not form stable nitride.

However, the spreading kinetics during the experiments listed in Table 7.6 were very different. They were relatively rapid for the experiments of Nicholas et al. (1990) and Naidich and Taranets (1995), in which the contact angles given in Table 7.6 were reached in a few minutes or tens of minutes. On the other hand, the initial contact angle at 1000°C measured by Tomsia et al. (1989) was 150° and it decreased very slowly through 90° after 7 hours to reach a stationary value of 53° after 30 hours. These slow kinetics cannot be controlled by deoxidization of Al drops because this occurs rapidly at 900°C and above in a high vacuum (see Section 6.4.2). They may, however, be controlled by segregation at the interface of an AlN impurity such as O and XPS analyses showed enrichment of AlN surfaces in O, in the form of Al_2O_3 or AlON (Jones and Nicholas 1989). However, the contact angle of Al on Al_2O_3 at 1000°C of 80° is significantly higher than the 50°–60° observed on sintered AlN, so the presence of Al_2O_3 or an oxynitride would be a barrier to wetting. Therefore, spreading kinetics may be controlled by the elimination of these oxide films on AlN, as in the case of SiC at temperatures below 1000°C. Liquid Al can activate the *in situ* "cleaning" of AlN surface by forming volatile Al suboxide (for example $4(Al) + \langle Al_2O_3 \rangle \Leftrightarrow 3[Al_2O]$). Thus, it can be concluded that the good wetting observed ultimately in the Al/AlN system is not due to an effect of impurities but rather to intrinsic interactions between Al and AlN. The interface between Al and AlN should be similar to that between Si and SiC from energetic point of view.

Figure 7.20 shows the change with temperature of the stationary contact angle for the Al/AlN system, measured by the sessile drop technique in high vacuum. Between 900 and 1150°C, the contact angle decreases slowly as for non-reactive couples. Two important changes occur in the low- and high-temperature ranges. A dramatic increase in contact angle is observed at low temperatures, indicating that the kinetics of Al drop and AlN surface deoxidization are too slow to be effective in times typical of sessile drop experiments. A contact angle close to zero is observed at high temperatures, but the fundamental significance of this observation is unclear because the intense evaporation of liquid Al associated with surface roughness of the substrate can produce an apparent decrease in θ (see Figure 3.3).

This analysis of the literature on the Al/AlN system reveals the complexity of the surface chemistry of AlN that is influenced strongly, and probably dominated, by oxygen even though the results do not exclude possible effect of other impurities, for instance carbon. A detailed study by spectroscopic techniques, for instance XPS, of the surface chemistry of AlN after heat treatment at different temperatures is needed to clarify the wettability of this ceramic by liquid metals.

Literature data for the wetting behaviour of low or moderate melting point metals on AlN in a high vacuum are reported on Table 7.7. Large non-wetting contact angles are observed, usually in the range 130-150° (it is noteworthy that

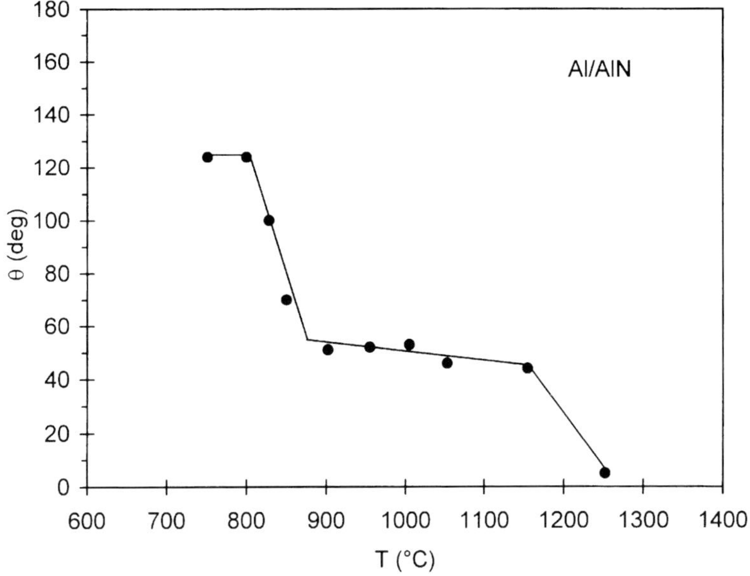

Figure 7.20. Contact angle versus temperature plot of sessile drops of Al on AlN in a high vacuum. Data from work reported in (Naidich and Taranets 1995).

Table 7.7. Wetting by pure metals on sintered AlN under high vacuum.

Metal	T_{min} - T_{max} (°C)	$\theta(T_{min})$ - $\theta(T_{max})$ (deg)
Sn	250-1150	130-125 [1]
	900	142 [2]
In	200-700	132-125 [1]
Ga	400-1000	120-110 [1]
Pb	350-750	148-137 [1]
	420-540	154-143 [3]
Ge	950-1200	122-114 [1]
Ag	1000	134 [1]
	1000	135 [4]
Au	1100-1250	138-134 [1]
Cu	1100-1200	123-118 [1]
	1150	135 [5]
	1150	134 [2]
	1150	135 [4]

[1] (Naidich and Taranets 1995), [2] (Tomsia et al. 1989), [3] (Rhee 1970b), [4] (Rhee 1970c) [5] (Nicholas et al. 1990)

these values are the same as those observed on polycrystalline oxides). These values suggest the occurrence of weak, van der Waals, interactions between AlN surfaces and the metals, including Cu. However, these results are in disagreement with those, both experimental and theoretical, of a comparative study of the bonding of Cu with Al_2O_3 and AlN (Ohuchi and Kohyama 1991). Both ceramics were fabricated *in situ* in an ultrahigh vacuum. The Al_2O_3 surface was obtained by thermal oxidation of Al using microleaks of O_2 and the AlN surface was formed by direct nitridation of clean Al using a nitrogen ion beam. Monitoring of the photoemission spectra during Cu deposition was used to follow the evolution of chemical bonding. In the case of Cu deposition on Al_2O_3, no significant reduction of either Cu or substrate energy levels was observed, indicating that interaction between Cu and Al_2O_3 is weak. On the contrary, very different photoemission features were observed during Cu deposition on AlN with a shift in the binding energy of Cu and an increased intensity of characteristic peaks. These results were interpreted by Ohuchi and Kohyama as signs of the formation of a chemical bond between Cu and AlN. This conclusion agrees with theoretical results of electronic structure and bonding of Cu on AlN obtained by first-principle calculations (Ohuchi and Kohyama 1991). Indeed, the total energy of the system decreased with the development between Cu and Al atoms of a metal-like interaction. The disagreement between the wetting behaviour and the electronic structure of the Cu/AlN interface can be explained by assuming that surfaces of sintered AlN used in the wetting experiments were oxidized.

Literature reports of the wetting behaviour of the Cu/AlN system (Table 7.7) do not mention any significant reactivity between Cu and AlN. For instance, Rhee (1970c) observed no severe chemical reaction and reported that a flat interface boundary was maintained throughout the experiment. In view of the data presented in Figure 7.19, this is rather surprising. A more precise calculation of X_{Al}^{eq} resulting from AlN dissociation into Cu (reaction 7.11), carried out using experimental values of activity of Al in Cu-Al alloys (Hultgren et al. 1973) and taking $P_{N2} = 10^{-9}$ atm, leads to $X_{Al}^{eq} = 0.25$ at $1100°C$ (for $P_{N2} = 10^{-8}$ atm, $X_{Al}^{eq} = 0.17$). These values indicate a considerable reactivity so there is again a disagreement between calculations and experimental findings, suggesting that the barriers of wetting which seem to exist on the AlN surface are also barriers of reactivity as for SiC.

Naidich and Taranets (1995) reported contact angles of ferrous metals on AlN for temperatures in the range 1500–1600°C. The values were 95° and 105° for Ni and Pd much lower than those in Table 7.7 for low or moderate melting point metals. Since the reactivity barriers existing at low temperatures (i.e., oxide films on AlN) are unlikely to be present at temperatures as high as 1500°C, a strong reactivity between Ni or Pd and AlN is expected (Figure 7.19). Interfacial reactions can affect contact angles both by changing the geometry at the triple line and by modifying the composition of the liquid.

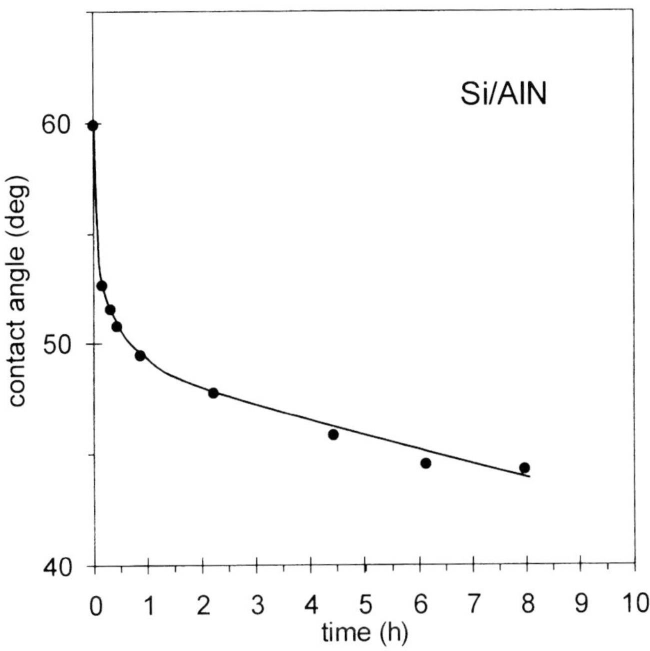

Figure 7.21. Contact angle versus time for Si on AlN at 1430°C in an H_2/H_2O flowing buffer gas under a total pressure of one atmosphere ($P_{O2} = 8.7 \times 10^{-21}$ atm). Data from work reported in (Barsoum and Ownby 1981).

The only other element that wets AlN for which data are available is Si (Barsoum and Ownby 1981, Naidich and Taranets 1995). In the investigation by Barsoum and Ownby (1981), AlN was in the form of layers produced by a variant of the CVD process and a H_2/H_2O flowing buffer gas at a pressure of 1 atm was employed to keep the O_2 partial pressure at a level close to 10^{-20} atm. No precision was given for P_{N2} in the gas but N_2 is generally present in gas mixtures as an impurity at the ppm level, which would lead to $P_{N2} = 10^{-5}$–10^{-6} atm. This value is much higher than P_{N2} in high vacuum environments, thus increasing the stability of AlN. At the melting temperature of Si, the contact angles are close to 60°, and then decrease to about 45° in 5 hours (Figure 7.21). Such contact angle variations with time can be due either to an *in situ* modification of the surface chemistry of AlN, like for Al on AlN, but cleaning of AlN surfaces is expected to occur rapidly at 1430°C, or to reactions between molten Si and AlN (since the calculated equilibrium molar fraction of Al dissolved in Si at 1700K and for $P_{N2} = 10^{-5}$ atm is 0.04). Some of this decrease may result also from evaporation at high temperature in the flowing buffer gas, that produces receding contact angles. Whatever the reason for the slow kinetics, Figure 7.21 indicates the formation of a strong Si/AlN interface. This conclusion is similar to that drawn for molten Si on covalent SiC, for which the contact angle was also close to 40°.

Improved wetting by the metals of Table 7.7 on AlN can be obtained by Al additions, as suggested by the low contact angle and the high work of adhesion observed in the Al/AlN system. Naidich and Taranets (1998) found that the wetting of Sn on AlN at 1100°C in a high vacuum was greatly improved by Al additions. For instance, with 60 at.% Al in Sn, a contact angle of 20° was observed after 35 min and the θ-X_{Al} isotherm passes through a minimum as does that for Sn-Al/Al$_2$O$_3$ (see Figure 6.32).

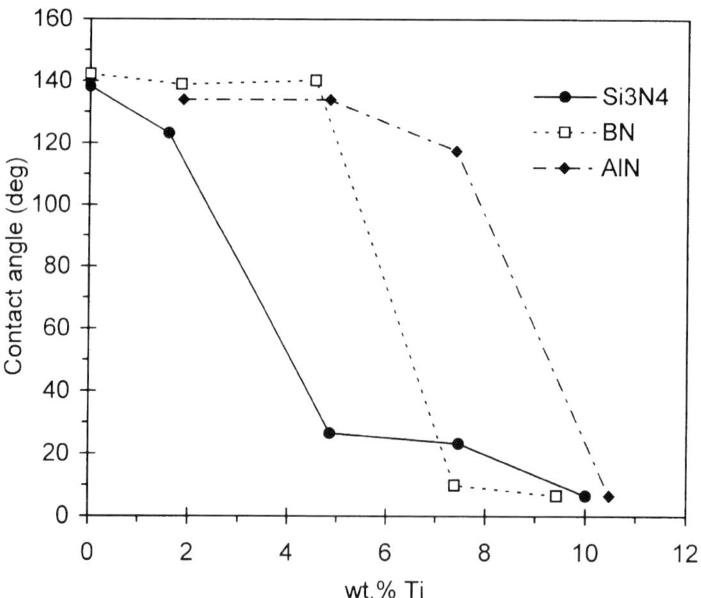

Figure 7.22. Contact angles of Cu-Ti alloys on Si$_3$N$_4$, BN and AlN at 1150°C in a vacuum (Nicholas et al. 1990) [2].

Naidich and Taranets (1995) attempted to improve wetting of Cu, Ge and Sn on AlN at 1150°C in a high vacuum by additions of up to 6 at.% of Cr, Ta, Nb, V and Ti which have a high affinity for N. Although some improvement in wetting was observed, contact angles lower than 90° were achieved only with Ti. The addition of 10 wt.% Ti in Cu at 1150°C produced a decrease of the contact angle from 135° to close to zero (Figure 7.22) with formation at the interface of a few microns thick continuous layer of wettable metal-like TiN (see Section 7.2). Similarly, very low contact angles have been achieved at temperatures close to 900°C in a high vacuum with Ag-Cu alloys containing 1 to 5 wt.% of Ti (Tomsia et al. 1989, Nicholas et al. 1990, Loehman and Tomsia 1992). It was found that a continuous layer of hypostoichiometric TiN ($TiN_{0.7}$) formed at the interface in accord with the reaction:

$$(Ti)_{Ag-Cu} + \langle AlN \rangle \Leftrightarrow \langle TiN \rangle + (Al)_{Ag-Cu} \tag{7.13}$$

Al liberated by this reaction can react with the excess Ti to form the Al_3Ti intermetallic (Tomsia et al. 1989).

Finally, contact angles less than 30° were measured for Ni alloyed with 2 wt.% of Ti, Zr or Hf on sintered AlN at 1500°C in gettered Ar (Trontelj and Kolar 1978). In all cases, dense reaction product layers of nitrides, up to several tens of microns in thickness, were formed at the interface.

7.1.2.2 Wetting of Si_3N_4. Si_3N_4 is similar to AlN in its structure (hexagonal), type of bonding (predominantly covalent), complex chemistry (it usually contains several percent of sintering aids, mainly Y_2O_3 and Al_2O_3) and oxidizability. Even after a few minutes exposure in air at room temperature, it forms 0.5–3 nm thick layers which appear to be Si_2N_2O (Maguire and Augustus 1972, Raider et al. 1976, Ljungberg and Warren 1989). These layers are relatively stable in temperature and several hours at 1140°C in an ultrahigh vacuum are needed to remove oxygen (Maguire and Augustus 1972). Of the three nitrides in Table 7.5, Si_3N_4 is the least stable and its stability is even lower when in contact with liquid metals with a high affinity for Si, such as Cu and Au (Figure 7.23). In sessile drop experiments carried out in a high vacuum at temperatures of 1100°C or below, no evidence of reactions was found (Ljungberg and Warren 1989, Tomsia et al. 1989). This confirms the presence of barriers to reactivity on the surface of Si_3N_4 resulted from a reduced atomic mobility i.e., a low value of the evaporation coefficient α (Table 7.5). These barriers can be enhanced by oxide or oxynitride layers, as for AlN. The presence of such layers can explain also the high contact angles measured for low or moderate melting point metals and alloys which do not form stable nitrides. As for AlN, these

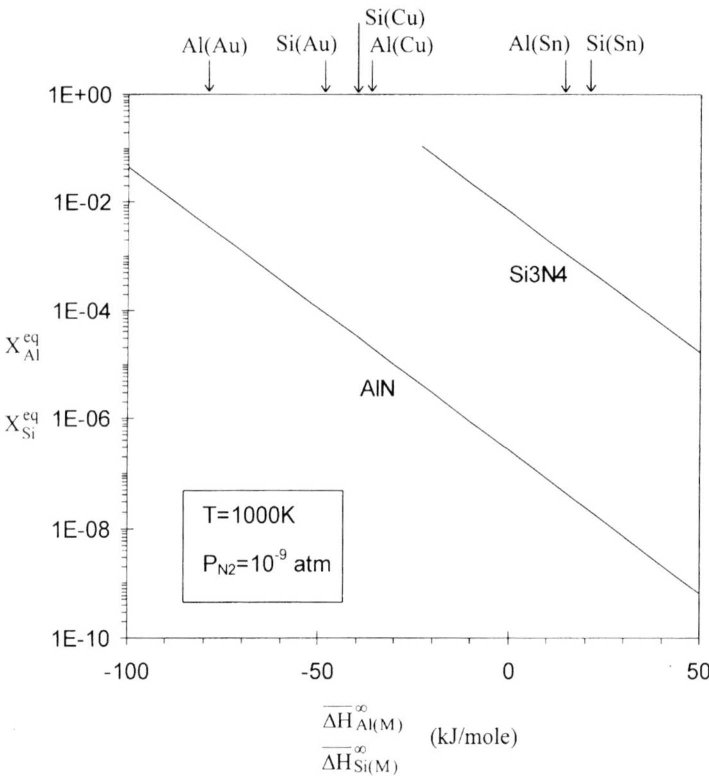

$$X^{eq}_{Al}$$
$$X^{eq}_{Si}$$
$$\overline{\Delta H}^{\infty}_{Al(M)}$$
$$\overline{\Delta H}^{\infty}_{Si(M)}$$
(kJ/mole)

Figure 7.23. Calculated values of molar fractions of Al and Si in liquid metal M in equilibrium with AlN and Si_3N_4 respectively as a function of the partial enthalpy of mixing of the nitride metal at infinite dilution in M. Arrows indicate values of $\overline{\Delta H}^{\infty}_{Al(M)}$ and $\overline{\Delta H}^{\infty}_{Si(M)}$ for some metals M which do not form stable nitride.

values lie between 120° and 170° at temperatures of no more than 1100°C (Table 7.8) and are close to those obtained for non-reactive metals on polycrystalline oxides which can have very rough surfaces. However, in the case of the Cu/Si_3N_4 system at 1150°C, some mobility of Cu drops on the substrate surface was detected (Tomsia et al. 1989) ; such drop mobility is often observed in systems in which interfacial reaction leads to formation of bubbles. Moreover, an exceptional decrease in contact angle was observed at a temperature between 1100°C and 1200°C by Allen and Kingery (1959) for Sn (Table 7.8). This result suggests that the barriers at low temperatures consist of oxide or oxynitride layers (contact angles of various metals on Si_3N_4 and Si_2N_2O are very close at 900°C (Tomsia et al.

1989)), but these disappear at high temperature to permit a direct contact and reactions between Sn and Si_3N_4.

Table 7.8. Wetting of Si_3N_4 in a high vacuum by pure metals and alloys which do not form stable nitrides.

Liquid	T (°C)	θ (deg)	Si_3N_4	Reference
Sn	1100	144	HIP	(Ljungberg and Warren 1989)
Cu	1100	131	1% Y_2O_3	
Au	1100	157		
Ag	1100	155		
Au-42.5 at.%Ni	1050	120	pressureless sintered with Y_2O_3 and Al_2O_3	(Paulasto et al. 1996)
Sn	900	153	no additives	(Tomsia et al. 1989)
Cu	1150	115-130 [*]		
Ag-28 wt.%Cu	900	142		
Cu	1150	135	6 wt.% La_2O_3 4 wt.% Al_2O_3	(Nicholas et al. 1990)
Sn	800	168	sintered 95.1 % purity	(Allen and Kingery 1959)
	1100	154		
	1200	29		

[*] mobile sessile drops

In a low P_{O2} atmosphere generated by a flowing buffer gas, pure Si wets Si_3N_4 produced by CVD (Barsoum and Ownby 1981). On melting, the initial contact angle is close to 70° and a decrease to 60° is observed after 1 hour. Although some dissociation of Si_3N_4 occurs at the melting point of Si, this should not modify the nature of solid/vapour or solid/liquid interfaces, and thus the Si/Si_3N_4 couple can be considered as non-reactive from the point of view of wettability. Slow wetting kinetics and stationary contact angles close to 50° have been observed in Ar (Li and Hausner 1992) and He (Duffy et al. 1980) atmospheres. The achievement of wetting in this system indicates the formation of a chemical bond localized between Si and Si_3N_4, as in the similar non-reactive couples Al/AlN ($\theta \simeq 55°$) and Si/SiC ($\theta \simeq 40°$).

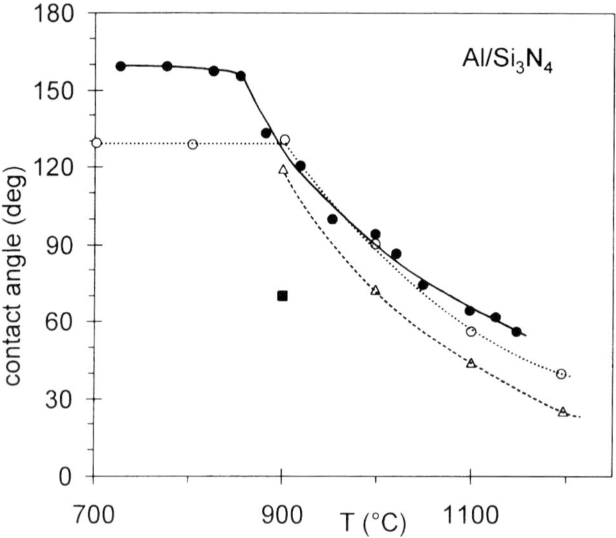

Figure 7.24. Contact angle versus temperature for Al on Si$_3$N$_4$. Full circles: (Nicholas et al. 1990), open circles: (Ljungberg and Warren 1989), triangles: (Naka et al. 1987), square: (Tomsia et al. 1989).

Because AlN has a higher thermodynamic stability than Si$_3$N$_4$ (Table 7.5), Al may react with Si$_3$N$_4$ forming AlN. The contact angle versus temperature curves of Figure 7.24 for Al on Si$_3$N$_4$ show a non-wetting to wetting transition, similar to transitions occurring for oxidizable metals, as discussed in Section 6.4.2 (see Figures 6.20 and 6.25). At low temperatures, large non-wetting contact angles are observed because both Al and Si$_3$N$_4$ are oxidized. At about 1100°C, contact angles much lower than 90° are obtained that indicate the establishment of direct contact between Al and Si$_3$N$_4$. The transition from non-wetting to wetting appears to occur at 950–1000°C in the experiments of Figure 7.24, but the transition temperature depends strongly on the furnace atmosphere and the substrate purity. For instance, a stationary wetting contact angle ($\theta = 71°$) was reached in 40 minutes at 900°C in a high vacuum (Tomsia et al. 1989). It is possible, and even probable, that wetting in this system can be achieved at even lower temperatures provided the Al and Si$_3$N$_4$ surfaces are cleaned. Characterization of Al/Si$_3$N$_4$ interfaces of sessile drop specimens by SEM or microprobe analysis did not reveal any evidence of reaction between Al and Si$_3$N$_4$ at the micrometer scale (Tomsia et al. 1989). However, higher spatial resolution techniques revealed the formation of submicronic layers with a complex chemistry, consisting of sialons (Ning et al. 1987) and/or a mixture of SiO$_2$ and Al$_2$O$_3$ (Tomsia et al. 1989).

In view of the high work of adhesion of Si on Si_3N_4 and the comparatively low surface energy of molten Si, an improvement in the wetting by inert pure metals can be expected by the addition of Si. Figure 7.25 shows the spreading kinetics at 1200°C in a high vacuum of a Ni-50 at.% Si alloy on Si_3N_4 containing Y_2O_3 and Al_2O_3 as sintering aids. A wetting contact angle ($\theta = 75°$) is achieved after several thousands of seconds (van Dal 1997). Because no chemical reaction was detected by SEM and microprobe analysis, the slow spreading kinetics may result from an *in situ* cleaning of Si_3N_4 surface by Si, acting as Al in the Al/AlN system (see Section 7.1.2.1).

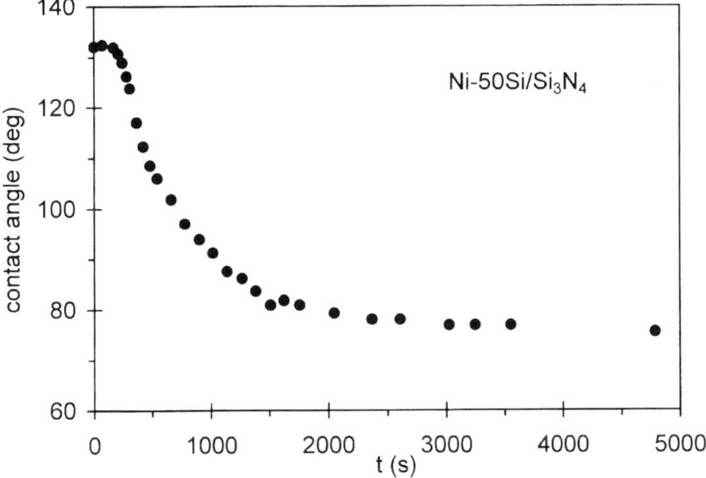

Figure 7.25. Contact angle versus time for Ni-50 at.% Si on Si_3N_4 at 1200°C in a high vacuum (van Dal 1997).

Even lower contact angles ($\theta = 40°$) have been obtained at a temperature close to 1200°C in a high vacuum using a Cu-10 wt.% Ni-7 wt.% Cr alloy. They were attributed to the formation of a continuous layer of the wettable Cr_2N compound that was several tens of microns thick (Xiao and Derby 1995). Similarly, while the Au-Ni alloy does not wet Si_3N_4 (Table 7.8), additions of a few at.% of V to the alloy leads to contact angles of 40–55° with formation of VN_x at the interface (Paulasto et al. 1996). Additions of Ti and Nb to pure Cu or Cu-Sn alloys produce a strong improvement in wetting (Figures 7.22 and 7.26). In the Cu-Ti/Si_3N_4 system, continuous layers of TiN up to 10 to 20 microns thick were found (Nicholas et al. 1990). A few at.% Ti in Ag-Cu alloys lead to very low contact angles in the range 0–20° at 850°C and above, with formation at the interface of a bilayer consisting of TiN, formed by the reaction $(Ti) + \langle Si_3N_4 \rangle \Leftrightarrow \langle TiN \rangle + (Si)$ and of Ti silicides formed by reaction between excess Ti in the alloy and Si liberated by the previous reaction (Tomsia et al. 1989, Nicholas et al. 1990, Ljungberg 1992, Nomura et al. 1998, 1999).

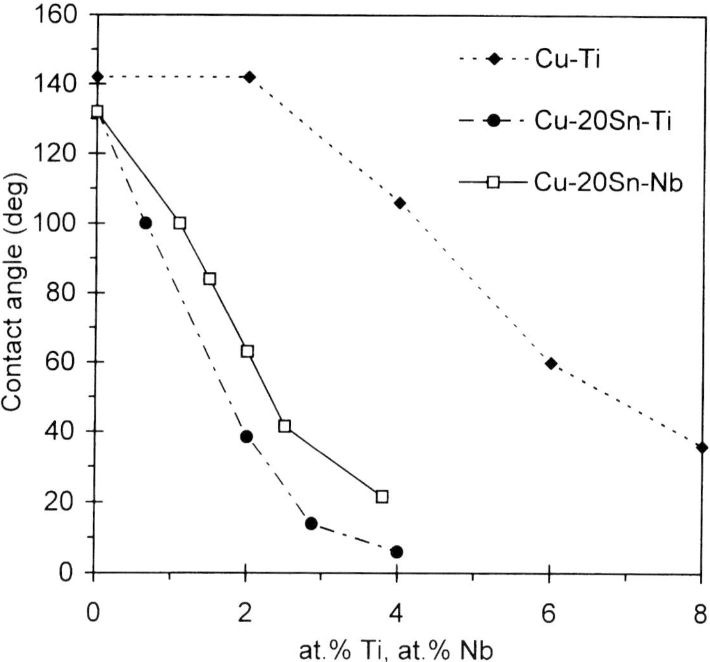

Figure 7.26. Contact angle on Si_3N_4 at 1150°C (holding time 30 min) as a function of concentration of active element (Ti or Nb) in pure Cu and Cu-Sn alloy. Data from work reported in (Zhuravlev et al. 1998).

7.1.2.3 Wetting of BN. At equilibrium, the partial pressure of N_2 for BN dissociation lies between the values for AlN and Si_3N_4 (Table 7.5). In practice, even in a high vacuum the stability of BN is much higher than that predicted by equilibrium thermodynamics because the evaporation coefficient is as low as 10^{-3} (Table 7.5). BN is also an oxidizable ceramic but boron oxide, B_2O_3, is volatile with a partial pressure at 1000°C $P_{B_2O_3} = 10^{-7}$ atm (Chase et al. 1985). Moreover, because B_2O_3 is liquid at temperatures higher than 450°C, no retarded evaporation of this oxide is expected. Use of equations (6.26) and (6.27) predicts an evaporation rate of B_2O_3 of several nm/s at 1000°C, indicating a rapid cleaning of BN (and also of B_4C) surfaces at this temperature. In practice, the surface chemistry of BN and its evolution with temperature may be complex due to the presence of O and C impurities in nitride ceramics (Perevertailo et al. 1998). A detailed chemical characterization of BN surfaces in a high vacuum at different temperatures is needed to understand the interaction between BN and liquid metals.

Few studies of wetting of metals on BN exist, the only systematic investigation known to us being that of Naidich (1981) which was performed in the range 1000–1500°C using both cubic and hexagonal BN. No oxidation of BN surfaces is expected under these conditions. For metals with a negligible (Ag, Au, Sn) or weak (Ge, Ga, Cu) affinity for both N and B, contact angles lie in the range 135–150° at 1000–1100°C for both structures of BN.

Ferrous metals form contact angles lower than 90° (Naidich 1981). However, these metals have a high affinity for B at 1500°C (for instance liquid Ni dissolves up to 60 at.% B (Massalski 1990)), so a strong reactivity between these metals and BN is expected and the contact angles are only apparent (see Figure 2.18).

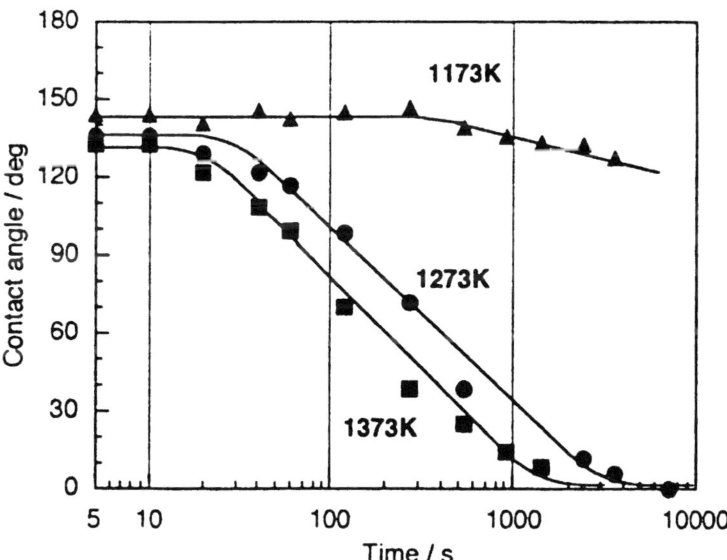

Figure 7.27. Contact angle versus time for Al on sintered hexagonal BN in an He-3% H_2 gas. Data from work reported in (Fujii et al. 1993) [24].

Al reacts with BN forming a continuous layer of AlN at the interface and liberating B which reacts with Al to form borides precipitates (Fujii et al. 1993). Figure 7.27 shows results obtained for Al on sintered hexagonal BN (99.7 wt.% purity). In this experiment, the BN substrate was first cleaned in a high vacuum at 1423K for 1 hour before purified He-3%H_2 gas was introduced and the temperature was lowered to the experimental value. A pure Al drop was placed *in situ* on the substrate by a dispensed drop device (Figure 3.7.c) allowing the oxide layer on Al to be removed. The initial contact angle was close to 140° as for non-reactive metals on BN but decreased after a few minutes with a spreading rate that was strongly dependent on temperature. Very low contact angles were obtained at 1273K and 1373K in 10^3 seconds, and nearly perfect wetting was obtained in 7×10^3 seconds at 1273K. Other investigators obtained higher stationary contact angles (in the range 35–60° at 1373K (Naidich 1981, Nicholas et al. 1990)), that are consistent with those for Al on AlN (Figure 7.20) i.e., for Al on the reaction product at the Al/BN interface.

Silicon has an affinity both for B (at 1500°C Si dissolves up to 17 at.% B (Massalski 1990)) and N (although Si_3N_4 is less stable than BN). Naidich (1981) achieved stationary contact angles of 95° and 110° for Si on cubic and hexagonal BN at 1500°C in a high vacuum in a few minutes. These values are much higher than those observed for Al, but significantly lower than the 140° or so observed for non-reactive metals. A detailed interpretation of these values is not possible in the absence of information on interfacial reactions which could occur in the Si/BN system.

Additions of Ti to Cu and Ag-Cu allowed very low contact angles to be achieved on BN at 1150°C (Figure 7.22) and 950°C (Nicholas et al. 1990). Electron probe micro-analysis revealed the formation of a Ti-rich layer several microns thick at the interface but the nature of Ti compounds (TiN or TiB_2 or both) was not identified.

The wetting behaviour of metals on B_4C is very similar to that on BN (Naidich 1981). For metals with a negligible or weak affinity for both B and C (Cu, Ag, Ge, Sn), contact angles in the range 130°–140° were observed in a high vacuum at temperatures close to 1100°C. The addition of metals such as Cr and Ti that form metal-like carbides strongly improved wetting. Again, Al and Si, which form carbides more stable than B_4C, wet well B_4C at 1150°C and 1430°C respectively. As on BN, ferrous metals that can dissolve large quantities of B and C wet B_4C well and form an interface that is expected to be non-planar.

7.1.3 Concluding remarks

This review of the experimental results for metal/covalent ceramic systems shows that the intrinsic wetting properties in these systems are not well established at present. Further work is needed using combined studies of the wetting and surface chemistry of ceramics, similar to those carried out by Rado (1997) for SiC. In future studies, to use monocrystalline ceramics is obviously desirable, and when they are not available, CVD layers are preferable to sintered materials.

For metals on "real" materials in neutral gas or high vacuum environments, good wetting is seldom observed at temperatures of 1000°C or less owing to pollution of the ceramic surfaces (SiC, Si_3N_4, AlN) by oxygen. At higher temperatures, cleaning of surfaces can occur but the wetting observed in many cases is accompanied by reactivity. Although BN and B_4C surfaces seem not to be oxidized at temperatures close to 1000°C, non-wetting is still observed for metals such as Pb, Sn and Ag that have a negligible affinity for B, N and C.

At temperatures of 1000°C and above, Al and Si wet covalent ceramics rather well with contact angles close to 50° for both non-reactive (Al/AlN and Si/SiC) and reactive systems (Al/SiC and Al/BN). This behaviour relates well to theoretical studies indicating the formation of metallic or covalent chemical bonds at the interfaces between Al or Si and covalent ceramics. The ability of Al and Si to bond strongly with ceramic surfaces appears to correlate with the degree of covalence (or, equivalently, with the degree of ionicity) of the ceramic, as shown by the data in Table 7.9 for Si on non-reactive solids. A similar tendency is observed for Al on various solids, including solid Al considering the metallic bond in solid Al as a homopolar Al-Al bond (Table 7.9).

Table 7.9. Contact angles of Si and Al on solids of variable degree of ionicity.

Liquid/solid	Degree of ionicity (%)	T (°C)	θ (deg)	W_a/W_c = $(1+\cos\theta)/2$
Si/SiO_2	51	~1410	87 [1]	0.53
Si/Si_3N_4	30	~1410	55 ± 5	0.79 ± 0.04
Si/SiC	11	~1410	40 ± 10	0.87 ± 0.05
Si/Si	0	~1410	14 [2]	0.98
Al/CaF_2	89	700-1000	130-100 [3]	0.18-0.41
Al/Al_2O_3	63	700-1000	90-80 [3]	0.50-0.59
Al/AlN	43	1000	53	0.80
Al/Al	0	660	~0 [4]	~1

[1] (Sangiorgi et al. 1988), [2] (Naidich et al. 1975), [3] (Naidich et al. 1983), [4] (see Section 5.1)

7.2. METALS ON METAL-LIKE CERAMICS

Refractory transition metal carbides, nitrides and borides (e.g., Table 7.10) are of great interest for both practical purposes and theoretical studies. This class of solids exhibits very high hardnesses and melting points, as indeed do many *covalent* compounds. However, most monocarbides (MeC) and mononitrides (MeN) crystallize in the sodium chloride structure, generally found with *ionic* compounds. Further, these ceramics also have aspects of *metallic* behaviour such as good thermal and electrical conductivities. This partly metallic bonding has repercussions on interfacial and adhesion properties with metals and we have seen in Sections 6.3 (Table 6.5) and 7.1.2 (Figure 7.22) that improvement in wetting and adhesion in metal/ionocovalent ceramic systems is obtained by creating metal-like compounds at the metal/ceramic interface.

Table 7.10. Wetting by Cu of metal-like carbides in a high vacuum at 1100°C (Ramqvist 1965). The equilibrium molar fraction of carbide metal Me dissolved in Cu, $X^{eq}_{Me(Cu)}$, is calculated from equation (7.16). Data for ΔG^0_f (reference state: pure liquid metal Me) come from (Rosenqvist 1983) except for Mo_2C and HfC (Kubaschewski and Alcock 1979). Data for $\overline{\Delta H}^{\infty}_{Me(Cu)}$ come from (Niessen et al. 1983).

Carbide Me_nC_m	θ (deg)	W_a / W_c	ΔG^0_f (kJ/g atom C)	$\overline{\Delta H}^{\infty}_{Me(Cu)}$ (kJ/mole)	$X^{eq}_{Me(Cu)}$
Mo_2C	18	0.98	-84	82	2×10^{-5}
WC	20	0.97	-57	101	10^{-6}
Cr_3C_2	47	0.84	-59	51	3×10^{-3}
$NbC_{0.97}$	70	0.67	-140	12	2×10^{-6}
TaC	78	0.60	-163	9	3×10^{-7}
TiC	112	0.31	-173	-40	8×10^{-6}
ZrC	127	0.20	-179	-110	2×10^{-3}
HfC	134	0.15	-216	-81	7×10^{-6}

In fundamental studies of ceramic wetting, many difficulties can arise from the lack of monocrystalline high-purity materials. In most cases, sintered materials have been used that, even when free of sintering additives, contain typically 0.1 wt.% and more of oxygen and other impurities that can strongly affect wettability (Figure 7.28). Because of the high affinity of transition metals of Groups IVB and VB for O, as compared to C or N, these compounds are oxidizable and this exerts a

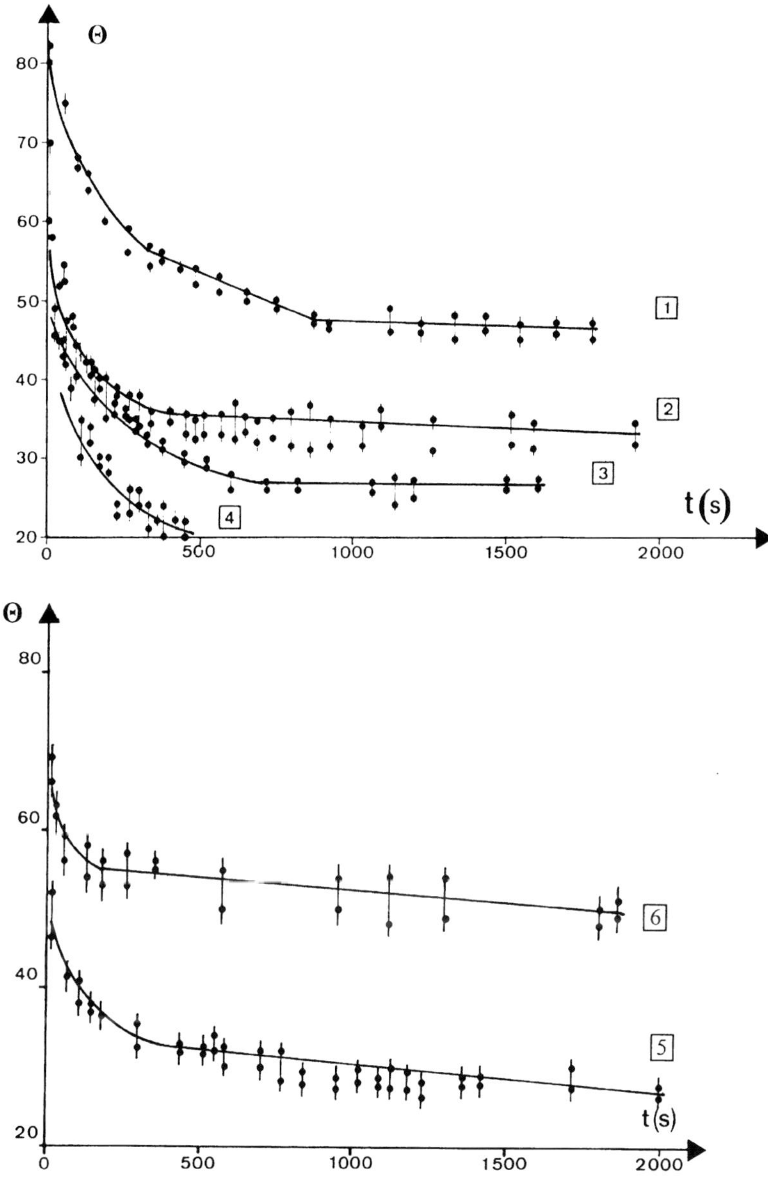

Figure 7.28. Variation of contact angle with time of Fe presaturated with TiB_2 on different TiB_2 substrates: sintered TiB_2 with 0.68 wt.% oxygen (1), 0.33 wt.% oxygen (2), 0.26 wt.% oxygen (3), CVD TiB_2 (4), sintered TiB_2 with 0.30 wt.% oxygen (5), the same as (5) after surface oxidation (6). From (Ghetta et al. 1992) [25].

strong influence on wetting and adhesion. Finally, several of these substances possess a large range of non-stoichiometry. Bulk non-stoichiometry can be easily controlled during processing of the material but it is difficult to control surface non-stoichiometry, which may be very different (Lequeux 1997). Note that studies of the wettability of metal-like ceramics by liquid metals were performed in attempts to produce new cermets, mainly in the period 1960-75, and hence did not benefit from modern techniques for chemical analyses of surfaces and interfaces.

Table 7.10 presents results obtained for pure Cu on different nearly stoichiometric carbides. These results come from the extensive study performed by Ramqvist (1965) on the wetting by metals on various sintered metallic carbides carried out in a high vacuum. The contact angles given on Table 7.10 are equilibrium values obtained generally in less than 15 minutes. However, the change in θ between t = 0 and t = 15 minutes was only a few degrees in many cases (Figure 7.29) indicating that a nearly equilibrium state was reached very rapidly on melting.

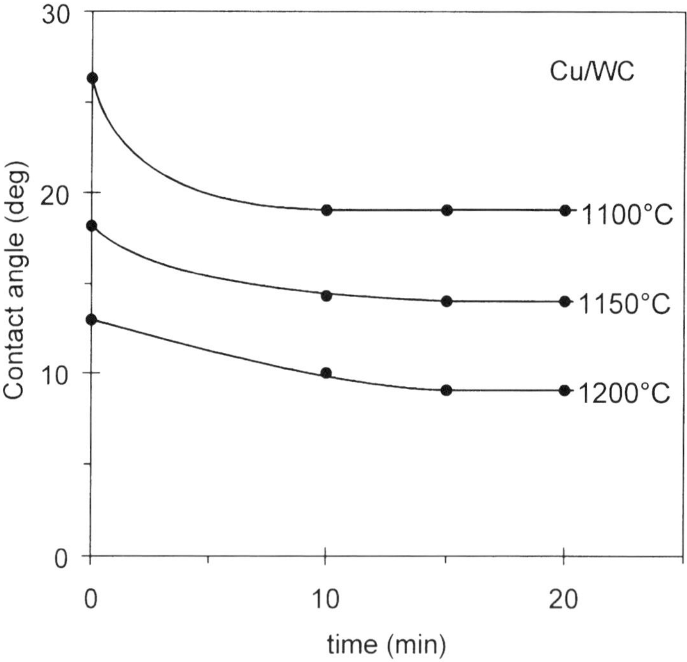

Figure 7.29. Contact angles of Cu on WC plotted as a function of time. Data from work reported in (Ramqvist 1965).

The results in Table 7.10 show contact angles varying from values typical of metal/oxide systems with van der Waals interactions at the interface ($\theta \approx 130°$, $W_a/W_c \approx 0.15$) to values typical of metal/metal systems ($\theta \approx 20°$, $W_a/W_c \approx 0.95$). It is noteworthy that the degree of wetting of Cu on metallic W (see Section 5.2) is nearly equal to that on WC.

In Section 6.1, we discussed a correlation between wettability and reactivity in metal/oxide systems: for a molar fraction of dissolved oxygen greater than 10^{-5}, the contact angle decreases from about 130° to 0° (Figure 6.2). To check if a similar correlation holds for metal/metallic carbide systems, the reactivity between Cu and different carbides at 1100°C can be calculated for the dissolution of carbide Me_nC_m in Cu:

$$\frac{1}{m}\langle Me_nC_m \rangle \Leftrightarrow \frac{n}{m}(Me)_{Cu} + (C)_{Cu} \tag{7.14}$$

As the stoichiometric carbide is also in equilibrium with free graphite, meaning that the activity of C in the carbide is equal to that in pure graphite i.e., to unity (except for Mo_2C for which according to the Mo-C phase diagram (Massalski 1990) the activity of C is lower but close to unity), this yields:

$$\frac{1}{m}\langle Me_nC_m \rangle \Leftrightarrow \frac{n}{m}(Me)_{Cu} + \langle C \rangle \tag{7.15}$$

For reaction (7.15), the equilibrium molar fraction of carbide metal Me dissolved in Cu is:

$$X_{Me(Cu)}^{eq} = \exp\left[\frac{m}{nRT}\left(\Delta G_f^0 - \frac{n}{m}\overline{\Delta H}_{Me(Cu)}^{\infty}\right)\right] \tag{7.16}$$

In this equation, ΔG_f^0 is the standard Gibbs energy of formation of the carbide Me_nC_m from graphite and pure liquid metal Me and $\overline{\Delta H}_{Me(Cu)}^{\infty}$ is the partial enthalpy of mixing of Me at infinite dilution in Cu. Results in Table 7.10 show that, although some correlation exists between contact angles and Gibbs energies of formation of carbides, there is no correlation between θ and $X_{Me(Cu)}^{eq}$. This means that the element Me is not active at Cu/carbide interfaces at a level of 0.1 at.% or less. Thus, the high values of W_a observed for carbides lying between Mo_2C and

TaC in Table 7.10 have to be explained by the nature of the interfacial bonding between Cu and the carbide and not by an effect of dissolved carbide metal.

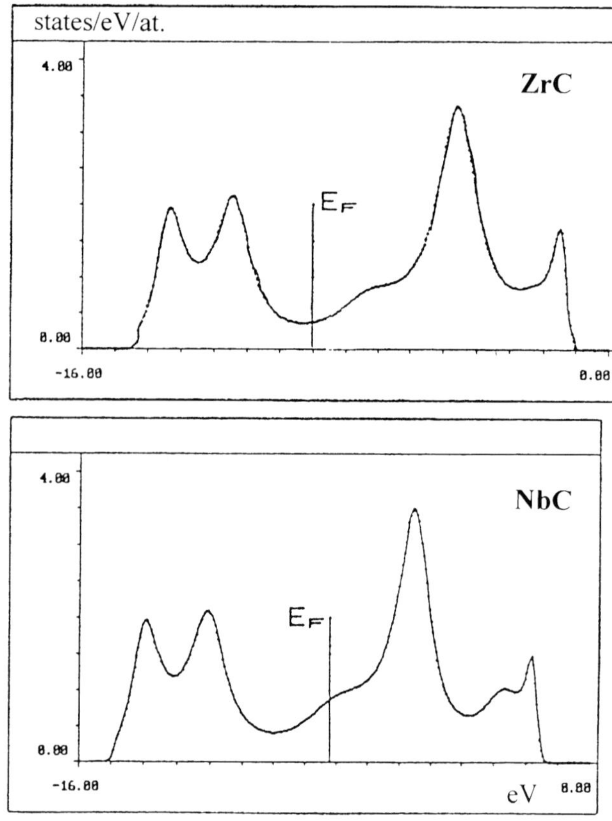

Figure 7.30. Density of electronic states (resulting from overlapping of d-orbitals of the metal and p-orbitals of carbon) and location of the Fermi level (E_F) for ZrC and NbC (Le 1990).

Electronic structure calculations for transition metal carbides (Neckel 1990, Le 1990, Le et al. 1991) reveal significant contributions to cohesion by all three main types of chemical bonding. Covalent bonds are due to the formation of molecular orbitals by combining atomic d-orbitals of the metal with p-orbitals of C. Ionic bonds result from charge transfer from the metal to the non-metal. Metallic bonds are due to s electrons and also to a non-vanishing density of d-p electronic states (DOS) existing at the Fermi level (Figure 7.30). The main difference between the DOS curves calculated for stoichiometric ZrC, TiC or HfC and NbC, TaC or VC is the position of the Fermi level. For carbides of transition metals of Group IVB (such as ZrC), the Fermi level is located at the minimum of DOS curve, while for carbides of metals of Group VB (such as NbC), it is shifted to the right of the minimum of the DOS curve i.e., towards antibonding states. This results in a low thermodynamic stability of carbides such as NbC compared to carbides such as ZrC.

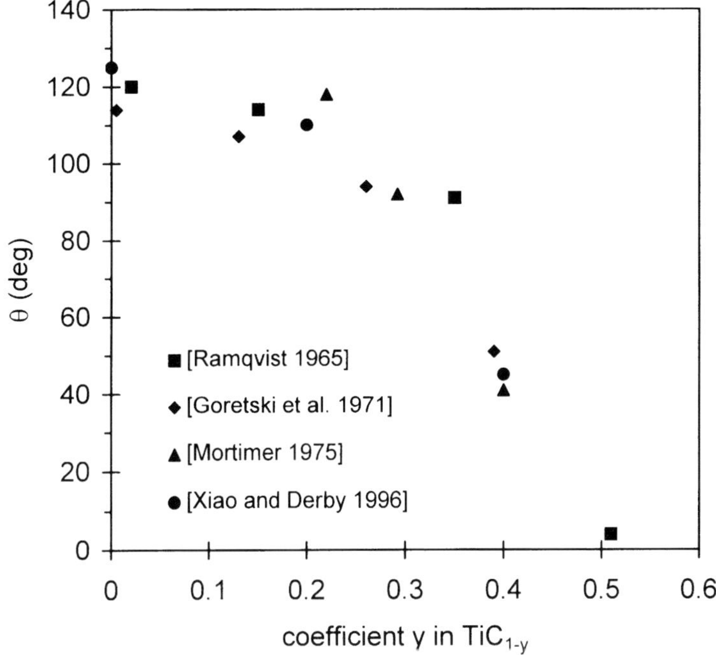

Figure 7.31. The influence of stoichiometry of TiC_{1-y} on the wetting behaviour of Cu at 1100–1150°C.

The good adhesion and wetting observed between Cu and carbides of Mo, W, Cr, Nb and Ta can be explained by considering that metallic bonding can be established at the interface, although covalent and other types of interactions cannot be excluded. More surprising are the experimental results on TiC, ZrC and HfC for which W_a values are several times lower than those found on NbC and TaC, despite a significant DOS at the Fermi level. However, the experimental results in Table 7.10 for TiC, ZrC and HfC are questionable. This appears clearly in the case of TiC and TiN. Indeed, Ramqvist (1965), as well as other authors

(Livey and Murray 1956, Rhee 1970c, Xiao and Derby 1996), found non-wetting of Cu on stoichiometric TiC ($\theta = 110$–$125°$ at the melting temperature of Cu, see Figure 7.31 for y = 0). However, other experiments carried out in very low P_{O2} atmospheres led to much lower contact angles ($\theta \cong 60°$ (Whalen and Humenik 1960, Li 1993)). Similar differences have been found for Cu on stoichiometric TiN in high vacuum ($\theta = 110°$ (Xiao and Derby 1996) and $\theta = 60°$ (Lequeux et al. 1998)).

There are at least two possible explanations for the different results obtained in the Cu/TiC and Cu/TiN systems: (i) the non-wetting observed on stoichiometric TiC (or TiN) is due to pollution of the ceramics by oxygen (Li 1993), and (ii) the low final contact angles observed by some authors on initially stoichiometric substrates are due to the development at the substrate surface of an hypostoichiometric zone. This could have been caused, in the case of TiC, by C losses as CO produced by reaction with the O present in these materials at a level of 0.1 wt.% or by N_2 evaporation in the case of TiN. Moreover, SIMS and nuclear microprobe analyses performed by Lequeux (1997) on *nominally* stoichiometric TiN substrates before being used in wetting experiments revealed a hypostoichiometric surface layer several tens of nanometers thick, probably formed during the cutting and polishing of the substrates. Therefore, the contact angles close to 60° found by Lequeux et al. (1997, 1998) were obtained in fact on hypostoichiometric TiN surfaces.

To eliminate these uncertainties about surface chemistry, further exeriments were performed with TiN in a $N_2 + H_2$ atmosphere. The role of H_2 was to suppress or minimise oxidation caused by the reaction $2TiN + (3/2)O_2 \Leftrightarrow Ti_2O_3 + N_2$ by decreasing P_{O2}. The role of N_2 was mainly to ensure a stoichiometric TiN surface. In these conditions, the contact angle of Cu on TiN was found to be $96 \pm 2°$ (van Deelen et al. 1998). This value corresponds to a $W_a/(2\sigma_{LV})$ ratio of 0.45 higher than for pure Cu on ionocovalent oxides (0.20) but lower than for Cu on WC or Mo_2C (≈ 0.98, Table 7.10).

In the case of TiC, further work on substrates with well characterised bulk and surface compositions is needed to elucidate the diverging results of the literature. Whatever the explanation, it has to be recognized that the wettability of TiC and TiN, and probably of other stable and oxidizable carbides and nitrides such as ZrC, ZrN, HfC and HfN, is very sensitive to the furnace atmosphere.

When y varies for TiN_{1-y} and TiC_{1-y}, the activities of both Ti (Figure 7.32) and the non-metallic element vary, while the products $a_{Ti}.a_C$ and $a_{Ti}.a_N$ ($= a_{Ti}.P_{N2}^{1/2}$) are nearly constant at a given temperature (Hultgren et al. 1973). If, instead of pure Cu, a Cu-Ti alloy is used, the activity of Ti at the interface and the stoichiometry parameter y are fixed by the Ti concentration in the alloy (y increases when the molar fraction of Ti in Cu increases, see Figure 7.33 taking Me = Ti). The results in Figure 7.31 for TiC and Table 7.11 for various carbides

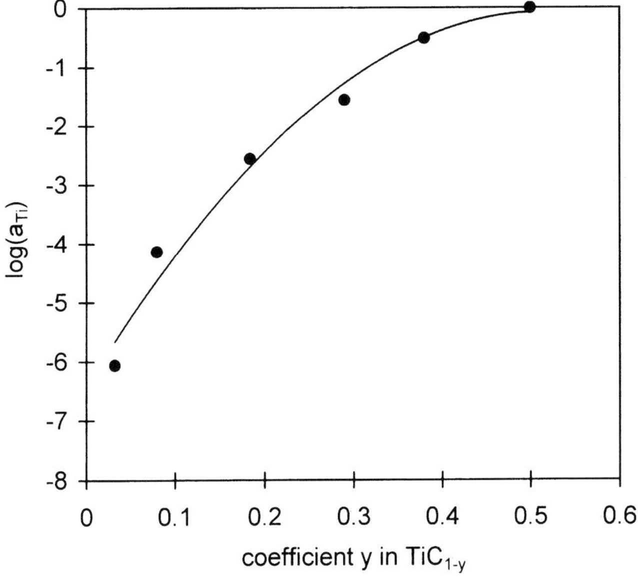

Figure 7.32. Activity of Ti in equilibrium with TiC_{1-y} at 1400K (Nicholas 1986).

show that hypostoichiometry favours wetting. This means that Ti additions to Cu are expected to improve wetting, which is indeed observed for both TiC and TiN (Xiao and Derby 1996, Lequeux et al. 1998). For instance, in the case of a Cu-5 at.% Ti alloy on TiN, for which the corresponding value of y is 0.20, the contact angle is only 12° (Lequeux 1997). Such a low contact angle is due at least partly to a modification of the interface at the solid-side (hypostoichiometry) but it is also possible that the liquid-side of the interface is modified by Ti adsorption, because Ti-TiN interactions are stronger than Cu-TiN interactions.

Ni, Fe and Co dissolve large amounts of carbides because of the relatively strong interactions between these metals and both carbon and the metal component of the carbide. For instance, at 1500°C, the molar fraction of carbide dissolved in Co lies between 0.05 for TaC and 0.23 for WC (Warren 1980). When a small piece of Co is placed on a MeC carbide substrate, the reaction between Co and MeC depresses the melting temperature of Co by 100 to 200°C because of the presence of eutectics in ternary Me-C-Co systems. Therefore, the final contact angle measured during such an experiment is for a Co alloy saturated with carbide. Furthermore, because of the dissolution process, the measured contact angle is only apparent and is lower than the true Young contact angle (see Figure 2.18). Despite this difficulty, the experimental results obtained by Ramqvist (1965) and other authors (Whalen and Humenik 1960, Warren 1980) show very good wetting by ferrous metals on all carbides, including the high-stability carbides of Ti, Zr and Hf (Table 7.12).

Ferrous metals are also known to wet stable nitrides with contact angles reported of 49° for Fe at 1590°C and 7° for Co at 1540°C on ZrN in vacuum (Kotsch 1967).

Table 7.11. Wetting by Cu of non-stoichiometric metal-like carbides in a high vacuum at 1100°C (Ramqvist 1965).

carbide	y	θ (deg)
TiC_{1-y}	0	112
	0.15	108
	0.34	88
	0.51	0
TaC_{1-y}	0	78
	0.06	75
	0.10	34
VC_{1-y}	0.12	50
	0.17	38

Table 7.12. Contact angles of initially pure Ni, Co and Fe on different carbides (Ramqvist 1965).

Carbide	Ni at 1380°C θ (deg)	Co at 1420°C θ (deg)	Fe at 1490°C θ (deg)
TiC	23	25	28
ZrC	24	36	45
HfC	23	40	45
VC	17	13	20
NbC	18	14	25
TaC	16	13	23
Cr_3C_2	0	0	0
Mo_2C	0	0	0
WC	0	0	0
$TiC_{0.49}$	0	21	—

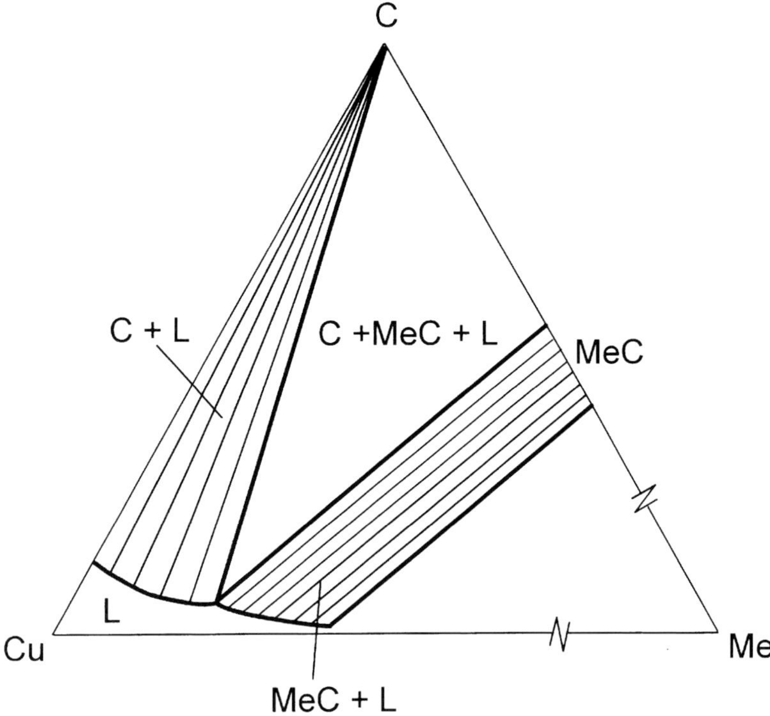

Figure 7.33. Schematic representation of isothermal section of Cu-Me-C phase diagram when the carbide MeC exhibits a range of non-stoichiometry.

For metal-like borides, the selected contact angles given in Table 7.13 assumed by Cu and Ni exemplify weak and high reactivities. Ni dissolves several mole % of many borides and wets in high vacuum or neutral gas environments. Results obtained by the same investigators in different atmospheres show significant effects of furnace atmosphere on the contact angle for the more oxidizable borides ZrB_2 and TiB_2. It may be noted that the results for Fe and Co are very similar to those for Ni.

In the case of relatively unreactive Cu, the dispersion of θ values is even higher, so that for TiB_2 both wetting contact angles and very marked non-wetting ones have been found by different investigators. However, for CrB_2, excellent wetting has been observed, as for Cu/Cr_3C_2 (see Table 7.10). Note that for CrB_2, the furnace atmosphere has less influence on θ than for TiB_2, probably owing to the lower stability of chromium oxide and to the formation of volatile chromium suboxides that promotes deoxidation at high temperatures.

Table 7.13. Wetting by Cu and Ni of different borides.

Boride	Cu			Ni		
	T (°C)	Atm.	θ (deg)	T (°C)	Atm.	θ (deg)
TiB_2	1100	Ar	158 [1]	1480	He	38.5 [1]
TiB_2	1120	vacuum	58 [2]	1500	vacuum	0 [2]
TiB_2	1100	Ar	143 [3]	1480	vacuum	20 [4]
				1480	Ar	72 [4]
ZrB_2	1120	vacuum	142 [2]	1500	vacuum	42 [2]
ZrB_2	1100	Ar	135 [3]	1480	vacuum	65 [4]
ZrB_2				1480	Ar	78 [4]
TaB_2	1100	Ar	77 [1]	1500	Ar	43 [1]
CrB_2	1430	He	50 [1]	1480	He	11 [1]
CrB_2	1100	Ar	26 [3]	1450	vacuum	20 [4]
CrB_2				1450	Ar	21 [4]

[1] (Eremenko and Naidich 1959), [2] (Merz and Kotsch 1963), [3] (Samsonov et al. 1973), [4] (Samsonov et al. 1973b)

Finally, consider the wetting of Al on metallic ceramics. In this case, experimental difficulties in the determination of wetting properties arise not only from substrate oxidizability and impurities but also from oxidizability of Al. As we have seen in Section 6.4.2, in moderate vacuum or in neutral gas, spreading of Al at temperatures lower than 900°C is inhibited by an oxide layer covering its surface and this results in very large apparent contact angles (140–160°) of Al on various carbides, nitrides and borides. Much lower contact angles have been found by Rhee (1970) using a furnace evacuated to 2×10^{-7} mbar and less and sintered, low porosity, TiB_2, TiC and TiN containing 0.2-0.5 wt.% of oxygen as the main impurity. For these ceramics, wetting was observed at temperatures higher than 750°C (Figure 7.34) while no reactivity at the micron scale was found at the Al/ceramic interface. The results of Rhee show an important improvement in wetting by temperature, perfect wetting of Al on TiC, TiB_2 and TiN being predicted at temperatures higher than 1000°C. This has been indeed observed for Al on TiC_{1-y} at temperatures close to 1100°C for y values between 0 and 0.5 (Frumin et al. 1997). However, at these temperatures, wetting is accompanied by a significant reactivity which, depending on y, consists either of some dissolution of TiC_{1-y} into molten Al or of Al_4C_3 formation. As in the case of Cu on TiC, hypostoichiometry of TiC_{1-y} greatly favours wetting of Al, even at temperatures close to the melting temperature of Al (Frumin et al. 1997).

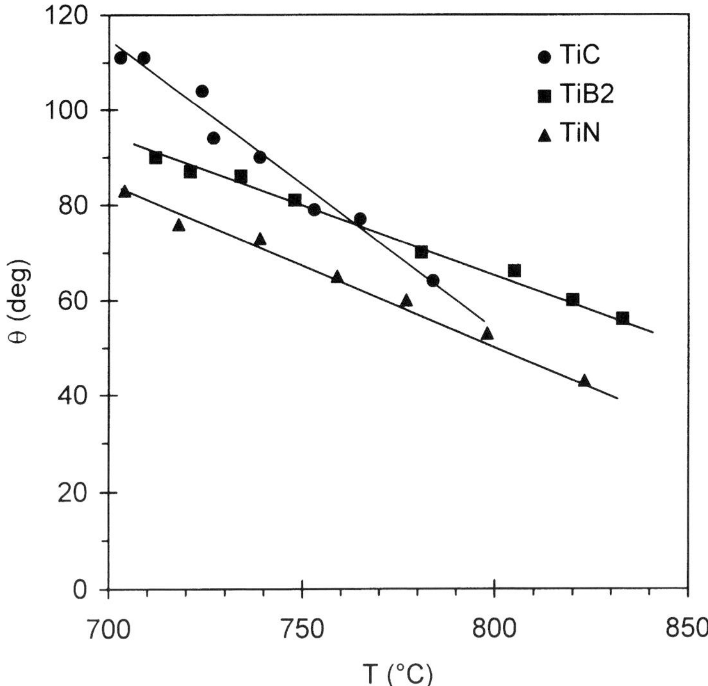

Figure 7.34. Change of stationary contact angle with temperature for times of no more than 30 minutes for Al on TiC, TiB_2 and TiN. From data reported in (Rhee 1970).

To sum up, despite difficulties arising from the lack of monocrystalline high-purity substrates for wetting studies, some experimental facts on the wetting behaviour of non-reactive metals on metal-like carbides emerge that are similar to those of nitrides and borides. Thus the wetting and adhesion of metals on carbides of Mo, W and Cr are good and very similar to those observed in non-reactive metal/metal systems. Compared to Mo_2C, WC and Cr_3C_2, stoichiometric TiC, ZrC and HfC are less wetted by liquid metals but the intrinsic contact angles on these carbides are not well-known because of their high oxidizability. Hypostoichiometry of carbides or nitrides favours wetting and adhesion and this can be controlled by adding Ti to a liquid metal matrix because this element fixes *in situ* hypostoichiometry at the alloy/carbide (or nitride) interface. For reactive couples such as Ni (or Co, Fe)/TiC, the presence of Ti in the alloy caused by dissolution of the substrate explains the excellent wetting observed. At present, there is no satisfactory fundamental understanding of wetting and adhesion in non-reactive metal/metal-like ceramic systems, as discussed by Xiao and Derby (1996). Such an understanding needs more work, both experimental and theoretical. Wetting studies must be combined with a detailed characterisation of surface chemistry and hypostoichiometry. Moreover, calculations of electronic structure and bonding, which exist at present only for bulk compounds, must also be performed for surfaces, taking into account electronic and atomic relaxations.

REFERENCES FOR CHAPTER 7

Allen, B. C. and Kingery, W. D. (1959) *Trans. Metall. Soc. AIME*, **215**, 30

Bakovets, V. V. (1977) *Sov. Phys. Crystallogr.*, **22**, 125

Barsoum, M. W. and Ownby, P. D. (1981) in *Surfaces and Interfaces in Ceramic and Ceramic-Metal Systems*, ed. J. Pask and A. Evans, Plenum Press, New-York, p. 457

Chase, M. W., Davies, C. A., Downey, J. R., Frurip, D. J., McDonald, R. A. and Syverud, A. N. (1985) *JANAF Thermochemical Tables*, ed. D. R. Lide, 3rd ed., J. of Physical and Chemical Reference Data, vol. 14, American Chemical Society and the American Institute of Physics for the National Bureau of Standards

Chatillon, C. and Massies, J. (1990) *Materials Science Forum*, Transtec Publications, Switzerland, **59&60**, 229

Drevet, B. (1987) DEA report, LTPCM, INP Grenoble (France)

Drevet, B., Kalogeropoulou, S. and Eustathopoulos, N. (1993) *Acta Metall. Mater.*, **41**, 3119

Duffy, M. T., Berkman, S., Cullen, G. W., D'Aiello, R. V. and Moss, H. I. (1980) *J. Cryst. Growth*, **50**, 347

Eremenko, V. N. and Naidich, Y. V. (1959) *Russian J. Inorg. Chem.*, **4**, 931

Ferro, A. C. and Derby, B. (1995) *Acta Mater.*, **43**, 3061

Frumin, N., Frage, N., Polak, M. and Dariel, M. (1997) *Scripta Mater.*, **37**, 1263

Fujii, H., Nakae, H. and Okada, K. (1993) *Acta Metall. Mater.*, **41**, 2963

Gasse, A. (1996) Ph.D. Thesis, INP Grenoble, France

Ghetta, V., Gayraud, N. and Eustathopoulos, N. (1992) *Solid State Phenomena*, **25& 26**, 105

Goretski, H., Exner, H. E. and Scheuermann, W. (1971) *Fundamentals of Sintering*, **4**, 327

Hildenbrand, D. L. and Hall, W. F. (1964) in *Condensation and Evaporation of Solids*, ed. E. Rutner, P. Goldfinger and J. P. Hirth, Gordon and Breach, New York, p. 399

Hoekstra, J. and Kohyama, M. (1998) *Phys. Rev. B*, **57**, 2334

Hultgren, R., Desai, P. D., Hawkins, D. T., Gleiser, M. and Kelley, K. K. (1973) in *Selected Values of the Thermodynamic Properties of Binary Alloys*, American Society for Metals

Iwamoto, C. and Tanaka, S. (1998) *Acta Mater.*, **46**, 2381

Jones, L. M. and Nicholas, M. G. (1989) *J. Mater. Sci. Lett.*, **8**, 265

Kalogeropoulou, S., Baud, L. and Eustathopoulos, N. (1995) *Acta Metall. Mater.*, **43**, 907

Kita, Y., Van Zytveld, J. B., Morita, Z. and Iida, T. (1994) *J. Phys. : Condens. Matter*, **6**, 811

Kotsch, H. (1967) *Neue Hütte*, **12**, 350, cited by : Naidich, Y. V. (1981) in *Progress in Surface and Membrane Science*, vol. 14, ed. by D. A. Cadenhead and J. F. Danielli, Academic Press, New York, p. 353

Kubaschewski, O. and Alcock, G. B. (1979) *Metallurgical Thermochemistry*, 5th Ed., Pergamon

Lamy, M., CEA-Grenoble/DRFMC/SP2M, France, private communication

Landry, K. and Eustathopoulos, N., LTPCM, INP Grenoble, France, unpublished work

Laurent, V., Rado, C. and Eustathopoulos, N. (1996) *Mat. Sci. Eng. A*, **205**, 1

Le, D. H. (1990) Ph.D. Thesis, INP Grenoble, France

Le, D. H., Colinet, C. and Pasturel, A. (1991) *Physica B*, **168**, 285

Lequeux, S. (1997) Ph.D. Thesis, INP Grenoble, France

Lequeux, S., Le Guyadec, F., Berardo, M., Coudurier, L. and Eustathopoulos, N. (1998) in *Proc. 2nd Int. Conf. on High Temperature Capillarity*, Cracow (Poland), 29 June-2 July 1997, ed. N. Eustathopoulos and N. Sobczak, published by Foundry Research Institute (Cracow), p. 112

Li, J. G. and Hausner, H. (1992) *J. Europ. Ceram. Soc.* **9**, 101

Li, J. G. (1993) *Mater. Lett.*, **17**, 74

Livey, D. T. and Murray, P. (1956) in *Plansee Proc. Ind. Seminar*, Reutte (Tyrol), 1955, p. 375

Ljungberg, L. and Warren, R. (1989) *Ceram. Eng. Sci. Proc.*, **10**, 1655

Ljungberg, L. (1992) Ph.D. Thesis, Chalmers University of Technology, Göteborg, Sweden

Loehman, R. E. and Tomsia, A. P. (1992) *Acta Metall. Mater.*, **40**, suppl., S 75

Magaud, L. and Pasturel, A. (1998) in *Proc. 2nd Int. Conf. on High Temperature Capillarity*, Cracow (Poland), 29 June-2 July 1997, ed. N. Eustathopoulos and N. Sobczak, published by Foundry Research Institute (Cracow), p. 3

Maguire, H. G. and Augustus, P. D. (1972) *J. Electroch. Soc.*, **119**, 791

Massalski, T. B. (1990) *Binary Alloy Phase Diagrams*, 2nd edition, ASM International

Merz, A. and Kotsch, H. (1963) in *Transactions of a Conference on Powder Metallurgy*, Eisenach Institute of Special Materials, Berlin-Dresden, p. 68

Mortimer, D. A. (1975) AERE-R7941 cited by : Standing, R. and Nicholas, M. (1978) *J. Mater. Sci.*, **13**, 1509

Muehlhoff, L., Choyke, W. J., Bozack, M. J. and Yates, J. T. (1986) *J. Appl. Phys.*, **60**, 2842

Naidich, Y. V. and Nevodnik, G. M. (1969) *Inorganic Materials*, **5**, 1759

Naidich, Y. V., Perevertailo, V. M. and Obushchak, L. P. (1975) *Poroshkovaya Metallurgiya*, **7**, 63 (English translation p. 567)

Naidich, Y. V. (1981) in *Progress in Surface and Membrane Science*, vol. 14, ed. by D. A. Cadenhead and J. F. Danielli, Academic Press, New York, p. 353

Naidich, Y. V., Chuvashov, Y. N., Ishchuk, N. F. and Krasovskii, V. P. (1983) *Poroshkovaya Metallurgiya*, **246**, 67 (English translation : Plenum Publishing Corporation p. 481)

Naidich, Y. V. and Taranets, N. Y. (1995) in *Proc. Int. Conf. High Temperature Capillarity*, Smolenice Castle, May 1994, ed. N. Eustathopoulos (Reproprint, Bratislava) p. 138

Naidich, Y. V. and Taranets, N. Y. (1998) *J. Mat. Sci.*, **33**, 3993

Naka, M., Kubo, M. and Okamoto, I. (1987) *J. Mater. Sci. Lett.*, **6**, 965

Neckel, A. (1990) in *The Physics and Chemistry of Carbides, Nitrides and Borides*, ed. R. Freer, NATO ASI Series, Series E : Applied Sciences, Vol. 35, Kluwer Academic Publishers, p. 485

Nicholas, M. G. (1986) *Br. Ceram. Trans. J.*, **85**, 144

Nicholas, M. G., Mortimer, D. A., Jones, L. M. and Crispin, R. M. (1990) *J. Mater. Sci.*, **25**, 2679

Niessen, A. K., de Boer, F. R., Boom, R., de Châtel, P. F., Mattens, W. C. M. and Miedema, A. R. (1983) *Calphad*, **7**, 51

Nikolopoulos, P., Agathopoulos, S., Angelopoulos, G. N., Naoumidis, A. and Grubmeier, H. (1992) *J. Mater. Sci.*, **27**, 139

Ning, X. S., Suganuma, K., Morita, M. and Okamoto, T. (1987) *Phil. Mag. Lett.*, **55**, 93

Nogi, K. and Ogino, K. (1988) *Transactions of the Japan Institute of Metals*, **29**, 724

Nomura, M., Iwamoto, C. and Tanaka, S. (1998) in *Proc. 2nd Int. Conf. on High Temperature Capillarity*, Cracow (Poland), 29 June-2 July 1997, ed. N. Eustathopoulos and N. Sobczak, published by Foundry Research Institute (Cracow), p. 23

Nomura, M., Iwamoto, C. and Tanaka, S. (1999) *Acta Mater.*, **47**, 407

Ohuchi, F. S. and Kohyama, M. (1991) *J. Am. Ceram. Soc.*, **74**, 1163

Paulasto, M., Ceccone, G., Peteves, S. D., Voitovich, R. and Eustathopoulos, N. (1996) *Ceramic Transactions*, **77**, 91

Pauling, L. (1960) *The Nature of the Chemical Bond*, Cornell University Press, Ithaca, NY

Perevertailo, V. M., Smekhnov, A. A. and Loginova, O. B. (1998) in *Proc. 2nd Int. Conf. on High Temperature Capillarity*, Cracow (Poland), 29 June-2 July 1997, ed. N. Eustathopoulos and N. Sobczak, published by Foundry Research Institute (Cracow), p. 127

Rado, C., Kalogeropoulou, S. and Eustathopoulos, N. (1994) *Mat. Res. Soc. Symp. Proc.*, **327**, 319

Rado, C., Rocabois, P., Kalogeropoulou, S. and Eustathopoulos, N. (1995) in *Proc. Int. Conf. High Temperature Capillarity*, Smolenice Castle, May 1994, ed. N. Eustathopoulos (Reprint, Bratislava) p. 143

Rado, C., (1997) Ph.D. Thesis, INP Grenoble, France

Rado, C., Senillou, C. and Eustathopoulos, N. (1998) in *Proc. 2nd Int. Conf. on High Temperature Capillarity*, Cracow (Poland), 29 June-2 July 1997, ed. N. Eustathopoulos and N. Sobczak, published by Foundry Research Institute (Cracow), p. 53

Rado, C., Kalogeropoulou, S. and Eustathopoulos, N. (1999) *Acta Mater.*, **47**, 461

Raider, S. I., Flitsch, R., Aboaf, J. A. and Pliskin, W. A. (1976) *J. Electroch. Soc.*, **123**, 560

Ramqvist, L. (1965) *Int. J. Powder Metallurgy*, **1**, 2

Rhee, S. K. (1970) *J. Amer. Ceram. Soc.*, **53**, 386

Rhee, S. K. (1970b) *J. Amer. Ceram. Soc.*, **53**, 426

Rhee, S. K. (1970c) *J. Amer. Ceram. Soc.*, **53**, 639

Rocabois, P., Chatillon, C and Bernard, C. (1996) *J. Am. Ceram. Soc.*, **79**, 1351

Rosenqvist, T. (1983) *Principles of Extractive Metallurgy*, 2nd edition, Mc Graw-Hill Book Company, New York

Samsonov, G. V., Panasyuk, A. D. and Borovikova, M. S. (1973) translated from *Poroshkovaya Metallurgiya*, **5**, 61, Consultants Bureau, New York, 1973, p. 403

Samsonov, G. V., Panasyuk, A. D. and Borovikova, M. S. (1973b) translated from *Poroshkovaya Metallurgiya*, **6**, 51, Consultants Bureau, New York, 1973, p. 476

Sangiorgi, R., Muolo, M. L., Chatain, D. and Eustathopoulos, N. (1988) *J. Am. Ceram. Soc.*, **71**, 742

Shimbo, M., Naka, M. and Okamoto, I. (1989) *J. Mater. Sci. Lett.*, **8**, 663

Sobczak, N., Ksiazek, M., Radziwill, W., Morgiel, J., Baliga, W. and Stobierski, L. (1998) in *Proc. 2nd Int. Conf. on High Temperature Capillarity*, Cracow (Poland), 29 June-2 July 1997, ed. N. Eustathopoulos and N. Sobczak, published by Foundry Research Institute (Cracow), p. 138

Tomsia, A. P., Pask, J. A. and Loehman, R. E. (1989) *Ceram. Eng. Sci. Proc.*, **10**, 1631

Trontelj, M. and Kolar, D. (1978) *J. Am. Ceram. Soc.*, **61**, 204

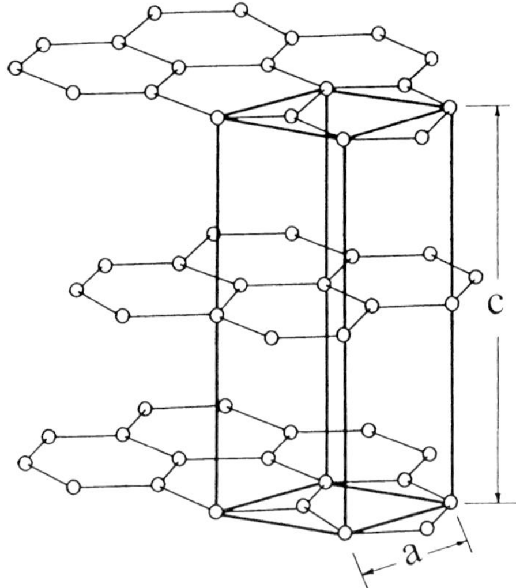

Figure 8.1. Structure of graphite.

A third allotrope, fullerene, has been identified (Kroto et al. 1985) but has not yet been used in high-temperature wetting studies. This Chapter, therefore, is concerned only with the two principal allotropes, diamond and graphite, which have physical properties that reflect their differing lattice structures and are sometimes markedly different as indicated by the typical values assembled in Table 8.2. Thus diamond has the highest known thermal conductivity and hardness, but graphite has a much higher electrical conductivity. The hardness, optical and thermal characteristics of diamonds have resulted in applications as cutting agents, as jewelry and as electronic heat sinks. The applications of graphite are very varied and depend on both inherent material characteristics as well as the effects of temperature and other processing parameters on microstructural features such as porosity and grain orientation. There are three main types of graphite of technical importance: *Bulk polycrystalline graphite* which has a coarse grain structure and contains several percent of gross open porosity. *Pyrolytic graphite* which is a dense and very anisotropic material formed by cracking hydrocarbons at 1400–3000°C. *Vitreous carbon* which has a very fine structure described as a tangle of graphitic ribbons only a few unit cells thick (Jenkins and Kawamura 1971) whose low density of 1.5 Mg/m^3 as against 2.26 Mg/m^3 for fully dense graphite is due to the presence of closed porosity at the nanometer scale. Because of its low density, vitreous carbon has a low dispersive (van der Waals) component of surface energy (32 mJ/m^2) compared to that of monocrystalline graphite (150 mJ/m^2, Section 4.2.3).

Table 8.2. Some physical properties of graphite and diamond.

Property	Graphite	Diamond
melting temperature ($^\circ$C)	3700	3550
density (Mg/m^3)	2.26	3.53
coefficient of thermal expansion (K^{-1})	[*]	0.8×10^{-6}
thermal conductivity at R.T. (W m^{-1} K^{-1})	100	1000–3000 [**]
electrical conductivity (Ω^{-1} m^{-1})	2.7×10^{6} [***]	10^{-15}
bulk modulus (GPa)	33	1100
hardness (Mohs scale)	1.5	10

[*] 10^{-6} K^{-1} parallel and 5×10^{-6} K^{-1} perpendicular to basal planes of oriented material. Greater anisotropy is displayed by single crystals.
[**] The precise value depends on the concentration of the ^{13}C isotope.
[***] in basal planes

Obviously, differences in bulk structure of the different types of graphite have repercussions on surface structure and properties. Thus effects on wetting behaviour can be expected but in studies with liquid metals, information given about carbon microstructure is usually insufficient. Further problems arise from the difficulty of preparing surfaces with a low roughness. This is particularly true for polycrystalline graphite, the form most used in published work, for which the average roughness R_a can be several hundreds of nanometers or even more even after careful polishing. This is due to the existence of some open porosity and especially to the weak cohesion between graphitic planes bonded only by van der Waals interactions, leading to detachment of particles by cleavage during polishing. Consequently, the resulting surface is rough and consists of microfacets of graphitic planes. The high R_a values can greatly affect equilibrium contact angles, especially in non-wetting systems for which deviations from Young contact angle as high as tens of degrees can be expected (see Section 3.3). Therefore, published data for this kind of system reveal only gross trends in contact angle and work of adhesion values.

In a limited number of studies, vitreous carbon (C_v) substrates have been used whose surfaces can be polished to an average roughness of less than 10 nm. On this type of surfaces, it can be expected that the departure of the equilibrium contact angle from the Young contact angle is only a few degrees. However, it must be emphasized that wetting properties obtained for both non-reactive and reactive metals on C_v may differ significantly from intrinsic properties on graphite because the density of these substrates is very different and also because in C_v there is no preferred orientation of graphitic hexagons with regard to the surface.

In the following, we shall consider successively non-reactive pure metals, i.e., metals in which the maximum solubility of carbon is lower than 1 ppm (such as Cu or Au), then metals that can form carbides (such as Si or Ti) and finally metals dissolving large amounts of carbon without forming stable carbides (such as Ni).

8.1. NON-REACTIVE SYSTEMS

These systems consist of metals which do not form carbides and do not dissolve carbon significantly: for instance, Cu dissolves only 1 ppm of carbon at its melting point (Oden and Gokcen 1992). The results in Table 8.3, obtained on vitreous carbon in high vacuum at a temperature close to the melting point of the metal, show that non-reactive metals do not wet carbon. As for non-reactive metal/oxide couples, contact angle values change only slightly with temperature (of the order of 1 deg for 100K), as shown in Figure 8.2.

Despite the difficulties in obtaining well-defined contact angles on polished surfaces of polycrystalline graphite mentioned in the introduction, the results of the systematic study of Naidich and Kolesnichenko (1965) provide clear evidence that, as for vitreous carbon, contact angles on polycrystalline graphite are much higher than 90° whatever metal was used. Then, values lying between 135° and 150° were found for metals such as Cu, Ag, In, Ge and Sn in high vacuum between 1100°C and 800°C. This non-wetting behaviour, associated with the absence of reactivity, is well-known to metallurgists and is exploited in the use of graphite crucibles.

Table 8.3. Contact angles of non-reactive metals on vitreous carbon in high vacuum at a temperature close to the melting point of the metal.

Metal	θ (deg)	Reference
Ga	138	(Naidich and Chuvashov 1983)
In	145	(Chizhik et al. 1985)
Sn	151	(Chizhik et al. 1985)
Bi	142	(Chizhik et al. 1985)
Pb	142	(Chizhik et al. 1985)
Au	138	(Chizhik et al. 1985)
	135	(Dezellus and Eustathopoulos to be published)
Cu	139	(Dezellus and Eustathopoulos to be published)
	145	(Standing and Nicholas 1978)
Ni-Pd [*]	137	(Grigorenko et al. 1998)

[*] composition with congruent fusion

Large non-wetting contact angles have also been found on diamond (111) faces (Naidich 1981). However, the different faces of diamond exhibit structural modifications with temperature which can lead at high temperatures to surface graphitization and changes of several tens of degrees in contact angle values (Figure 8.3). Despite these transformations, the contact angles remain in all cases higher than 90°.

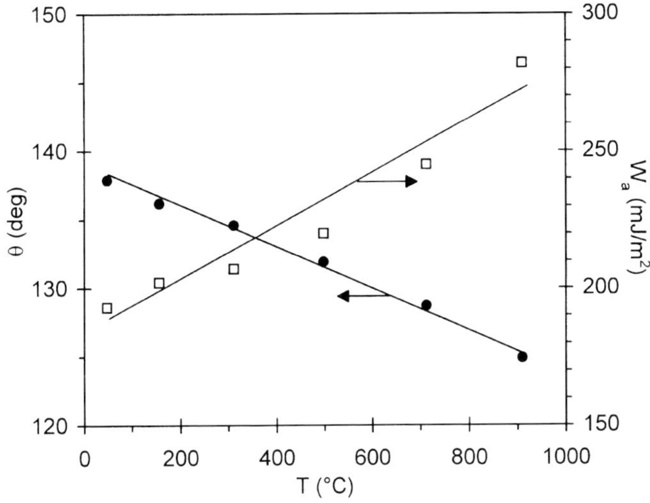

Figure 8.2. Contact angle and work of adhesion of Ga on vitreous carbon versus temperature. A variation of 900K leads to a contact angle change of only ten degrees. From data reported by Naidich and Chuvashov (1983).

The values of the work of adhesion corresponding to the contact angles of Table 8.3 are of the order of 100 mJ/m^2 and are attributed to van der Waals interactions resulting from dispersion forces (Naidich 1981). According to the equation given in Figure 6.8.b for the van der Waals interactions between two condensed phases, the interaction energy is proportional to the atomic densities of the two phases in contact. The higher the atomic density of the carbon substrate, the larger is the dispersive (van der Waals) contribution to interaction energy, resulting in a higher adhesion energy and lower contact angle. Therefore, if the predominant interactions at non-reactive metal/carbon interfaces are van der Waals interactions, different values of contact angle would be expected for the same metal on *vitreous carbon* (C$_v$) and *monocrystalline graphite* (C$_{mg}$) which have very different atomic densities

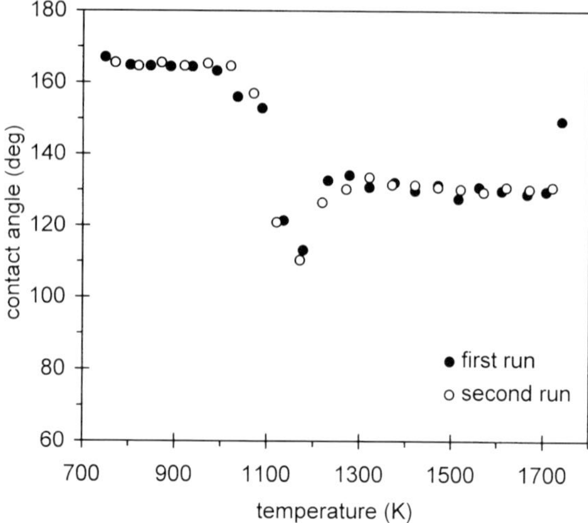

Figure 8.3. Contact angles of liquid Sn on (100) plane of diamond versus temperature in an Ar-10% H_2 atmosphere. Data from work reported in (Nogi et al. 1998) [7].

(Table 8.4). In contrast to polycrystalline graphite, these types of carbon allow very smooth surfaces to be obtained with an average roughness of a few nanometers either by polishing for C_v or by cleavage for C_{mg}. Such surfaces of C_{mg} are basal planes while no specific orientation is expected for polished C_v surfaces. For such substrates, the measured contact angles are expected to be equal to the Young contact angle to within 2–3°. The results in Table 8.4 for Cu and Au, as well as the initial contact angle measured for Al before it reacts (see Figure 8.7), confirm predictions based on van der Waals interactions that the contact angle decreases with increasing density of C substrates. Note that if adhesion in this kind of system resulted from chemical interactions, the basal planes of high atomic density present on the surfaces of graphite single crystals would be less wetted than the low atomic density surfaces of vitreous carbon (see Section 5.1).

Further evidence of the effects of van der Waals interactions on metal/carbon bonding has been offered by Nogi et al. (1995) who used an atomic force microscope to measure the force resulting from van der Waals interactions between three differing faces of diamond and the probe and found these scaled well with experimental values of the work of adhesion for the interfaces formed by Bi, Pb or Sn.

It is interesting to note that Heyraud and Metois found that small, micron-size, solid particles of Au (Heyraud and Metois 1980) and Pb (Heyraud and Metois 1983) assumed equilibrium shapes at high temperatures on monocrystalline graphite, with their (111) faces parallel to the graphite surface. The (111) plane is the most densely packed plane of a fcc metal so this orientation maximises the number of metal atoms per unit area in nearest-neighbour interactions with C atoms and this is consistent with the predominance of van der Waals interactions at the interface. In this study (Heyraud and Metois 1980), the contact angle of nearly spherical particles of Au on graphite was found to be 127°, a value close to that of liquid Au (Table 8.4). This concordance is expected because the changes of density and surface energy of a pure metal at melting are small and have opposing effects on θ.

Table 8.4. Contact angle and work of adhesion of different metals on vitreous carbon (C_v) and pseudomonocrystalline graphite (C_{mg}). From (Dezellus and Eustathopoulos to be published) except for Al (Landry et al. 1998).

Substrate	Cu at 1373K		Au at 1373K		Al at 1100K	
	θ (deg)	W_a (mJ/m^2)	θ (deg)	W_a (mJ/m^2)	$\theta^{(*)}$ (deg)	W_a (mJ/m^2)
C_v ($\rho = 1.5\,\mathrm{Mg/m^3}$)	139	319	135	338	139	206
C_{mg} ($\rho = 2.26\,\mathrm{Mg/m^3}$)	122	611	119	595	123	383

$^{(*)}$ initial contact angle before reaction

Effects of non-reactive alloying elements on contact angle and work of adhesion values can be discussed in terms of the thermodynamic model initially developed for non-reactive metal/oxide systems (see Section 6.5.1). This model allows prediction of the shape of θ and W_a isotherms for binary alloys from the values of σ_{LV} and W_a of pure metals and the regular solution parameter λ defined in equation (4.3). As in metal/oxide systems, contact angles are very large and vary little between the different metals (Table 8.4) so the ratio W_a/σ_{LV} is approximately constant. It can be readily shown that binary alloys developing van der Waals interactions with carbon produce type (b) isotherms of Figure 6.29. For such systems, addition of B to A when $\sigma_{LV}^B \ll \sigma_{LV}^A$ will decrease the work of adhesion of A and increase or decrease the contact angle depending on whether the contact angle of pure A, θ_A, is higher or lower than a critical value θ^* given by equation (6.35). An example is Cu-Sn on vitreous carbon at 1150°C for which, according to results of Standing and Nicholas (1978), $W_a^{Cu} = 245$ mJ/m^2 and $W_a^{Sn} = 160$

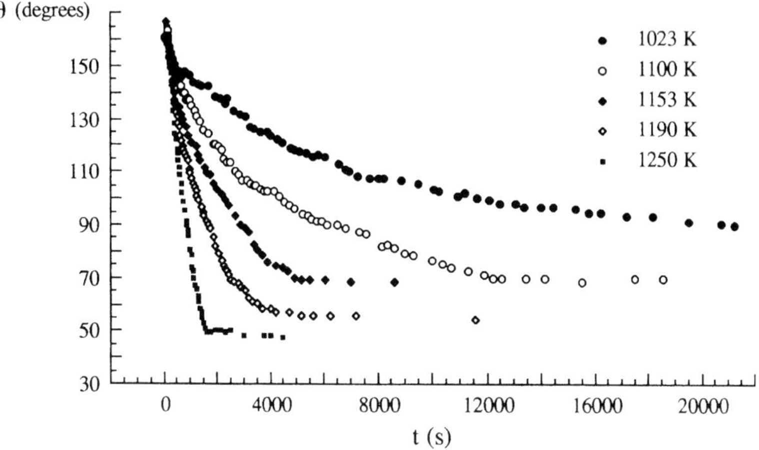

Figure 8.5. Contact angle kinetics in the Al/vitreous carbon system at different temperatures (Landry et al. 1998) [17].

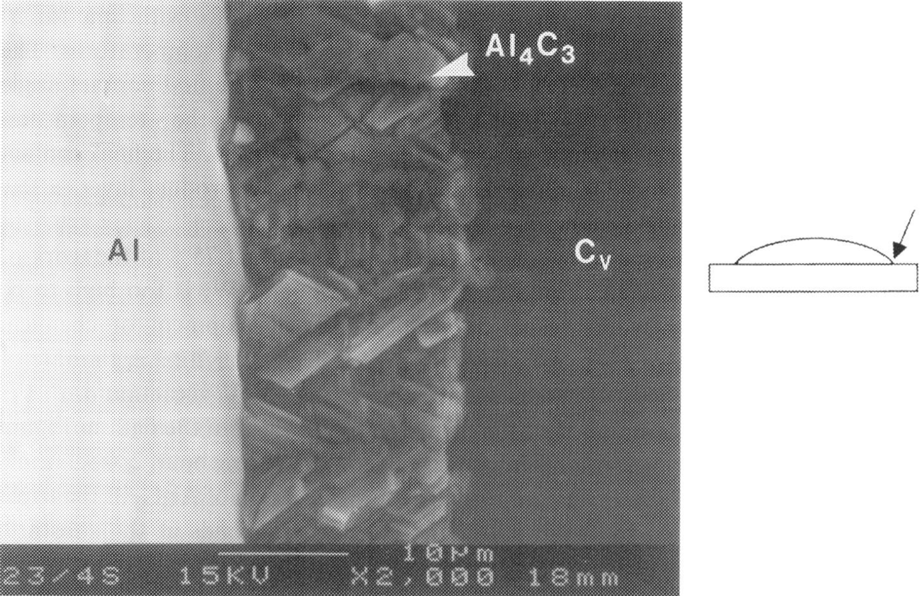

Figure 8.6. SEM micrograph taken from above of an Al/vitreous carbon specimen cooled from 1100K showing an Al_4C_3 layer close to the triple line (Landry and Eustathopoulos 1996b).

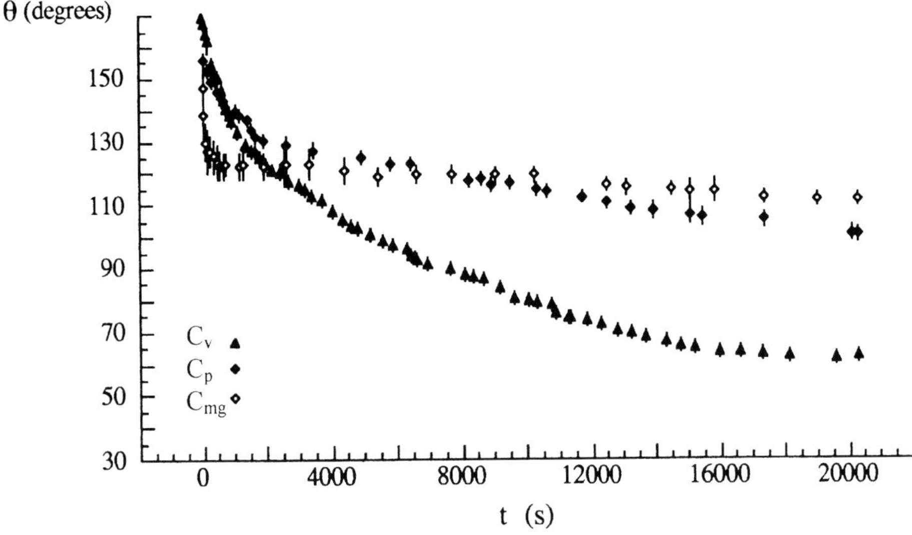

Figure 8.7. Contact angle kinetics of Al on vitreous carbon (C_v), pyrocarbon (C_p) and monocrystalline graphite (C_{mg}) at 1100K (Landry et al. 1998) [17]. For C_p and C_{mg}, no stationary contact angle was obtained after 2×10^4 seconds.

reaction and spreading kinetics on vitreous carbon are much faster than those on monocrystalline graphite and pyrocarbon. Note that in the case of C_{mg}, θ remains nearly constant for about 10^4 seconds after the end of the first stage (deoxidation of Al), reflecting the very low reactivity between molten Al and basal planes of graphite.

On polycrystalline graphite, a material of interest for many applications, several tens of minutes at 1000°C are needed to obtain a contact angle lower than 90° (Mori et al. 1983, Nogi et al. 1988, Weirauch et al. 1995, Sobczak et al. 1996) due to the high activation energy of the wetting process. Extremely slow kinetics are observed at temperatures close to the melting point of Al, sessile drop experiments performed with deoxidized Al droplets on polycrystalline graphite showing no contact angle variations for several hours at 700°C (Mori et al. 1983, Nogi et al. 1988, Sobczak et al. 1996). Therefore, in this range of temperature, Al behaves as a non-reactive and non-wetting metal interacting with graphite by weak van der Waals dispersion forces.

Liquid Si dissolves little C at its melting point (the molar fraction of C in Si at saturation is less than 10^{-3}) but reacts with C to form predominantly covalent SiC. This product is well wetted by molten Si, the contact angle of Si on (0001) basal planes of α-SiC being 30–40° (see Section 7.1.1). The wetting kinetics of Si on

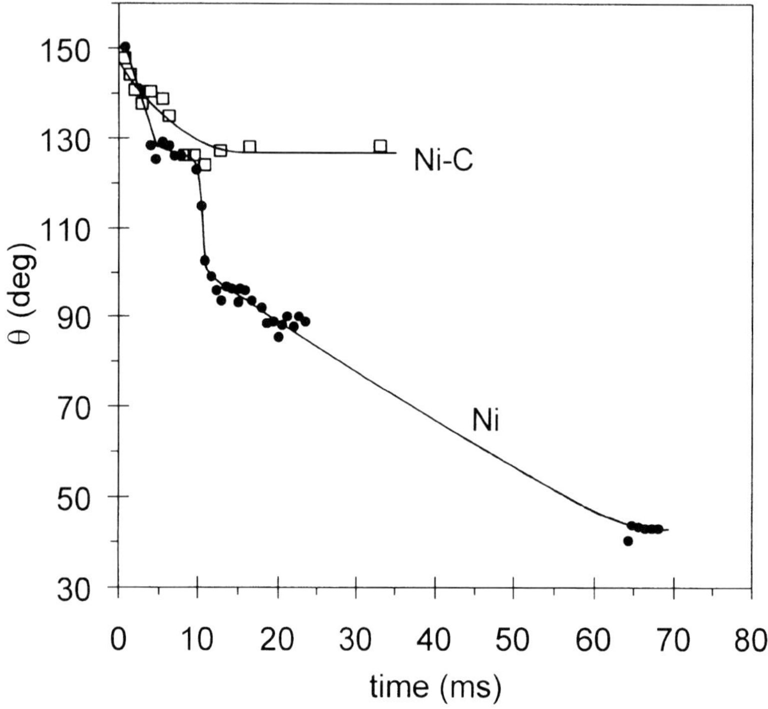

Figure 8.8. Spreading kinetics for C-presaturated Ni (squares) and pure Ni (full circles) on polycrystalline graphite at 1460°C. From data reported in (Naidich et al. 1972).

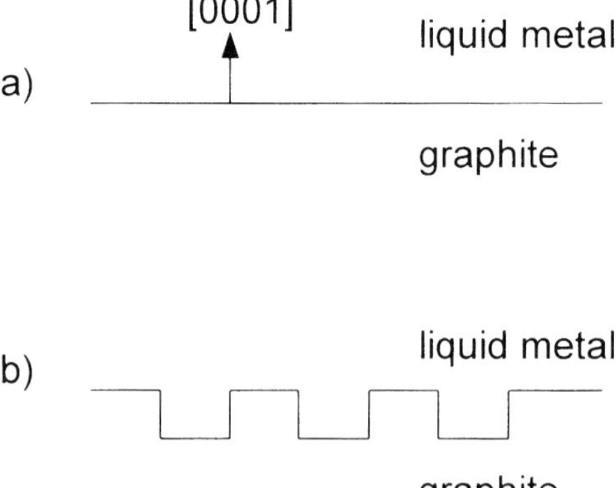

Figure 8.9. A possible effect of dissolution of graphite on wetting is the change of the crystallographic nature of the metal/graphite interface. a) Initial interface, b) interface during dissolution.

To sum up, wetting by ferrous metals can be described in terms of a contact angle measured on a nearly planar interface that is achievable only when using C-saturated metals. These contact angles are high but not as high as those observed for non-reactive metals. The formation of the interface for initially pure ferrous metals is a complex process due to both a slight lateral extension of the liquid on the substrate and deep attack of the substrate. Consequently, published contact angles of 50°–70° given for ferrous metals on graphite are only apparent values which are not governed by the Young equation.

8.2.3 Reactive alloying elements

Attempts to improve wetting by non-reactive pure metals such as Cu by the addition of ferrous metals such as Ni or Fe have been unsuccessful (Naidich 1981). In contrast, improvements in wetting have been achieved by adding carbide-forming elements such as Cr or Ti. Additions of Cr to Cu above a critical value of the molar fraction of Cr, X_{Cr}^*, produce a sharp wetting transition (Figure 8.10) owing to the formation of a continuous layer of wettable Cr carbides in accord with the reaction:

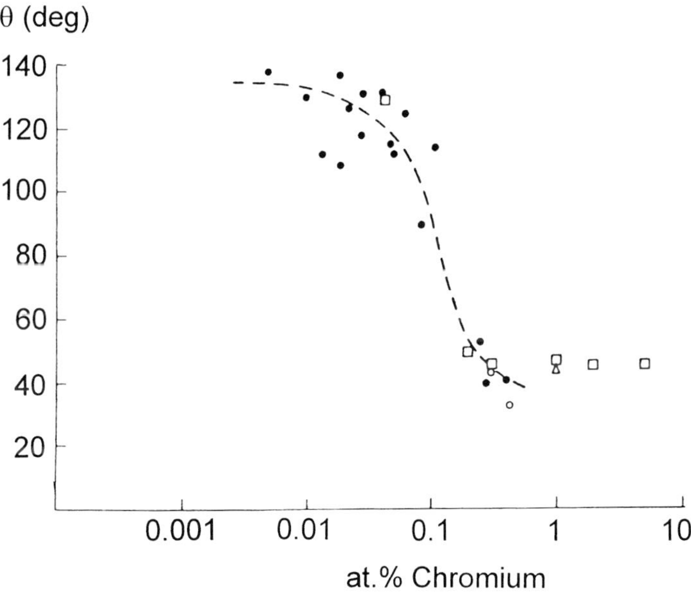

Figure 8.10. Contact angles of Cu-Cr alloys on C at 1150°C. Full circles: diamond (Scott et al. 1975). Open circles: diamond (Naidich and Kolesnichenko 1964). Squares: vitreous carbon (Mortimer and Nicholas 1973). Triangle: graphite (Mortimer and Nicholas 1973) [2].

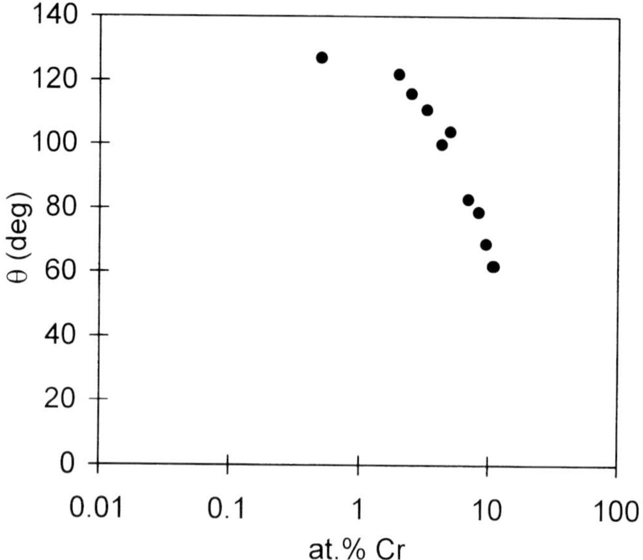

Figure 8.11. Contact angles of Ga-Cr alloys on graphite at 900°C. From data reported in (Naidich and Chuvashov 1983).

$$3(Cr)_M + 2\langle C\rangle \Leftrightarrow \langle Cr_3C_2\rangle \tag{8.1}$$

when M is Cu. The critical value X_{Cr}^* depends not only on the molar Gibbs energy of carbide formation, ΔG_f^0, but also on the activity coefficient of Cr in the matrix M, γ_{Cr}:

$$X_{Cr}^* = \frac{1}{\gamma_{Cr}}\exp\frac{\Delta G_f^0}{3RT} \tag{8.2}$$

As Cu-Cr interactions in the liquid alloy are very weak, γ_{Cr} is much higher than unity and for dilute alloys is about 50 so values of X_{Cr}^* are small, as indicated by the wetting transition in Figure 8.10. In matrices such as Ga (Naidich and Chuvashov 1983) or Ni-Pd (Kritsalis et al. 1991), much lower values of γ_{Cr} cause the wetting transition to occur at much higher values of X_{Cr}^* (see Figure 8.11 for Ga-Cr alloys). It is interesting to note that contact angles lower than those obtained for Cu-Cr and Ga-Cr binary alloys have been observed with ternary

Cu-Ga-Cr alloys on graphite (Figure 8.12). Adding Cu to Ga while keeping X_{Cr} constant increases the activity of Cr and above a critical value causes formation of Cr carbide to induce wetting. On the other hand, adding Ga to Cu causes the surface energy of the liquid to diminish and the contact angle decreases in accord with the Young-Dupré equation (for instance, the addition of 20 at.% Ga to Cu decreases the surface energy of Cu by 30%). As a result, the contact angle isotherm of Ga-Cu-Cr alloys with constant X_{Cr} passes through a minimum (Figure 8.12).

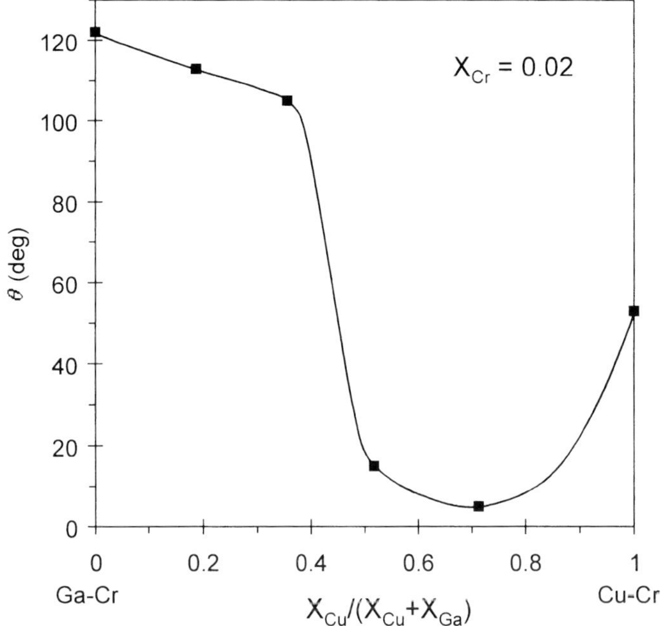

Figure 8.12. Variation of contact angle for Ga-Cu-Cr alloys ($X_{Cr} = 0.02$) on graphite at 1173K (except for Cu-Cr: 1423K). From data reported in (Naidich and Chuvashov 1983).

Ti additions to Cu and also to Sn, Ag (Naidich and Kolesnichenko 1968), Ga (Naidich and Chuvashov 1983) or Ni-Pd (Kritsalis et al. 1991) also produce a non-wetting/wetting transition due to the formation of a wettable reaction product (TiC). Note that although there is some uncertainty about the wettability of TiC by pure non-reactive metals, it is clearly established that TiC is wetted by these metals when they contain even low concentrations of Ti leading to the formation of hypostoichiometric TiC at the interface (see Section 7.2, Figure 7.32). An example

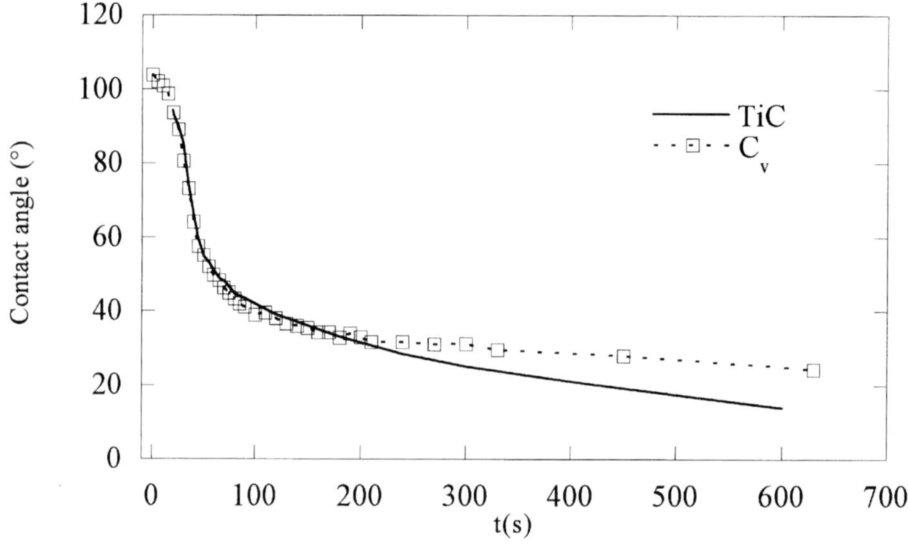

Figure 8.13. Variation of contact angle with time for CuAg-2.7 at.% Ti alloys on vitreous carbon and TiC at 830°C in high vacuum (Ljunberg et al. 1999).

is given in Figure 8.13 for CuAg-Ti alloys on vitreous carbon. Similar results have been obtained on graphite and diamond by alloying non-reactive metals such as Sn, Cu, Au or Ag with Ti, V, Nb or Ta which also form metal-like carbides (Naidich 1981).

Improvements in wetting by non-reactive metals can be produced also by adding Al or Si which form covalent carbides, as shown for Cu-Si alloys on vitreous carbon in Figure 8.14. In the Cu-Si/C system at 1100°C, formation of SiC occurs for $X_{Si} > X_{Si}^J \cong 0.15$ (see equation 7.4). This relatively high value of X_{Si}^J is due to the low value of $\Delta G_{f(SiC)}^0$, compared to $\Delta G_{f(TiC)}^0$ for instance, and to the small value of γ_{Si} in Cu, compared to γ_{Cr} in Cu for instance. For all Cu-Si alloys, the initial contact angle is that of a nearly unreacted alloy on C which is much higher than 90° (Figure 8.14). The kinetics of spreading are controlled by the formation of SiC at the alloy/C/vapour triple line, as for pure Si on C. However, because both the experimental temperature and the Si activity are much lower than for pure Si, the kinetics are 2 to 3 orders of magnitude slower than the 1 minute for pure Si (Section 8.2.1).

Several attempts have been made to improve wetting of Al on different kinds of C substrates by use of alloying elements, but no promising results have been obtained (Manning and Gurganus 1969, Nicholas and Mortimer 1971,

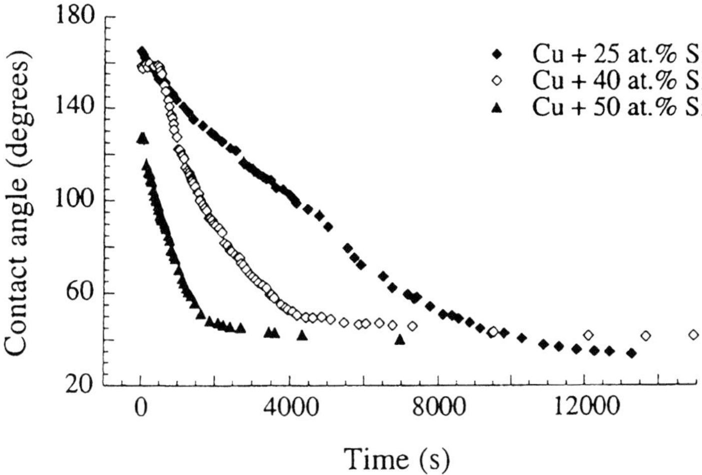

Figure 8.14. Contact angle kinetics for Cu-Si alloys on vitreous carbon at 1423K (Landry et al. 1996) [26].

Eustathopoulos et al. 1974). In certain cases, limited improvement in wetting has been observed due to an effect of the alloying element on the stability of the oxide layer covering Al rather than to an effect on intrinsic interfacial energies of the system. A typical example of this is provided by Mg additions (Eustathopoulos et al. 1974).

Landry et al. (1998) attempted to improve wetting of vitreous carbon by deoxidized Al by adding 0.8 at.% Ti. At the experimental temperature of 1223K, this concentration is close to the maximum solubility of Ti in liquid Al but neither the spreading kinetics nor the final contact angle were improved (Figure 8.15). Al_4C_3 was observed on the carbon side of the interface and TiC on the liquid side at the centre of the drop. However, only Al_4C_3 was observed in the vicinity of the triple line (Figure 8.15), due to the consumption of Ti to form a micron-thick layer of TiC in the central region during the initial spreading. For this reason, TiC formation does not affect the final contact angle of the system. Sobczak et al. (1996) avoided Ti exhaustion by using Al-Ti alloys supersaturated in Ti and observed contact angles as low as 20°.

8.3. CONCLUDING REMARKS

Non-reactive systems of graphite and Ag, Au, Cu, Pb, Sn, etc. are non-wetting with contact angles of about 120° due to the action of only van der Waals dispersion forces between the metal and C atoms. As for non-reactive metal/ionocovalent oxide systems, contact angles and work of adhesion of metals on graphite change only slightly with temperature. An improvement in wetting is obtained by alloying with elements that form wettable metal-like carbides, such as Cr, V and Ti, or covalent carbides, such as Si and Al. In this case, spreading kinetics depend critically on temperature.

The ferrous metals Fe, Ni, Co, etc. dissolve graphite and the resultant wetting behaviour is very complex and not yet understood. More work is needed to quantify wetting in terms of true, Young, contact angles and to identify the effect on wetting of dissolved C and the crystallographic orientations of the graphite surface.

Pure Al and Si react with graphite substrates to form wettable Al and Si carbides. However, the reaction kinetics are so slow at temperatures close to the melting point of Al that Al behaves as a non-reactive and non-wetting metal.

These conclusions are also valid for vitreous carbon and diamond although some differences exist in the final contact angle for non-reactive metals and the spreading kinetics for reactive metals.

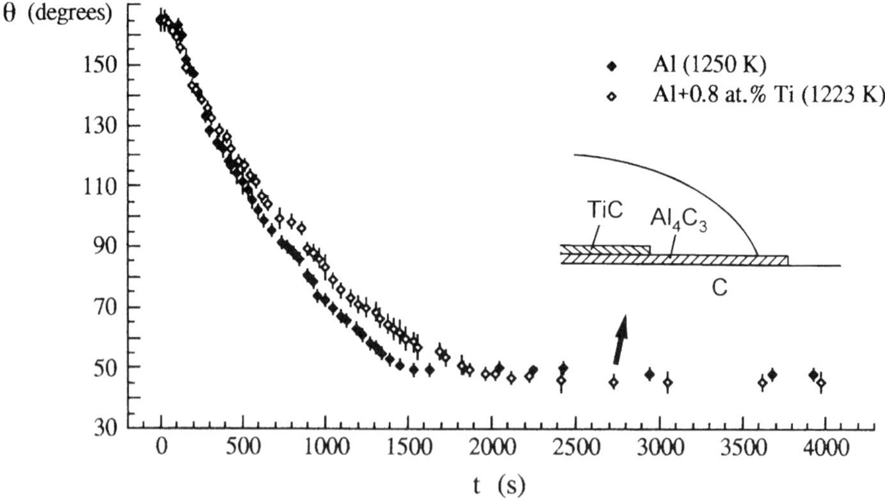

Figure 8.15. Contact angle kinetics for Al and Al-0.8 at.% Ti on vitreous carbon at 1250K and 1223K respectively (Landry et al. 1998) [17].

REFERENCES FOR CHAPTER 8

Barbangelo, A. and Sangiorgi, R. (1992) *Mater. Sci. Eng. A*, **156**, 217

Chizhik, S. P., Gladkikh, N. T., Larin, V. I., Grigor'eva, L. K., Dukarov, S. V. and Stepanova, S. V. (1985) *Poverkhnost*, **12**, 111 (in Russian)

Dezellus, O. and Eustathopoulos, N., to be published

Drevet, B., Kalogeropoulou, S. and Eustathopoulos, N. (1993) *Acta Metall. Mater.*, **41**, 3119

Eustathopoulos, N., Joud, J. C., Desre, P. and Hicter, J. M. (1974) *J. Mat. Sci.*, **9**, 1233

Ferro, A. C. and Derby, B. (1995) *Acta Metall. Mater.*, **43**, 3061

Gangopadhyay, U. and Wynblatt, P. (1994) *Metall. Mater. Trans. A*, **25**, 607

Grigorenko, N., Poluyanskaya, V., Eustathopoulos, N. and Naidich, Y. V. (1998) in *Proc. 2nd Int. Conf. on High Temperature Capillarity*, Cracow (Poland), 29 June-2 July 1997, ed. N. Eustathopoulos and N. Sobczak, published by Foundry Research Institute (Cracow), p. 27

Hara, S., Nogi, K. and Ogino, K. (1995) in *Proc. Int. Conf. High Temperature Capillarity*, Smolenice Castle, May 1994, ed. N. Eustathopoulos (Reproprint, Bratislava), p. 43

Heyraud, J. C. and Metois, J. J. (1980) *Acta Met.*, **28**, 1789

Heyraud, J. C. and Metois, J. J. (1983) *Surface Science*, **128**, 334

Hill, C. C. and Holman, J. S. (1981) *Chemistry in Context*, Nelson, Walton-on-Thames, UK

Hoekstra, J. and Kohyama, M. (1998) *Phys. Rev. B*, **57**, 2334

Jarfors, A. E. W., Svendsen, L., Wallinder, M. and Fredriksson, H. (1993) *Metall. Trans. A*, **24**, 2577

Jeffrey, G. A. and Wu, V. Y. (1966) *Acta Cryst.*, **20**, 538

Jenkins, G. M. and Kawamura, K. (1971) *Nature*, **231**, 175

Kritsalis, P., Coudurier, L., Parayre, C. and Eustathopoulos, N. (1991) *J. Less-Common Metals*, **175**, 13

Kroto, H. W., Heath, J. R., O'Brien, S. C., Curl, R. F. and Smalley, R. E. (1985) *Nature*, **318**, 162

Landry, K. (1995) Ph.D. Thesis, INP Grenoble, France

Landry, K., Rado, C. and Eustathopoulos, N. (1996) *Metall. Mater. Trans.*, **27A**, 3181

Landry, K. and Eustathopoulos, N. (1996b) *Acta Mater.*, **44**, 3923

Landry, K., Kalogeropoulou, S. and Eustathopoulos, N. (1998) *Mat. Sci. Eng. A*, **254**, 99,

Li, S., Arsenault, J. and Jena, P. (1988) *J. Appl. Phys.*, **64**, 6246

Li, J. G., Coudurier, L. and Eustathopoulos, N. (1989) *J. Mater. Sci.*, **24**, 1109

Ljunberg, L., Dezellus, O., Jeymond, M. and Eustathopoulos, N. (1999) in *Proceedings of 9th CIMTEC*, vol. "Ceramics : Getting into the 2000's - part C", Florence (Italy), 14-19 June 1998, in press

Manning, C. R. and Gurganus, T. B. (1969) *J. Am. Ceram. Soc.*, **52**, 115

Mori, N., Sorano, H., Kitahara, A., Ogi, K. and Matsuda, K. (1983) *J. Japan Inst. Metals*, **47**, 1132

Mortimer, D. A. and Nicholas, M. (1973) *J. Mater. Sci.*, **8**, 640

Naidich, Y. V. and Kolesnichenko, G. A. (1964) *Poroshkovaya Metallurgiya*, **3**, 23 (English translation p. 191)

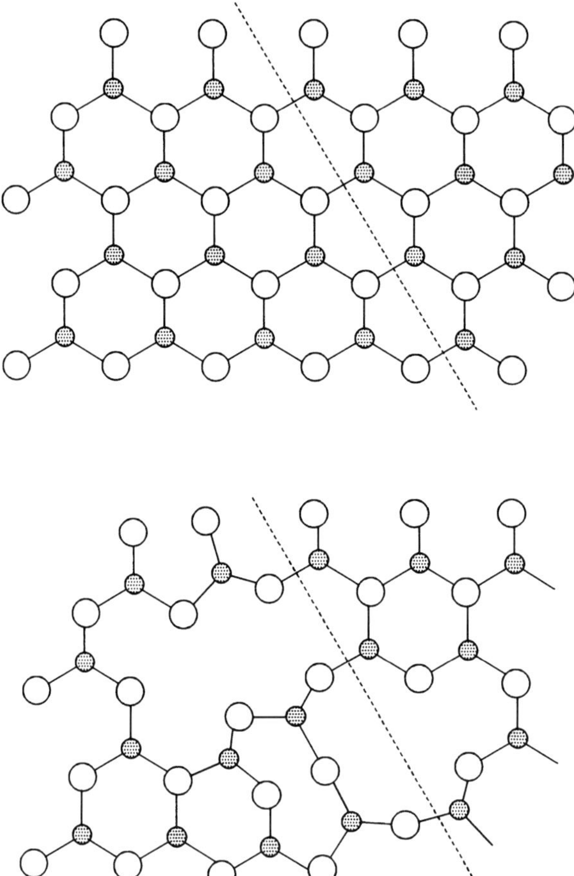

Figure 9.1. The two-dimensional structures of a hypothetical compound J_2O_3 as, top, a crystalline solid and, bottom, a glass. The J^{3+} ions are shown as small shaded circles and the O^{2-} ions as open circles. The dashed lines represent the traces of surface planes. Because of the numerous defects contained in the glass, the surface energy change corresponding to the reversible creation of a new surface is lower than that of the corresponding crystalline solid.

of surface creation. For these reasons, the surface energy cannot be related in a simple way to pair energies in the bulk.

The differences between the values of viscosity, η, and liquid surface energy, σ_{LV}, of metals and glasses are of importance because the ratio σ_{LV}/η is of major significance in determining spreading rates (see equation (2.4)) and flow rates in capillaries (see equation (10.1)). For glassy materials, the ratio is less, and sometimes much less, than 0.1 m.s^{-1}, as opposed to several hundred m.s^{-1} for liquid metals and alloys.

Table 9.1. Properties of some glassy silicates, SiO_2 (Turkdogan 1983) and Cu. The values of density, ρ, viscosity, η and liquid surface energy, σ_{LV}, are for T_F.

Material	T_F (K)	T_G (K)	ρ (Mg.m^{-3})	η (Pa.s)	σ_{LV} (J.m^{-2})
$Na_2O\text{-}3SiO_2$	1043	775	2.14	7.8	0.28
$Na_2O\text{-}2SiO_2$	1147	725	2.25	2.8	
$K_2O\text{-}2SiO_2$	1318	763	2.20	8.0	0.22
$Na_2O\text{-}Al_2O_3\text{-}3SiO_2$	1391	1088	2.28	$>10^3$	
$CaO\text{-}Al_2O_3\text{-}2SiO_2$	1826	1086	2.45	4.0	0.40
SiO_2	1993	1373	2.08	$10^{5.6}$	0.31
Cu	1356	NA	8.0	4.3×10^{-3}	1.35

9.2. WETTING BEHAVIOUR

As oxide materials, glasses can be expected to, and do, wet *oxide components* exceedingly well, Figure 9.2. Further, the growth of oxide films on metal substrates has been used to promote wetting, (Nicholas 1986a, Tomsia and Pask 1990). The basic reason for this behaviour is the similarity of the binding forces of the glass and the solid oxide and hence the high degree of electronic continuity across the solid/liquid interface. This reason explains why ionic non-oxide ceramics can be well wetted by glasses, as shown by the $20°$ contact angle of a lead zinc borate glass on MgF_2 at $600°C$, (Freeman et al. 1976).

The excellent wetting of oxides by glasses is exploited in the bonding of Si_3N_4 and other high temperature structural ceramics which, as discussed in Section 7.1.2, are usually covered by oxide, oxynitride or oxycarbide layers. The greatest success has been achieved with glass compositions that resemble those naturally present as intergranular phases, (Peteves et al. 1996). Particularly effective wetting has resulted from the use of compositions based on mixtures of the sintering additives Al_2O_3, Y_2O_3 and MgO and of SiO_2 which is present as an impurity in Si_3N_4, (Mecartney et al. 1985). Other compositions found to be successful with Si_3N_4 include reactive mixtures that dissolve some of the ceramic or form Si_2N_2O at the glaze/ceramic interface as a bonding phase. Thus compositions used to join Si_3N_4 include $ZrO_2\text{-}ZrSiO_4$, $MgO\text{-}SiO_2\text{-}Al_2O_3$, $Y_2O_3\text{-}Al_2O_3\text{-}SiO_2$, $CaO\text{-}SiO_2\text{-}TiO_2$

and SiO_2-MgO-CaO, (Becher and Halen 1985, Iwamoto et al. 1985, Peteves et al. 1996).

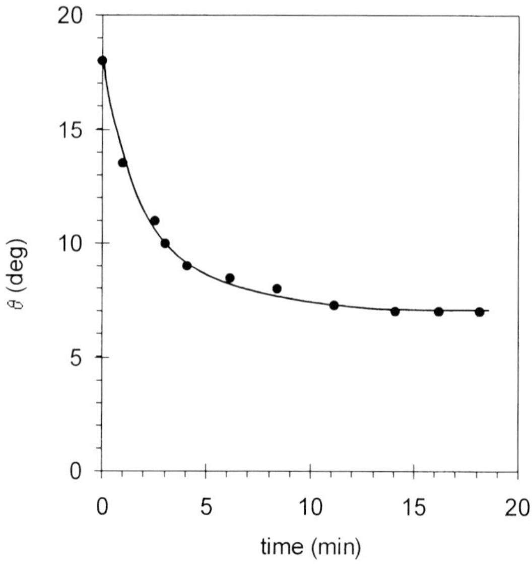

Figure 9.2. Variations of contact angle values with time observed for a Al_2O_3-32.4SiO_2-30.6CaO (wt.%) slag on Al_2O_3 at 1415°C, (Towers 1954) [27].

The wetting behaviour of *glass/metal systems* was first examined in the pioneer work of Ellefson and Taylor (1938) in the case of a non-reactive system consisted of a sodium silicate glass (with a composition close to $2Na_2O$-$3SiO_2$) on solid Au (see also the study performed by Pask and Tomsia (1981) in the same system). As Au has no affinity for O and the adsorption of O on Au is negligible, no reaction can be expected and the contact angle measured in a vacuum, as well as in N_2 and O_2 atmospheres, is 55–60°. These values correspond to a work of adhesion of only 0.45-0.5 $J.m^{-2}$, which is typical of those obtained for liquid metal/solid oxide systems when only van der Waals forces act across the interface (see Section 6.2.2) even though the contact angles are much higher. According to the Young-Dupré equation $\cos\theta = (W_a/\sigma_{LV}) - 1$, the similar values of W_a imply that the differing wetting behaviour of liquid Au on a solid oxide and glass on solid Au, Figure 9.3, is due in the main to the differing values of σ_{LV}: 1.14 $J.m^{-2}$ for Au and about 0.3 $J.m^{-2}$ for oxide glasses, Table 9.1. Thus despite the low value of W_a, non-reactive glasses may wet metal surfaces because of their low surface energies.

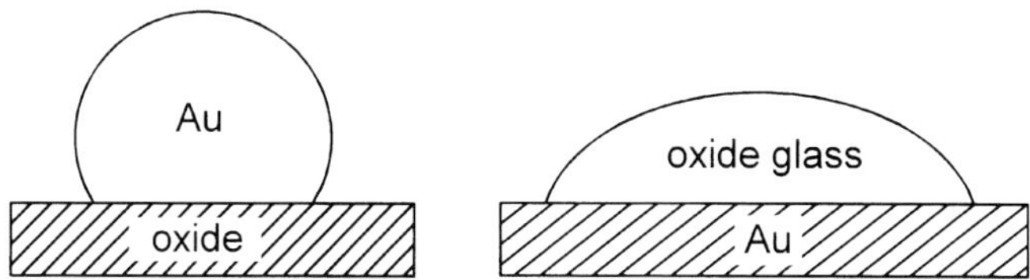

Figure 9.3. Schematic representation of the configurations that can be assumed by Au-oxide systems. Molten Au does not wet ionocovalent oxides ($\theta \cong 135°$) while molten glasses wet solid Au ($\theta \cong 60°$).

In a N_2 atmosphere and in vacuum, the sodium silicate glass wets Pt, the contact angle being 60-65°. However, in this case, a significant decrease of the contact angle was observed when the experiments were repeated in a O_2 atmosphere. Regarding spreading rates, Ellefson and Taylor (1938) found that the spreading rate of sessile drops scales with the inverse of the viscosity of the glass.

Good wetting with contact angles in the range 40-60° has also been observed for several slags containing chemically stable oxides such as Al_2O_3, SiO_2, CaO or MgO when in contact with Pt, Table 9.2. A systematic study by Towers (1954) showed that the contact angle values for non-reactive slag/Pt couples varied little with slag composition or temperature. The effect of the partial pressure of O_2, P_{O2}, when comparing measurements made in air or N_2, was weak but significant, Table 9.2. Similarly, the effect of P_{O2} on the wetting of W by a SiO_2-20Al_2O_3-10CaO-10MgO (wt.%) mixture has been studied using gas buffer systems (CO/CO_2 and H_2/H_2O), (Ownby et al. 1995). Wetting was found to be independent of temperature in the range 1300-1500°C but the contact angle values were found to decrease when P_{O2} was raised, Figure 9.4. Adsorption of oxygen can occur at both the metal/vapour and glass/metal interfaces and this will result in the formation of W-O bonds with a partially ionic character and to the development of ionic interactions between the modified W surface and the oxide glass, Figure 9.5.b. As with liquid metals in contact with solid oxides, the energy decrease of the metal/vapour surface is less than that of the glass/metal interface, so that W_a is increased and wetting is enhanced.

A further enhancement in wetting and bonding can occur if the metal surface has been oxidised. Thus a twofold increase in W_a was achieved when Fe and steel surfaces were oxidised before being contacted by silicate slags, Figure 9.5.c, (Perminov et al. 1961). Enhancement can be achieved also by the *in situ* oxidation of the metal by reaction with a slag oxide that has a low thermodynamic stability. Extensive studies of interfacial reactions and bonding in glass/metal systems have been performed by Pask and Tomsia, (1981, 1987, 1993), and these have shown that *in situ* oxidation of the metal and enhancement of wetting occur when CoO is added to Na-silicate glasses that are brought into contact with Fe, (Pask and

Table 9.2. Contact angle and work of adhesion values for slag/Pt systems, (Towers 1954).

Slag composition (wt.%)	Temperature (°C)	Atmosphere	θ (deg)	σ_{LV} (J.m^{-2})	W_a (J.m^{-2})
SiO_2-40CaO	1500	air	49.5	0.37	0.60
SiO_2-38CaO-20Al$_2$O$_3$	1300	N$_2$	50	0.42	0.68
	1300	air	43	0.42	0.72
CaO-34SiO$_2$-27Al$_2$O$_3$	1410-1550	air	42.5	0.45	0.77
SiO_2-20MgO-18Al$_2$O$_3$	1410	N$_2$	64.5	0.33	0.47
	1410	air	47	0.33	0.55
MnO-30SiO$_2$ [*]	1400	N$_2$	2	0.49	0.99

[*] there is a chemical reaction between the slag and Pt

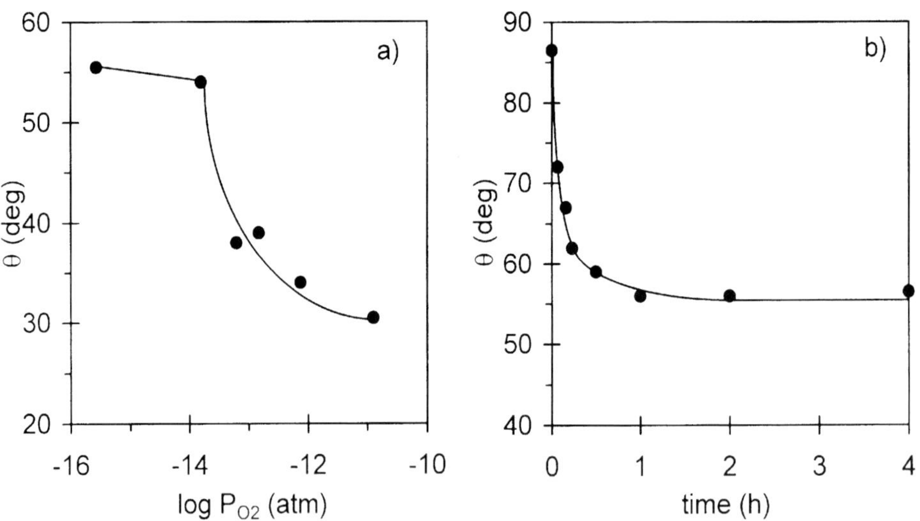

Figure 9.4. Contact angle values for SiO$_2$-20Al$_2$O$_3$-10CaO-10MgO (wt.%) on W at 1400°C plotted as (a) a function of logP$_{O2}$ and (b) as a function of time for logP$_{O2}$ = −15.5, (Ownby et al. 1995).

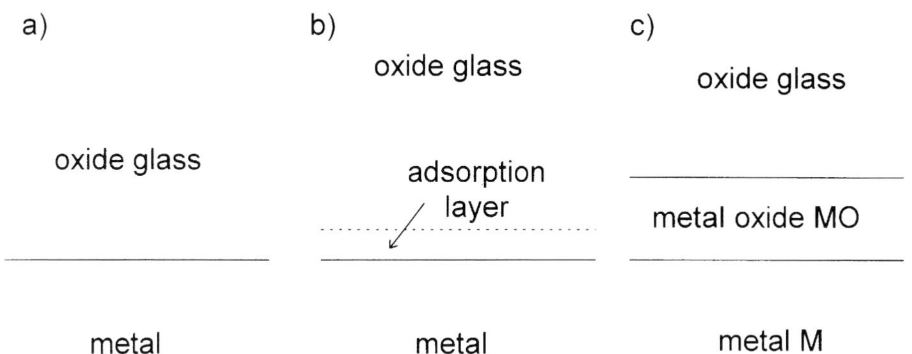

Figure 9.5. A schematic representation of glass/metal interactions: (a) a simple non-reactive oxide glass/metal interface across which only van der Waals interactions are developed; (b) a more intense interaction between the metal and the oxide glass is developed by the presence of an adsorption layer with a typical thickness of ~ 1 nm that is rich in oxygen; (c) improved wetting and bonding produced by preoxidation or *in situ* oxidation of the metal that leads to the creation of a glass/MO interface.

Tomsia 1981). Similarly, the presence of MnO in the glass seems to be the cause of the almost perfect wetting of Pt by MnO-$30SiO_2$, Table 9.2.

The viscosities of glasses are several orders of magnitude higher than those of metals so that spreading to assume an equilibrium contact angle can continue for minutes or even hundreds of minutes, Figure 9.4.b. Such spreading curves can be described by the de Gennes-Tanner model of viscous spreading for non-reactive couples (Section 2.1.2). However, marked departures from the predictions of this model can be observed in systems where spreading is accompanied or caused by chemical reactions at the liquid/solid interfaces, (Nicholas 1986b).

It is generally accepted that mixtures of molten oxides (glasses, slags) do not wet *graphite*. For instance, Ellefson and Taylor (1938), in their measurements of surface tension of molten silicates in N_2 by the sessile drop technique, used graphite substrates that produce contact angles as large as 130-160°. As these values were obtained on electrode graphite with a high porosity and a high surface roughness, they are only rough estimates of Young contact angles. However, these values suggest that interfacial bonding is ensured by weak, van der Waals, interactions. The non-wetting of graphite by oxide slags is regarded as one of the advantages of carbon bricks in the blast-furnace (Towers 1954). However, for certain slags, the contact angle appears to decrease with time towards values lower than 90° as a result of interfacial reactions between the slag components and carbon. For instance, initial contact angles close to 160° were measured on polished surfaces of graphite in N_2 at temperatures in the range 1330-1450°C for a slag SiO_2-$40CaO$-$20Al_2O_3$ (wt.%). Thereafter, the contact angle decreased slowly

Chapter 10
Wetting when joining

Brazing and soldering are widely used methods for joining metals and ceramics that employ a liquid metal or alloy as the bonding material. The term "brazing" arose from the original use of brass as the joining material but nowadays a wide range of alloys are used and brazing is differentiated from soldering by virtue of the joining materials having melting temperatures in excess of 450°C, (Schwartz 1995).

The process depends on a liquid metal flowing over surfaces to form a fillet between components and into the gap between the components, and then solidifying to form a permanent bond. Thus it is essential that the braze experiences high temperature capillary attraction. Without such attraction, solid braze material placed between components will flow out of the gap, "sweat", when it melts. Any residue of non-wetting liquid that remains within the gap will not conform to the microscopic features of the component surfaces but form an array of voids, as illustrated schematically in Figure 10.1, that is mechanically deleterious and should be avoided if at all possible. The size of such voids can be decreased if an external pressure is used to confine a non-wetting liquid braze into a gap but cannot eliminate them because the pressure needed to shrink voids increases as they become smaller.

Successful brazing depends not only on the characteristics of the liquid and the solid materials and of the vapour phase but also on the geometry and size of the joint. In the first Section, flow of liquid metals into idealised capillary gaps will be discussed. Thereafter, since the techniques used to join them differ somewhat, the role of wetting during the brazing of metal and of ceramic components will be considered separately. Finally, the role of wetting behaviour during joining by brazing-related techniques and of its influence on joint strengths will be discussed.

10.1. FLOW INTO CAPILLARY GAPS

10.1.1 Predicted penetration rates

The idealised flow of non-reactive liquids in smooth channels has been considered by founding physicists such as Newton and Poiseuille and is well understood. Slow flow of a liquid is laminar or streamlined while rapid flow is turbulent with the transition occurring when the Reynolds number - the dimensionless parameter $(Ud\rho/\eta)$ in which U is the velocity of the liquid, ρ and η are the density and

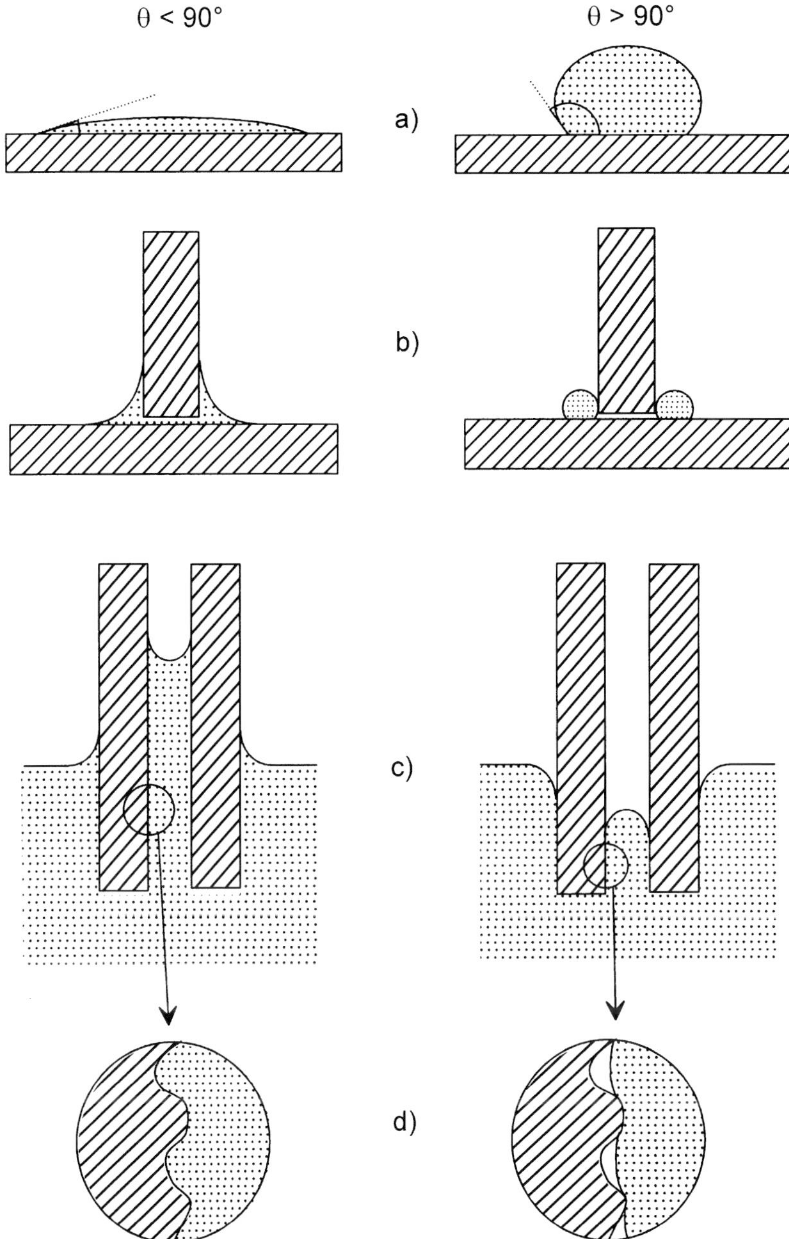

Figure 10.1. Schematic illustration of the influence of contact angles on the profiles of (a) sessile drops, (b) fillets between T configuration components, (c) the entry of liquid into capillary gaps and (d) the microscopic contact between the liquid and the solid components.

viscosity of the liquid and d is the width of a parallel sided channel - exceeds one to two thousand, (Bird et al. 1960). Strictly the density term, ρ, is the difference between the advancing liquid and the fluid it displaces, but this refinement can be ignored when the displaced fluid is a gas. Laminar flow is characterised by a velocity profile that varies right across the channel width from zero at its walls to a maximum in its centre, while the velocity profile of turbulent flow is almost uniform except in thin boundary layers at the walls. Note that even at low flow rates when laminar behaviour should prevail, entry into a channel and flow to a distance of a few widths is turbulent. Even if the liquid near the front in a partially filled tube also experiences turbulent flow, so that it moves from the centre to the walls of the tube, the front still advances at a rate determined principally by the laminar flow of the bulk of the liquid behind it, (Milner 1958).

The Reynolds numbers for the flow of a molten metal along a capillary braze gap are usually less than 1000 as will be shown later, and the theoretical laminar flow rates for such configurations have been calculated by Milner (1958) for both horizontal and vertical joints. He regarded the flow into a horizontal joint induced by capillary attraction as being impeded only by viscous drag, and derived a simple parabolic expression to describe such behaviour,

$$t = \frac{3\eta l^2}{\sigma_{LV} \cos \theta_Y d} \tag{10.1}$$

in which l is the penetration distance, t is the time, σ_{LV} is the liquid surface energy and θ_Y is the Young contact angle. This expression predicts that flow will continue indefinitely. However, penetration of a vertical capillary is helped or impeded also by gravitational forces, and this complicates analysis of the penetration kinetics, leading to the relationship

$$t = \left[\frac{12\eta}{\rho^2 g^2 d^3}\right]\left[-\rho g l d - 2\sigma_{LV} \cos \theta_Y \log_e\left(1 - \frac{\rho g l d}{2\sigma_{LV} \cos \theta_Y}\right)\right] \tag{10.2}$$

where g is the acceleration due to gravity. This equation describes an asymptotic approach to the equilibrium penetration, but expansion of the logarithm shows that the early stages of ascent can be described by the parabolic relationship presented as equation (10.1).

The predicted behaviour of individual liquid metals depends on their specific physical properties, such as those listed in Table 10.1 for Hg and a number of solder and braze metals and alloy solvents. Substitution of these values into equation (10.1) predicts that penetration time of completely wetted horizontal gaps by various metals should be virtually identical because of the similarity of their σ_{LV}/η ratios, as illustrated in Figure 10.2.a. However predictions of the penetration kinetics of vertical gaps are more complex: penetration should approach an equilibrium height, l_e, that depends on the liquid density as well as its surface energy and which is defined by the equation:

$$l_e = \frac{2\sigma_{LV}\cos\theta_Y}{\rho g d} \qquad (10.3)$$

Note that equation (10.3) is similar to equation (1.55) for a gap of circular geometry. For completely wetted gaps, l_e differs up to one order of magnitude for the different metals of Table 10.1. For similar values of l_e, the penetration time decreases slightly with increasing σ_{LV}/η, as illustrated in Figure 10.2.b for Pb and Hg.

Table 10.1. Physical properties of some liquid metals (Iida and Guthrie 1988) and equilibrium rise, l_e, in a vertical perfectly wetted 0.1 mm wide parallel sided gap. Note that the σ_{LV}/η ratio for H_2O is 70 and for a glass at a typical glazing temperature falls in the range of 10^{-3} to 10^{-4}.

Metal	T_F (°C)	σ_{LV} (J.m^{-2})	η (mPa.s)	ρ (Mg.m^{-3})	σ_{LV}/η (m.s^{-1})	l_e (m)
Hg	−39	0.50	2.0	13.7	250	0.07
Sn	232	0.56	1.8	7.0	311	0.16
Pb	328	0.46	2.6	10.7	176	0.09
Zn	420	0.78	3.5	6.6	222	0.24
Al	660	0.91	2.7	2.4	337	0.76
Ag	961	0.97	4.3	9.3	225	0.21
Au	1065	1.17	5.4	17.4	217	0.14
Cu	1083	1.30	4.3	8.0	302	0.32
Ni	1455	1.78	5.5	7.9	323	0.45

Differentiation of equation (10.1) to obtain a value of dl/dt, U, permits the Reynolds number to be rewritten as equal to $(\sigma_{LV}\cos\theta_Y\rho d^2/(6\eta^2 l))$. Assuming the workpieces are completely wetted, substitution of Table 10.1 data yields Reynolds numbers for a typical braze joint 0.1 mm wide and 10 mm long, (Schwartz 1995), ranging from 49.5 for Al to 117 for Au and even smaller numbers are obtained if the workpieces are not completely wetted so that $\cos\theta_Y$ is less than 1. These values are much smaller than that needed to induce a transition in flow behaviour, and therefore it can be predicted that even exceptionally wide or short braze joints will be filled by predominantly laminar flow.

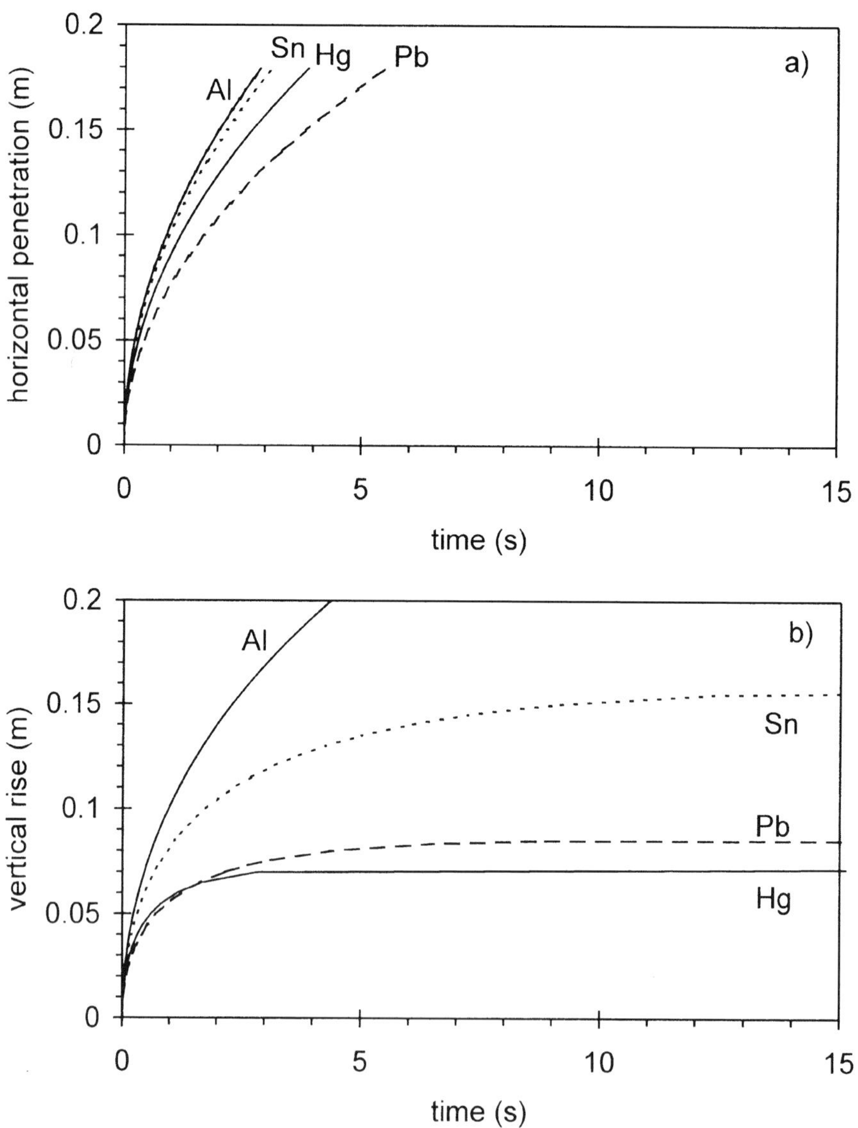

Figure 10.2. Predicted flow of liquid metals into completely wetted 0.1 mm wide parallel sided gaps: (a) horizontal gaps, (b) vertical gaps.

10.1.2 Experimental results

Quantitative descriptions of the flow or spreading behaviour of real systems of relevance to high temperature capillarity are not numerous but the published observations do illustrate the importance of material and process factors. Most published work describes the spreading of sessile drops because of the easier experimental techniques (see Chapter 2).

The flow along horizontal capillary channels was observed by Latin (1946) who found that penetration of parallel sided channels made from tinned Cu by fluxed Sn was predicted precisely by equation (10.1) when $\cos\theta_Y$ was equated to 1. However, flow along channels that had not been tinned was only half as fast even though $\cos\theta_Y$ for Sn on Cu was about 0.9. This discrepancy implies that the rate of advance was controlled by a Sn-Cu interaction at the liquid front.

When analysing the behaviour of fluxed soldering or brazing systems, attention must be paid to the characteristics of the flux as well as the liquid metal so that, for example, the density term, ρ, should be replaced by the difference in density between the two fluids.

Vertical penetration of capillaries in sintered metal compacts was examined by Semlak and Rhines (1958) who found the rates of rise of molten Cu in Fe compacts and of Pb in Cu compacts up to heights of about 20% of the calculated equilibrium ascent to be in excellent quantitative agreement with the predictions of equation (10.2). However, the observed rate of rise of Ag in Cu compacts was twice that predicted by equation (10.2) because structural rearrangement of the compact during penetration increased the effective radius of the capillaries.

Fillet formation when brazing Al alloys with Al-Si brazes has been studied using a hot stage fitted in a scanning electron microscope, (Nicholas and Ambrose 1989). Joint formation is slow, a 2 mm high fillet taking minutes to form rather than the millisecond or less suggested by modelling ideal systems. This retardation is because the main constraint to braze flow and fillet formation is not viscosity but the Al_2O_3 films present on the liquid and solid surfaces. That on the braze has to be disrupted by evolution of volatile Mg additions from the liquid, while that on the solid is displaced by "tunnelling" i.e., by penetration of the braze at the oxide/Al alloy interface through naturally occurring flaws of the oxide.

10.2. JOINING METAL COMPONENTS

Braze fluidity can be assured only if the joining temperature is higher than the braze liquidus. The viscosities of liquid metals are relatively low, only a few mPa.s, Table 10.1, and decrease slowly as the temperature is raised. A dramatic increase in viscosity can occur, however, if the temperature drops below the liquidus so that the braze structure becomes a liquid-solid mixture. Other factors affecting the selection of a suitable joining temperature are the need for the braze liquidus to be lower than the component solidus, and often to be a lot lower if detrimental microstructural changes are to be avoided. Similarly, it is necessary for the braze solidus to be sufficiently high to withstand any possible service temperatures.

Brazes with very varied melting temperatures have been developed for joining metal components of steel, Ni base alloys or Al alloys and braze compositions have been codified in national and international standards. Al alloy brazes are recommended only for the joining of Al alloy components, but brazes based on Ag or Cu or their alloys can be used for the joining of a wide range of component materials including steels, Ni alloys, precious metals and refractory metals, while Ni alloy brazes find applications in the production of high temperature joints between steel or Ni alloy components. Such recommended brazes are generally a pure metal or are based on eutectic alloys that have singular melting temperatures, as can be seen in Table 10.2 which lists some widely used brazes. Selection of brazes with singular melting temperatures or sharp melting ranges enables the fabricator to specify very precisely defined joining temperatures and to avoid or at least minimise the possibility of "liquation", i.e., the premature flow of low melting temperature phases over surfaces or into gaps between components during the heating process as the temperature is raised through the solidus-liquidus range, which could leave refractory phases outside the joint and result in incomplete filling.

While the use of pure metals and eutectic alloys as brazes is generally good practice, exceptions do occur. Thus it may be necessary to use brazes that deviate from eutectic compositions to minimise the concentration of an embrittling temperature depressant, or to complicate the melting characteristics and accept degraded fluidity by introducing elements that enhance the corrosion resistance of the solidified braze. Thus recommended Al-Si brazes have hypo-eutectic compositions and Cr is introduced into Ni-P and Ni-Si-B braze alloys.

The capillary attraction exerted by clean metal surfaces on metallic brazes is high. Indeed the contact angles of liquid metals on clean metal substrates is invariably less than, and usually much less than, 90° even for chemically inert and mutually insoluble systems, as discussed in Chapter 5.

Of even more practical importance in determining the behaviour of systems is the tendency of many component materials, and some brazes, to form chemically stable and physically tenacious oxide films when their surfaces are exposed to even mildly oxidising environments. (Note that these oxides are in turn covered by absorbed layers of organic contaminants, but these can be removed relatively easily by washing in solvents, vapour degreasing or baking in a vacuum). Oxide films can

Table 10.2. Some brazes for metal components.

AWS code [*]	Alloy composition (wt.%)	Melting range (°C)	AWS code [*]	Alloy composition (wt.%)	Melting range (°C)
	Ag	960	BCu-1	Cu	1083
BAg-8	Ag-28Cu	780	BAu-1	Cu-37.5Au	990-1010
BAg-1a	Ag-15Cu-17Zn-18Cd	620-640	BCuP-2	Cu-7.5P	732
BAg-3	Ag-16Cu-16Zn-16Cd-3Ni	630-660	BCuP-4	Cu-6Ag-7P	643-813
BAg-5	Ag-30Cu-26Zn	690-775	BCuP-5	Cu-15Ag-5P	643-801
BAlSi-2	Al-7.5Si	565-625	BNi-2	Ni-7Cr-3B-4.5Si-3Fe	970-1000
BAlSi-3	Al-10Si-3Cu	550-570	BNi-3	Ni-3B-4.5Si-1.5Fe	980-1040
	Al-10Si-1.5Mg	555-577	BNi-4	Ni-2B-3.5Si-1Fe	980-1070
			BNi-5	Ni-19Cr-10Si	1080-1135
	Au	1065	BNi-6	Ni-11P	880
BAu-2	Au-20Cu	890	BNi-7	Ni-14Cr-10P	890
BAu-4	Au-19Ni	950			

[*] American Welding Society

be responsible for beneficial characteristics such as the corrosion resistance of Al, Fe-Ni-Cr and Ni-Cr components conferred by surface coverage by Al_2O_3 or Cr_2O_3, but other aspects of their presence are not beneficial. Oxides are wetted only by a few braze metals or alloys as described in Chapter 6, while the oxide films on molten brazes can inhibit liquid flow. Thus it is essential that oxide films on components and brazes should be destabilised, removed or at least disrupted if wetting is to be achieved. In practice this is achieved by using fluxes or controlled atmospheres.

The fluxes used when brazing are alkali halide and borate mixtures and compounds, and they have two main functions: first, to dissolve the oxide film on the component surface or at least to degrade its adhesion by penetration of naturally occurring flaws and electrolytic action at the oxide-substrate interface, and secondly to prevent formation on the liquid surface of oxide skins which would restrict braze flow. Fluxes can be contained in a bath held at the brazing temperature in which the, usually aluminium, component is placed or else applied as a paste to surfaces of the component or braze.

Fluxes for Al components based on $Na_2B_4O_7/KBF_4$ mixtures were developed with the aim of dissolving Al_2O_3 films from component surfaces and have been used successfully in practice for decades. However, this achievement is an illustration of serendipity because they do not dissolve the oxide films but merely undermine them, (Jordan and Milner 1956). Few systematic studies of flux action for other brazing processes have been reported, but McHugh et al. (1988) examined some physical properties of a range of commercial fluoride/borate fluxes used with Ag or Cu brazes. They found that the fluxes had a network structure similar to glass which was degraded by an increasing F content. Flux viscosities were higher than those of the brazes with which they were used, varying over their working temperature ranges from 10^3 to 5×10^{-2} Pa.s and they were effectively impermeable to atmospheric O_2.

Adjustment of brazing torch chemistries can be used to produce reducing flames that not only heat but also clear away the oxide films on some component surfaces. Similarly, furnace brazing using controlled atmospheres of inert gases (generally Ar or He), H_2 or reducing mixtures of N_2 (or Ar)-H_2 (or CO), and vacua of 10^{-4} mbar or less can be used to achieve environments in which flow is unrestrained by oxide skins and component surfaces are clean metals. Ag, Cu, Fe and Ni, but not Al, can destabilise their oxide skins at high temperatures by taking O into solution, (Massalski 1990), and Ag_2O, Cu_2O and NiO also dissociate in even moderately O active environments (see Table 3.1). However, it is in practice impossible to achieve controlled atmospheres with O_2 activities low enough to cause dissociation of stable oxides, thus dissociation of Cr_2O_3 would require a partial pressure of less than 10^{-45} bar at 900°C while that for Al_2O_3 would be less than 10^{-60} bar at 600°C. Nevertheless components covered with such oxides are brazeable. Thus the oxide films containing Cr_2O_3 that are formed on surfaces of Fe-Cr-Ni components can be reduced when vacuum brazing by the presence of trace amounts of C in the steel, (Lugscheider and Zhuang 1982), or even by the low O activity within a liquid braze that covers a component surface, (Cohen et al. 1981, Thorsen 1984). The oxide can be removed also by penetration of naturally occurring flaws and undermining by dissolution of its metallic substrate, (Ambrose and Nicholas 1996). Again, the presence of Mo in the steel also can result in ready disruption of the surface oxides even in moderately oxidising environments, (Kubaschewski and Hopkins 1953, Gale and Wallach 1991), due to volatilisation of MoO, and volatilisation of Mg from Al-Si-Mg brazes causes disruption of the Al_2O_3 films that otherwise encase the liquid braze and prevent its flow, (McGurran and Nicholas 1984).

The final treatment of the components before assembling to braze is often to create a particular surface finish. It is recognised that some roughening of a surface can result in enhanced wetting by inherently well wetting liquids in accord with the Wenzel equation (1.35). However, the use of very rough surfaces can degrade the strength of the braze/component interfaces. A British standard recommends that the roughness of a component surface should not be greater than 1.6 μm R_a if the finish is anisotropic, (BS1723 1986a), while Schwartz (1995) recommends the use of abrasion with clean metallic grit to produce anisotropic roughness of 0.8-2.0 μm

R_a. The effects of substrate texture on wetting behaviour have been described in Section 1.3.1 and these lead to the expectation that anisotropic roughening will promote the spreading of very well wetting brazes but will impede that of moderately wetting brazes.

Relating these observations to brazing practice is complicated by the fact that the scientific studies showed the steepness of surface asperities can be a more significant characteristic than their maximum heights, but this parameter is not usually considered or quantified by brazing engineers. Thus the recommendations about component roughening cannot be assessed in terms of capillarity effects, and indeed it should be recognised that abrasion not only changes the topography of surfaces but also their chemistry by removing mature oxide films.

When suitable environments and surface preparation techniques are used, wetting should be achieved. Few recent data are available even for technically important braze/component systems apart from the early area of spread tests of Feduska (1959) with mainly Ni brazes on high-temperature base metals in high vacuum and dry He or H_2 and the more recent work of Keller et al. (1990) with Ag and Au brazes on stainless steels in high vacuum. These showed that interdiffusion induced by the differing complex chemistries of braze/component systems can affect the wetting behaviour of some systems quite markedly but does not completely obscure all the effects of process parameters and compositions. Thus it is clear that wetting is better for brazes with high liquidus, and therefore high use temperatures. Since the substrate surfaces were at least partially covered by oxide films, this temperature effect can be plausibly related to an increase in the rapidity with which these barriers to wetting are undermined and dispersed. As a general trend, at a fixed temperature, wetting in vacuum was superior to that obtained in dry H_2. The data of Feduska and Keller et al. also show that further increasing the temperature at which a particular braze is used generally decreases the extent of wetting, and this can be attributed to the rapidity of braze/substrate interdiffusion resulting in premature solidification of the molten braze. Similar work by Amato et al. (1972) shows spreading of a Ni-13Cr-10P braze on a Ni alloy substrate at first increases and then diminishes as the temperature is raised progressively from 950 to 1050 and then to 1075°C.

Examination of the area-of-spread data of Keller et al. (1990) for AISI 304 stainless steel with a composition of a Fe-18Cr-10Ni-1Mn in wt.%, as are all compositions in this Chapter, shows that Au and Ag-Cu-Pd alloys wet extremely well and that the presence of a small amount of Ni or a large amount of Pd promotes wetting by Ag-Cu and Ag-Cu-Sn alloys. Similar effects on wetting have been observed for other Fe-Cr-Ni alloys, but the wetting of AISI 316 steel, (Fe-17Cr-12Ni-2.5Mo), was generally better than that of AISI 304 and this could be due to the volatilisation of MoO providing more opportunity for liquid metal/solid metal contact and the creation of low energy interfaces.

The beneficial effects of Ni and Pd imply a decrease in the liquid/solid interfacial energy values, as discussed in Chapter 5. Work by researchers using other families of braze alloys has shown that the wetting of both steel and Ni alloy components by Ni brazes is promoted by the presence of B and Si which can flux, that is cause chemical reactions to disrupt the surface oxide, (Amato et al. 1972).

Liquid flow into joints will occur if the braze wets the component materials, but the penetration achieved is also a function of the configuration of the joint and, for vertical joints, the density of the liquid braze. Having selected a braze and an environment that promotes wetting, the configuration of the joint must be selected. Widely accepted standards of good design practice can be found in national and international specifications. These recommended designs take into account factors other than capillary phenomena, such as the general need to avoid tensile stressing of the brazed joint as it is cooled and to maximise its area to enhance its shear strength. Additionally, it is desirable that any machining or metalworking needed to form the configuration should be simple and that the joint should be held firmly in place without the use of complicated jigs. Satisfying these desires has led to a preference for lap, sleeve and pocket joints, such as those shown schematically in Figure 10.3, and hence to the creation of the narrow, capillary, gaps between the components that increase the driving force for flow of the liquid braze from its original location outside the joint.

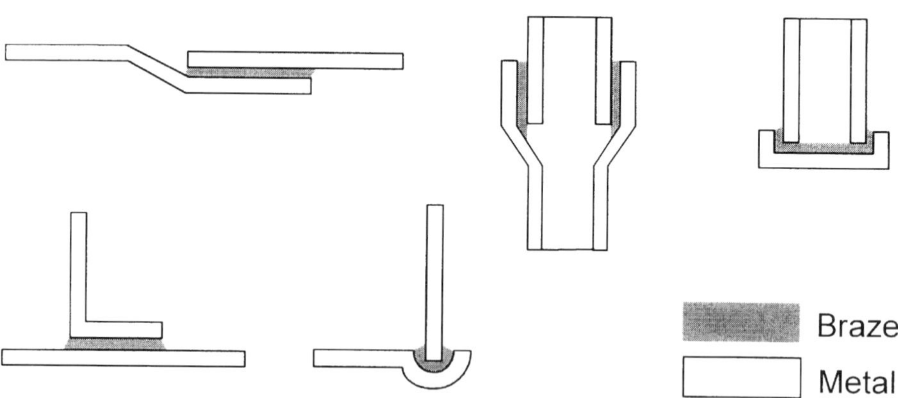

Figure 10.3. Some preferred joint configurations for brazed metal components.

Despite the variations in wetting behaviour displayed by differing braze alloys, the range of gap widths with which acceptable joints can be formed readily does not vary greatly, as shown by the recommendations in Table 10.3. It should be noted that the gaps recommended when joining with Ni brazes includes zero, meaning that the braze alloy flows along the microscopic channels left between the contacting asperities on the surfaces of the components. When joining with Cu brazes, even fewer and smaller residual microscopic channels required so BS1723 (1986a) is able to recommend even negative gaps produced by shrink fitting prior to brazing. In general, however, there is a consensus that a gap of about 0.1 mm is acceptable for joining, except possibly when an Al braze is used. In principle, it is possible to calculate the optimum vertical or horizontal gap width by quantifying the effect of variations in the joint gap on the equilibrium penetration and the rate of flow of the liquid braze by assuming the system to behave in an ideal manner. In practice this exercise is thwarted by the lack of relevant data for technically important systems and the presence of oxide films on the surfaces of the components and sometimes also on the surface of the braze. Nevertheless, insight into probable effects of gap widths on braze flow can be gained by reference to the behaviour of simpler braze/component material combinations.

Table 10.3. Recommended braze gaps.

| Braze | Typical widths (micron) | |
	(BS1723 1986a)	(Schwartz 1995)
Ag alloys	25 - 150	0 - 130
Al alloys	50 - 250	50 - 250
Au alloys	25 - 125	0 - 130
Cu alloys	−50 - 150	0 - 130
Ni alloys	0 125	0 - 130

Figure 10.2 showed the predicted capillary penetration by some liquid metals which were assumed to wet perfectly. The predictions for all the systems suggest that both horizontal and vertical 0.1 mm wide braze gaps will be filled to a depth of 10 mm, a typical braze depth, in much less than 1 second.

Significant differences due to the varying importance of gravity with orientation and the nature of the liquid metal become apparent only for untypical gaps longer than about 100 mm, and therefore the consensus about recommended joint gap widths is not surprising. The sometime exception to this rule, Al brazing of Al components, is probably due to the fact that the braze and component solvents are identical. The melting temperature of the braze is depressed by the addition of Si but this readily diffuses into solid Al and, therefore, premature isothermal resolidification of the Al-Si braze will occur if the joint is thin.

Rapid liquid flow into the capillary gaps should occur for simple liquid-solid systems. However, the predicted times of less than 1 second to fill a 10 mm deep gap shown in Figure 10.2 for some pure, perfectly wetting and non-reactive liquid metals are much less than the brazing temperature hold times of 100-1000 seconds that are typical of much brazing practice. These apparently excessively long hold times are necessary because the actual flow rates of liquid brazes are far less than those predicted by models of flow impeded only by viscous and gravitational forces. Thus hot stage microscope observations have measured flow rates of about 1 μm.s^{-1} for the formation of Al-Si-Mg brazed Al component joints and the spread of Ni-11P braze over Fe-Cr surfaces, (Ambrose et al. 1988). The slowness of these rates was attributed to difficulties in removing oxide films from the component surfaces by undermining the substrate metal, and to the confinement of flow to the narrow undermined region, thereby increasing the viscous drag experienced by the liquid braze, (Ambrose et al. 1988, 1993).

10.3. JOINING CERAMIC COMPONENTS : CERAMIC-CERAMIC AND CERAMIC-METAL JOINTS

The same capillary phenomena affect brazing practice for joining both ceramic and metal components, but the relative importance of the phenomena differs, and this makes it convenient to discuss their effects in a different sequence. Further, most joining of ceramics is to metals and the different thermal expansion and mechanical characteristics of these two families of materials, as exemplified in Table 10.4, have a profound effect on joint design that is not related to capillarity.

The capillary attraction for non-reactive metals exerted by ceramics is usually negative ($\theta > 90°$) as discussed in Chapters 6 to 8. For ceramic-ceramic joining, the braze must wet the ceramic surfaces but creating a ceramic-metal joint does not require the ceramic to be wetted but merely that the sum of the contact angles assumed by the ceramic and metal components should be less than 180°. This less demanding requirement arises because it is sufficient for the total energy for a liquid to advance while in contact with both component surfaces to be negative. Thus the energy change caused by the small advance δx shown in Figure 10.4 is δF where

$$\delta F = \delta x(\sigma_{SL} - \sigma_{SV})_{metal} + \delta x(\sigma_{SL} - \sigma_{SV})_{ceramic} \qquad (10.4)$$

Table 10.4. Physical properties of some ceramics and metals.

Material	Melting temperature (°C)	Thermal conductivity (W.m^{-1}.K^{-1})	Thermal expansion (10^6 K^{-1})	Elastic modulus (GPa)
Al$_2$O$_3$	2030	35	7.9	387
SiO$_2$	1720	1.5	3.0	354
ZrO$_2$	2960	19	7.5	140
Si$_3$N$_4$	1900 [*]	17	2.5	176
TiN	2900	17	8.1	
SiC	2700 [*]	50	4.3	211
TiC	3140	36	7.2	422
WC	2777	84	5.2	704
graphite	3650	25	[**]	
Ag	960	425	19.1	83
Al	660	238	23.5	71
Cu	1083	379	16.6	130
Fe	1535	78	12.1	211
Mo	2615	137	5.1	325
W	3387	174	4.5	411

[*] Does not melt but decomposes or vaporises
[**] 1 to 5×10^{-6} K^{-1} for varyingly oriented polycrystalline material

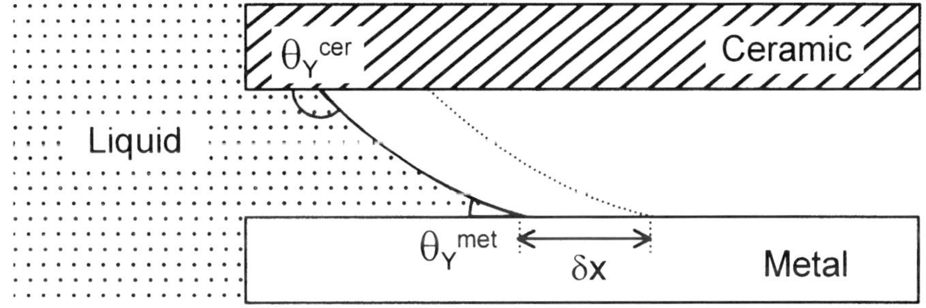

Figure 10.4. Infiltration of a liquid into a gap between metal and ceramic components.

Introducing the equilibrium contact angle of the liquid on the solid metal, θ_Y^{met}, and on the ceramic, θ_Y^{cer}, and taking into account the equilibrium condition $d(\delta F)/d(\delta x) = 0$ leads to the conclusion that $\theta_Y^{met} + \theta_Y^{cer} = 180°$. Thus the condition for the liquid to advance is $\theta_Y^{cer} < 180° - \theta_Y^{met}$. For clean metal surfaces,

Ceccone et al. 1995). However, this effort has not yet resulted in application of the alloys for the mass production of ceramic-metal components because of excessive formation of embrittling silicides by BNi-5 and unacceptably poor wetting by BNi-7. Recently, high melting-point brazes have been developed to join SiC based materials, with use temperatures up to 1000°C, that have good wetting, negligible reactivity with SiC and form strong braze/SiC interfaces (Moret and Eustathopoulos 1993, Moret et al. 1998).

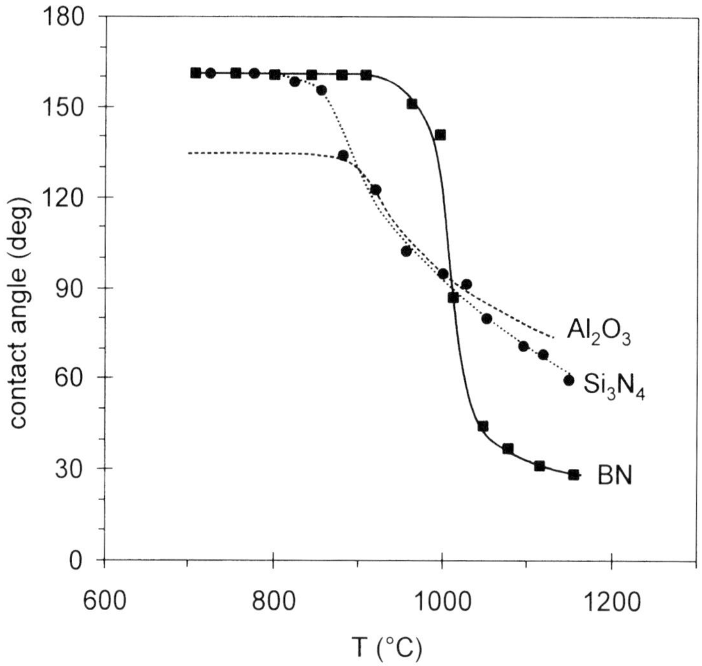

Figure 10.5. The wetting behaviour of sessile drops of Al on reactive ceramic substrates (Si$_3$N$_4$, BN) and Al$_2$O$_3$ during temperature rise at about 5 K/min in a high vacuum, (Nicholas et al. 1990) [2].

Liquid flow into ceramic-metal joints will be affected by their design. The joint gaps used when the ceramic component has been metallized will be similar to the 0.1 mm recommended for the brazing of most metal components, but usually the preferred gaps when using active metal brazes will be quite wide so that a reasonable reservoir of reactant is available. However, the principal consideration determining joint designs, regardless of whether the process involved is metallization followed by bonding with normal brazes or the active metal brazing of unmetallized ceramics, is the difference in physical properties of ceramics and metals. In particular, attention has to be paid to the smaller thermal expansivities of ceramics and to their brittleness. Thus it is desirable to ensure that the ceramic members are subjected to small compressive stresses as the assembly cools after being joined. Finite element modelling, experiment and experience have led to the development of a number of successful designs that possess the desired stressing characteristics and some of these are shown schematically in Figure 10.6.

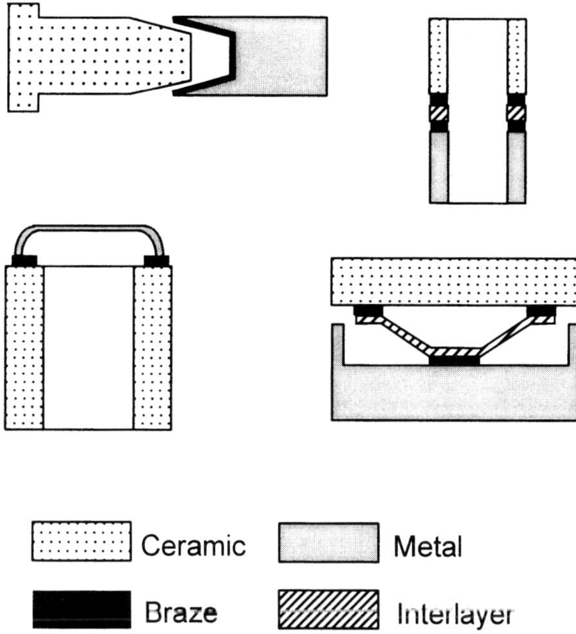

Figure 10.6. Designs for ceramic-metal joints.

Each of the applications shown in Figure 10.6 is different, but they all accommodate the mismatched thermal expansivities of ceramics and metals. The top left hand sketch shows a tapered joint between Si_3N_4 and a steel shaft in which the circumferential compressive stress on the ceramic is attenuated by the progressive thinning of the steel wall, (Mizuhara and Huebel 1986). Progressing clockwise, the next sketch shows a tubular Si_3N_4/steel device that uses a thick Mo interlayer to decrease thermal expansion mismatch stresses, (Schwartz 1995). In the bottom right hand corner is shown a ceramic crowned engine tappet that uses a flexible metal interlayer to accommodate mismatched thermal expansions,

(Bucklow et al. 1992), and the remaining sketch shows a ceramic tube capped by a flexible metal member, (Mizuhara and Huebel 1986).

Rapid fluid flow cannot be achieved with active metal brazes because of the need to form solid wettable reaction product layers for their liquid fronts to advance. Equations (10.1) to (10.2) relating liquid flow rates to the opposed effects of surface energy imbalances and of viscous drag are not relevant. Actual penetration rates are so slow, usually of the order of 1 μm.s^{-1}, that the usual practice is to place the active metal braze alloy within the joints rather than expecting it to fill them, and, as explained already, gap width is not the dominant consideration when designing ceramic-metal joints.

The thermal cycles used to braze ceramic-metal and ceramic-ceramic assemblies are usually prolonged with slow heating and cooling rates being used to minimise thermal gradients, and hence stresses, within the ceramics. Common practice is to heat at about 10°C.min^{-1} to just below the solidus temperature and then to hold for several minutes before continuing to heat at perhaps 3°C.min^{-1} to the brazing temperature. The dwell times at the brazing temperature are typically 10 to 30 minutes, fortuitously enabling a quasi-equilibrium joint microstructure to be achieved through braze/component chemical reactions and slow spreading of the liquid braze.

10.4. JOINING BY RELATED TECHNIQUES

Brazing is such a long established technique that is not surprising that many variations have been developed. Some of these such as active metal brazing have been referred to already, but this Section starts with comments on other techniques in which the liquid bonding phase is formed *in situ* by interdiffusion or chemical interaction with the environment. Finally, a small sub-section is devoted to "glazing", i.e., joining metals and ceramics by glasses.

10.4.1 Transient liquid phase bonding
This technique employs a thin and chemically different metal or metal alloy interlayer placed between the surfaces of metallic components so that contact results in interdiffusion and the transient production of a composition that is liquid and wets the solids, (Zhou et al. 1995).

The changes in the phase structure of a hypothetical eutectic system caused by interdiffusion of the interlayer and component materials are illustrated schematically in Figure 10.7 for contact between an interlayer of pure metal A and a component of pure metal B brought into contact at a temperature T_1 that is lower than the melting point of either A or B. Ingress of B into A and of A into B initially causes the A_{SS} and B_{SS} solid solutions to be formed, but ultimately the solubility limits identified as a and d in the figure are exceeded and a liquid phase is formed with a composition between b and c that wets and spreads over the whole solid surfaces. Formation of this liquid, therefore, increases the rate of interdiffusion and ultimately all the interlayer is consumed. However, interdiffusion does not cease and the liquid becomes richer in B and commences to resolidify isothermally when its composition becomes d.

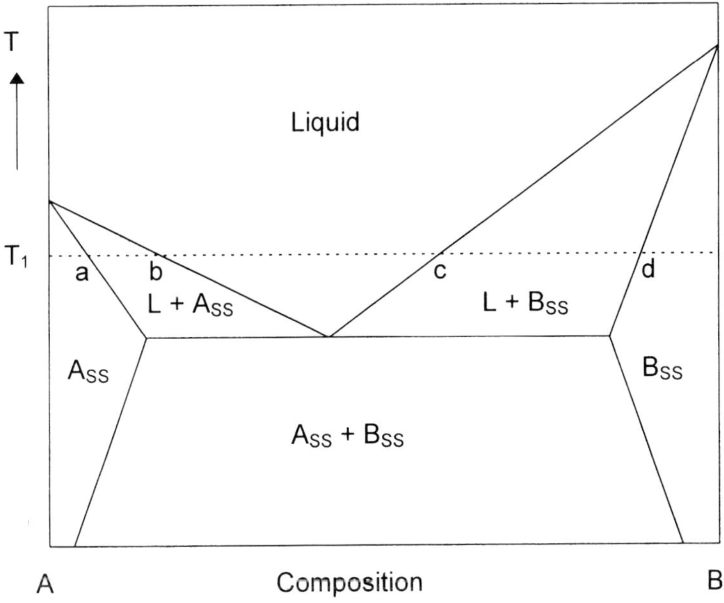

Figure 10.7. Phase diagram illustrating the changes caused by interdiffusion of A and B at a temperature of T_1.

Modelling the progress of a transient liquid phase bonding between components and an interlayer, therefore, is largely a matter of considering rates of interdiffusion and solubility limits. Wetting behaviour can be assumed to be excellent if the process progresses at all because the liquid can be formed only after metal-metal contact has been achieved. Thus the onset of transient liquid phase bonding may be delayed by the need to disrupt and disperse oxide films on substrate surfaces but once metal-metal contact has been achieved in some locations the oxide film residues can be undermined by the volume changes

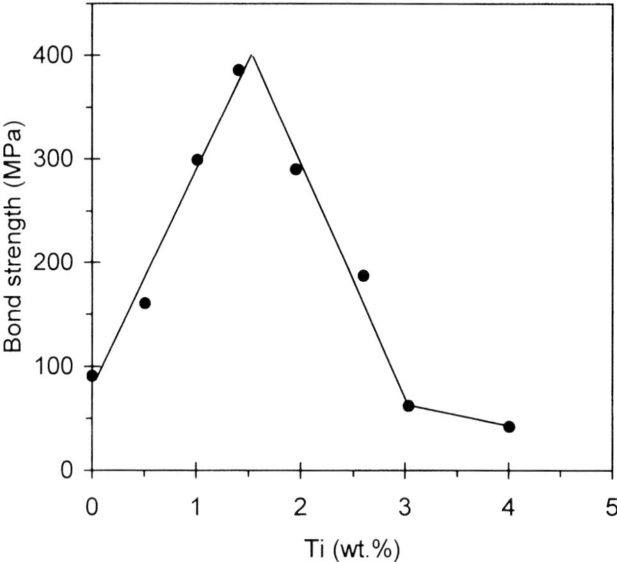

Figure 10.8. The room-temperature tensile strengths of solidified sessile drop interfaces formed between Al_2O_3 and Cu-Ti alloys plotted as a function of the Ti concentration, (Nicholas 1986).

Peaking of strength values has been observed for many reactive alloy-ceramic systems and attempts have been made to correlate the strength maxima with other system characteristics. Thus comparisons between peak strength values and the chemical affinity of the solute for O have revealed empirical correlations, such as that shown in Figure 10.9 for Al_2O_3 bonded by Ni alloys and by pure Ni. It is interesting to note that all solutes of Figure 10.9 are known (or expected) to improve wetting and thermodynamic adhesion on alumina. However, the detailed mechanism responsible for such a correlation has yet to be established and it has to take into account the fact that the extent of reaction and the morphology of the reaction product layers can vary significantly from system to system. For instance, in low P_{O2} environments, no reaction product forms between Al_2O_3 and pure Ni and Ni-Al and Ni-Cr alloys (only a slight dissolution of Al_2O_3 occurs in these alloys) while Ti in Ni reacts with Al_2O_3 to form different Ti oxides depending on Ti concentration.

The thermodynamic driving force for interface creation for clean metal-metal systems, defined in terms of the work of adhesion, is invariably large as shown in Chapter 5 and hence generates the expectation that the mechanical properties of metal-metal interfaces will be excellent. This is usually observed in practice when wetting produces complete and intimate contact of the braze and component materials, but there is not always complete interfacial contact and the presence of flaws can have a profound effect on mechanical properties of both metal-metal and metal-ceramic interfaces.

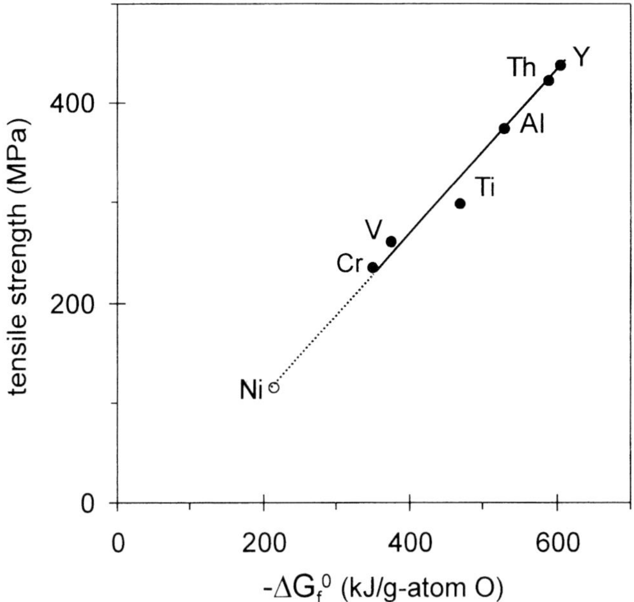

Figure 10.9. Peak room-temperature tensile strength values of the interfaces formed by Al_2O_3 with sessile drops of Ni or Ni-metal M alloys plotted as a function of the standard Gibbs energy of formation of M oxide, (Crispin and Nicholas 1976). Full circles: Ni alloys; hollow circle: Ni; from (Nicholas 1968).

10.5.2 Influence of interfacial flaws

Interfaces can exhibit both physical and chemical flaws, as exemplified by the presence of small voids in the valleys of roughened surfaces brought into contact with a non-wetting liquid ($\theta > 90°$), or retained islands of unwettable oxide on the surfaces of metal components or of surface contaminants on both metal and ceramic components. Such flaws can have detrimental effects on mechanical properties because at the very least their presence decreases the bonded area. Thus the mechanical preparation and cleaning of component surfaces are crucially important steps of the brazing process.

The formation of composite interfaces when very rough surfaces are contacted by liquids that do not wet, (Cassie and Baxter 1944), results in the presence of microvoids at the interface between the component and the solidified liquid (see Section 1.3.1 and Figure 1.27). This effect should seldom be of importance when brazing metal components because of their generally excellent wettability, but it can be of significance when joining ceramic components and similar populations of microvoids can be produced when metal components with rough surfaces are diffusion bonded, by bringing solid components surfaces into contact by applying pressure at a high but sub-solidus temperature, (Baker and Partridge 1991). The influence of such populations of flaws on mechanical properties depends on the type of stressing to which the interface is subjected as well as the characteristics of the joining and component materials. Thus diffusion bonded Ti-4Al-4Mo-2Sn samples can accommodate 10%, or more, of their interfaces being microvoids without any noticeable deterioration of tensile strength, but the fatigue strength is decreased by about 30% and the impact strength by 50-75%, (Baker and Partridge 1991).

The roughened surface topographies of ceramic components can have a particularly detrimental marked effect on mechanical properties both because of their influence on wetting behaviour and the likelihood of microvoid formation in valley bottoms and because the mechanical treatments may nucleate microscopic surface cracks that can propagate and cause premature failure when the interface is stressed. However, these microscopic cracks can be healed by subsequent heat treatment that promotes the flow of glassy binder phases. Thus the effect of refiring ground surfaces of 99.5% pure Al_2O_3 at 1650°C on its joining to 6.5×32 mm strips of Kovar using Ag-28Cu active metal brazes was to increase the peeling loads of the joints from 77 to 112N when 1 wt.% Ti alloy was used and from 32 to 120N when the Ti content was 3 wt.%, (Mizuhara and Mally 1985).

Of similar importance when preparing components prior to brazing is the need to clean their surfaces, (Schwartz 1995), so that microvoids and unbonded regions are not formed where the liquid metal contacts weakly bound and poorly wetted contamination such as grease. This is such standard practice that few data are available that exemplify the problems caused by failing to do so, but some of the effects of using different cleaning procedures can be realised by reference to the mechanical properties of diffusion bonded joints. Thus Table 10.7 summarises the effects of using different cleaning fluids to prepare the surfaces of Fe-0.45C samples prior to bonding by applying a pressure of 20 MPa for 10 minutes at 1200°C. Unfortunately, while some of the strength variations are quite marked, there is no supporting chemical evidence to explain why, for example, CCl_4 was a more effective cleaning agent than the other organic solvents.

Table 10.7. Effects of cleaning processes on the tensile strengths of diffusion bonded Fe-0.45C steel (Kazakov 1985).

Cleaning procedure	Interfacial strength (MPa)
none	320
degreasing with C_2H_5OH	460
degreasing with CH_2COCH_2	455
degreasing with CCl_4	570
pickling in acid	460

10.5.3 *Influence of interfacial interdiffusion*

By definition, brazes have a different composition from the components they are used to join and hence interdiffusion will occur during and after interface creation. Reference has been made already to the detrimental effects of the growth of thick reaction products at metal-ceramic interfaces and similar effects can occur with metal-metal systems. Thus it is not good practice to use Al brazes for the joining of steel or Cu components or to use Ni brazes containing Si for the joining of refractory metal components because of the rapid formation of fragile layers of intermetallic compounds.

However, some braze alloys are inherently brittle and interdiffusion with certain component materials can confer a degree of ductility. This effect is particularly important and often essential for the successful application of Ni brazes based on the Ni-11Si, Ni-11P or the Ni-3.6B eutectics such as those listed in Table 10.2 which wet steels and Ni super alloys well but solidify to produce continuous seams or dense clustered islands of brittle intermetallic along the braze centre lines unless interdiffusion causes egress of the temperature depressants. These solutes are quite mobile and hence significant egress from the joint can occur during a normal brazing cycle, but in practice it is usually necessary to use very prolonged braze dwells or subsequent heat treatment at a sub-solidus temperature to achieve complete egress. The effect of this chemical change is to convert the braze seam into a single-phase structure of a Ni-base solid solution that has greatly enhanced toughness and impact characteristics but is softer, as illustrated in Figure 10.10.

The detrimental effect of such intermetallic phases in Ni brazed joints is particularly severe when the microstructures are coarse, and this is true also of Si precipitated from Al-Si brazes that are slow cooled from the brazing temperature. Finally it should be noted that while egress of embrittling species can enhance the mechanical characteristics of the braze joints, their ingress can initially degrade

The enhancement of joint strength at room and elevated temperatures as the separation of the components is decreased has been observed with both metal-ceramic and metal-metal systems and data for a Ag brazed steel system are presented in Figure 10.11. The right hand branch of the curve in the figure was derived from equation (10.5) with w equal to 0.18 mm and an assumed value of 30.7 MPa for S_Y and fits the experimental data reasonably well. Extrapolation of the data lead to the expectation that the joint will be as strong as the steel when the gap is 0.007 mm or less but the most notable feature of the figure is the peaking of the joint strength at a value of about 105 MPa when the gap is about 0.06 mm. The weakening of the joint as the gap becomes even thinner is due to technical reasons rather than a failure of slipline field theory. It requires exceptional care to machine components to very precise dimensions so that the resultant capillary gaps are extremely narrow and of uniform width. Even if this is achieved, incomplete filling of the joints may occur because of the difficulty of sweeping oxide and flux debris out of very narrow gaps as the braze penetrates.

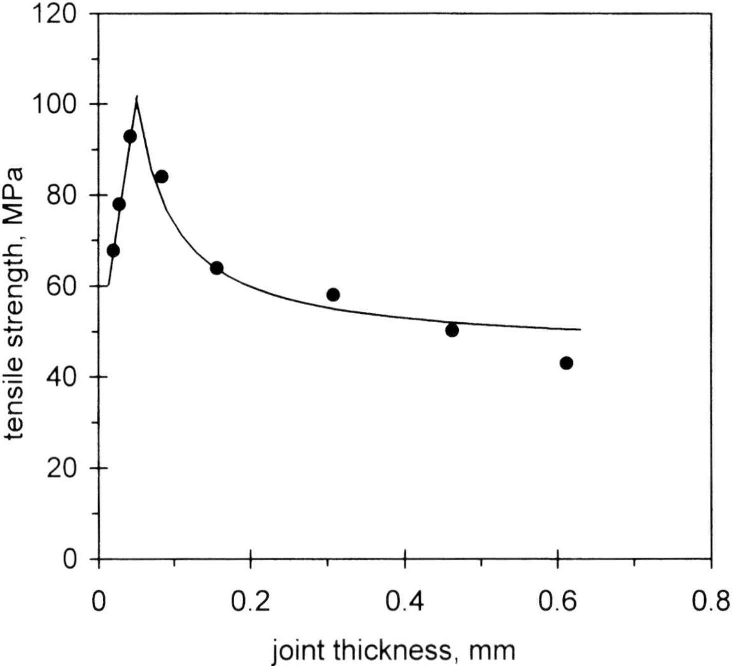

Figure 10.11. The room-temperature tensile strengths of stainless steel joints brazed with a fluxed Ag-Cu-Zn-Cd alloy. After (Udin et al. 1954).

Narrowing the joint gap affects not only strength but toughness characteristics. Thus both the toughness of low alloy steel samples brazed with Cu and the fracture energy of Al_2O_3 samples bonded with Au increased as the joint gap was widened, (Baker and Kavishe 1987, Reimanis et al. 1991). Therefore, while the thinnest filled joints may be the strongest, they can also have the greatest probability of failing prematurely because of easy crack propagation. It is interesting to therefore note that the maximum recommended joint gaps quoted in Table 10.3 for Ag and other braze families of 125-250 microns are significantly larger than the 60 microns suggested by the peak strength for Ag-brazed stainless steel in Figure 10.11.

REFERENCES FOR CHAPTER 10

Amato, I., Baudrocco, A. and Ravizza, M. (1972) *Weld. J.*, **51**, 341s

Ambrose, J. C., Jenkins, S. and Nicholas, M.G. (1988) *Brazing and Soldering*, **14**, 30

Ambrose, J. C., Nicholas, M. G. and Stoneham, A. M. (1993) *Acta Metall. Mater.*, **41**, 2395

Ambrose, J. C. and Nicholas, M. G. (1996) *Met. Sci. Technol.*, **12**, 72

ASTM F19-64 (1964) Standard Method for Tension and Vacuum Testing Metallized Ceramic Seals

Baker, T. J. and Kavishe, F. P. L. (1987) in *Proc. BABS 5th International Conference on High Technology Joining*, British Association for Brazing and Soldering, Abington, Cambridge, paper 21

Baker, T. S. and Partridge, P. G. (1991) in *Diffusion Bonding*, ed. R. Pearce, Cranfield Institute of Technology, Bedford, p. 73

Bird, R. B., Stewart, W. E. and Lightfoot, E. N. (1960) *Transport Phenomena*, Wiley International Edition, New-York

Bondley, R. (1947) *Electronics*, **20**, 97

Bredz, N. and Tennenhouse, C. C. (1970) *Weld. J.*, **48**, 189-s

BS1723 (1986a) *Brazing, Part 2 : Guide to Brazing*

BS1723 (1986b) *Brazing, Part 3 : Destructive Testing and Non-destructive Evaluation*

Bucklow, I. A., Dunkerton, S. B., Hall, W. G. and Chardon, B. (1992) in *4th Int. Symp. Ceramic Materials and Components for Engines*, ed. R. Carlsson, T. Johansson and L. Kahlman, Elsevier Applied Science, Stuttgart, Germany, p. 324

Burgess, J. F. and Neugebauer, C. A. (1973) *Direct Bonding of Metals with a Metal-Gas Eutectic*, US Patent 3744120

Burgess, J. F., Neugebauer, C. A. and Flanagan, G. (1975) *J. Electrochem. Soc.*, **122**, 6989

Cassie, A. B. D. and Baxter, S. (1944) *Transactions of the Faraday Society*, **40**, 546

Ceccone, G., Nicholas, M. G., Peteves, S. D., Kodentsov, A. A., Kivilahti, J. K. and van Loo, F. J. J. (1995) *J. Eur. Ceram. Soc.*, **15**, 563

Chang, Y. A., Goldberg, D. and Neuman, J. P. (1977) *J. Phys. Chem. Reference Data*, **6**, 621

Christensen, J. and Sheward, G.E. (1982) in *Behaviour of Joints in High Temperature Materials*, ed. T. G. Gooch, R. Hurst and M. Merz, Applied Science Publishers, London, p. 117

Cohen, J. M., Castle, J. E. and Waldron, M. B. (1981) *Metal Science*, **15**, 455

Courbière, M., Tréheux, D., Béraud, C., Esnouf, C., Thollet, G. and Fantozzi, G. (1986) *J. Physique*, Coll. C1, **47**, 187

Crispin, R. M. and Nicholas, M. G. (1976) *J. Mater. Sci.*, **11**, 17

Feduska, W. (1959) *Weld. J.*, **38**, 122-s

Gale, W. F. and Wallach, E. R. (1991) *Weld. J.*, **69**, 76-s

Guy, K. B., Humpston, G. and Jacobson, D. M. (1988) "Some Novel Developments in Aluminium Based Brazing Alloys", in *Problem Solving in Brazing and Soldering*, Proc. of Brit. Assoc. Brazing and Soldering Autumn Conference, Abington, Cambs., UK

Hey, A. W. (1990) in *Joining of Ceramics*, ed. M. G. Nicholas, Chapman and Hall, London, p. 56

Hill, R. (1950) *Mathematical Theory of Plasticity*, Oxford University Press

Holloway, D. G. (1973) *The Physical Properties of Glass*, Wykeham, London

Howe, J. M. (1993) *Int. Mater. Rev.*, **38**, 233 and 257

Iida, T. and Guthrie, R. I. L. (1988) *The Physical Properties of Liquid Metals*, Clarendon Press, Oxford

Ito, M. and Taniguchi, M. (1993) in *Designing Ceramic Interfaces II : Understanding and Tailoring Interfaces for Coating, Composites and Joining Applications*, ed. S. D. Peteves, Commission for the European Communities, Directorate-General XIII, Luxembourg

Johnson, S.M. and Rowcliffe, D. J. (1985) *J. Amer. Ceram. Soc.*, **68**, 468

Jordan, M. F. and Milner, D. R. (1956) *J. Inst. Metals*, **85**, 33

Kazakov, N. F. (1985) *Diffusion Bonding of Materials*, Pergamon Press, Oxford

Keller, D. L., McDonald, M. M., Heiple, C. R., Johns, W. L. and Hofmann, W. L. (1990) *Weld. J.*, **69**, 31

Kinloch, A. J. (1987) *Adhesion and Adhesive : Science and Technology*, Chapman and Hall, London

Kohl, W. H. (1967) *Handbook of Materials and Techniques for Vacuum Devices*, Reinhold, New York, p. 441

Kubachewski, O. and Hopkins, B. E. (1953) *Oxidation of Metals and Alloys*, Butterworths Scientific Publications, London

LaForge, L. H. (1956) *Bull. Amer. Ceram. Soc.*, **35**, 117

Latin, A. (1946) *J. Inst. Met.*, **72**, 265

Locatelli, M. R., Dalgleish, B. J., Tomsia, A. P., Glaeser, A. M., Matsumoto, H. and Nakashima, K. (1995) *Fourth Euro-Ceramic Conf.*, **9**, 109

Lugscheider, E. and Zhuang, H. (1982) *Schweissen u. Schneiden*, **34**, 490

Massalski, T. B. (1990) *Binary Alloy Phase Diagrams*, 2nd edition, ASM International

McDermid, J. R., Pugh, M. D. and Drew, R. A. L. (1989) *Met. Trans. A*, **20**, 1803

McGurran, B. and Nicholas, M. G. (1984) *Weld. J.*, **64**, 295-s

McHugh, G., Nicholas, M. G., Corti, C. W. and Notton, J. (1988) *Brazing and Soldering*, **15**, 19

Milner, D. R. (1958) *Brit. Weld. J.*, **5**, 90

Mizuhara, H. and Mally, K. (1985) *Weld. J.*, **64**, 27

Mizuhara, H. and Huebel, E. (1986) *Weld. J.*, **66**, 43

Moorhead, A. J. and Becher, P. F. (1987) *Weld. J.*, **66**, 26-s

Moret, F. and Eustathopoulos, N. (1993) *Journal de Physique IV*, Colloque C7, Supplement to Journal de Physique III, **3**, 1043

Moret, F., Sire, P. and Gasse, A. (1998) in *Proc. from Materials Conference '98 on Joining of Advanced and Specialty Materials*, 12-15 Oct. 1998, Rosemont (Illinois), ed. M. Singh, J. E. Indacochea and D. Hauser, published by ASM, p. 67

Nicholas, M. G. (1968) *J. Mater. Sci.*, **3**, 571

Nicholas, M. G. (1986) *Brit. Ceram. Trans.*, **85**, 144

Nicholas, M. G. and Ambrose, J. C. (1989) *DVS Berichte*, **125**, 19

Nicholas, M. G., Mortimer, D. A., Jones, L. M. and Crispin, R.M. (1990) *J. Mater. Sci.*, **25**, 2679

Reed, L., Wade, W., Vogel, S., McRae, R. and Barnes, C. (1966) *Metallurgical Research and Development for Ceramic Electron Devices*, AD 636950, Clearing House for Federal Scientific and Technical information, Washington, DC

Reid, C. G., Peteves, S. D. and Nicholas, M. G. (1994) *J. Mater. Sci. Lett.*, **13**, 1497

Reimanis, I. E., Dalgleish, B. J. and Evans, A. G. (1991) *Acta Metall. Mater.*, **39**, 3133

Schwartz, M. (1995) *Brazing for the Engineering Technologist*, Chapman and Hall, London

Scott, P. M., Nicholas, M. G. and Dewar, B. (1975) *J. Mater. Sci.*, **10**, 1833

Semlak, K. A. and Rhines, F. N. (1958) *Trans. Met. Soc. AIME*, **212**, 325

Shalz, M. L., Dalgleish, B. J., Tomsia, A. P. and Glaeser, A. M. (1992) *Ceram. Trans.*, **55**, 301

Thorsen, K. A., Fordsmand, H. and Praestgaard, P. L. (1984) *Weld. J.*, **62**, 339-s

Tomsia, A. P., Saiz, E., Dalgleish, B. J. and Cannon, R. M. (1995) in *Proc. 4th Japan International SAMPE Symposium*, Sept. 25-28, p. 347

Twentyman, M. E. (1975) *J. Mater. Sci.*, **10**, 765, 777 and 791

Udin, H., Funk, E. R. and Wulff, J. (1954) *Welding for Engineers*, Wiley, New York

van Houten, G. R. (1959) *Bull. Amer. Ceram. Soc.*, **38**, 301

Wan, C., Kritsalis, P., Drevet, B. and Eustathopoulos, N. (1996) *Mater. Sci. Eng. A*, **207**, 181

Yamada, T., Horini, M., Yokoi, K., Satoh, M. and Kohno, A. (1991) *J. Mater. Sci. Lett.*, **10**, 807

Zhou, Y., Gale, W. F. and North, T. H. (1995) *International Materials Reviews*, **40**, 181

Appendix A
The Laplace equation

Consider a spherical liquid (L) drop of radius r_d in a vapour V. The volume of this drop is increased slowly, for instance by using a syringe to inject fresh liquid inside the drop (Figure A.1). The increase of drop radius from r_d to $r_d + dr_d$ leads to an increase of the surface energy of the system equal to $d(4\pi r_d^2 \sigma_{LV}) = 8\pi r_d \sigma_{LV} dr_d$. If P_L is the pressure inside the drop and P_V the pressure in the vapour, the increase of r_d is associated with mechanical work to displace the surface by a distance dr_d, i.e. $(P_L - P_V)4\pi r_d^2 dr_d$. Equating the two amounts of work yields:

$$P_L - P_V = \frac{2\sigma_{LV}}{r_d} \qquad (A.1)$$

In the general case of a surface with a curvature $R_1^{-1} + R_2^{-1}$ where R_1 and R_2 are the principal radii (see Figure 1.8), equation (A.1) remains valid if $2/r_d$ is replaced by $R_1^{-1} + R_2^{-1}$.

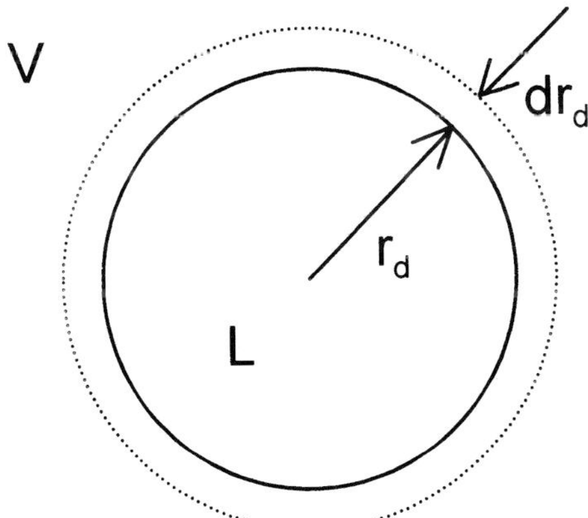

Figure A.1. Displacement of a liquid surface allowing derivation of the Laplace equation.

Appendix B

Free energy of formation of a meniscus on a vertical plate in the gravitational field

During the formation of a meniscus of height z^* (Figure 1.9) and for a triple line of unit length, the free energy change $\Delta F_{s,l}$ resulting from the substitution of a S/V surface by a S/L interface is:

$$\Delta F_{s,l} = z^*(\sigma_{SL} - \sigma_{SV}) \tag{B.1}$$

The meniscus rise z^* can be related to the instantaneous contact angle θ by means of the Laplace equation (1.20) relating the pressure difference across a curved surface at point Q to the curvature. The pressure inside the liquid is higher than that in the vapour phase for the configuration shown in Figure 1.8, but the opposite occurs in the case of a wetting liquid on a vertical plate (Figure 1.9.a) i.e., $P_L(z) < P_V(z)$. Moreover, because of the cylindrical symmetry, $1/R_2 \to 0$, so

$$P_L(z) - P_V(z) = \frac{\sigma_{LV}}{R_1} \tag{B.2}$$

and for the concave liquid surface in Figure 1.9.a, R_1 is negative. Since

$$P_L(z=0) - P_V(z=0) = 0 \tag{B.3}$$

if the differences in vapour pressure with z are neglected, $P_V(z=0) = P_V(z)$, then:

$$P_L(z) - P_L(z=0) = \frac{\sigma_{LV}}{R_1} \tag{B.4}$$

This difference is equal to the hydrostatic pressure of a column of height z. Accordingly,

$$\frac{\sigma_{LV}}{R_1} = -\rho g z \tag{B.5}$$

where ρ is the liquid density and g the acceleration due to gravity. The calculation of the radius of curvature at any point on the L/V surface and at the triple line leads to (Neumann and Good 1972):

$$z^* = \pm\left(\frac{2\sigma_{LV}}{\rho g}\right)^{1/2}(1 - \sin\theta)^{1/2} = \pm l_c(1 - \sin\theta)^{1/2} \qquad (B.6)$$

where a positive value of z^* corresponds to $0 \leq \theta \leq 90°$ and a negative value to $90 \leq \theta \leq 180°$ and $l_c = (2\sigma_{LV}/(\rho g))^{1/2}$. Substituting this expression for z^*, equation (B.1) becomes:

$$\Delta F_{s,1} = \pm(\sigma_{SL} - \sigma_{SV})l_c(1 - \sin\theta)^{1/2} \qquad (B.7)$$

The term $\Delta F_{s,2}$ accounts for the increase in the L/V area during meniscus formation and for a plate of unit width this can be related to the increase Δl in the length of the L/V line:

$$\Delta F_{s,2} = \sigma_{LV}\Delta l \qquad (B.8)$$

with:

$$\Delta l = \int_{x=0}^{\infty} (ds - dx) \qquad (B.9)$$

where the increment of L/V length, ds, is such that:

$$ds^2 = dx^2 + dz^2. \qquad (B.10)$$

Δl can be calculated by introducing equation (B.10) into equation (B.9) to obtain:

$$\Delta F_{s,2} = \sigma_{LV}l_c[2^{1/2} - (1 + \sin\theta)^{1/2}] \qquad (B.11)$$

The variation of the potential energy ΔF_b can be calculated by considering a small column of liquid of rectangular cross section 1.dx (we recall that the liquid column in the y-axis is of unit length). The column is composed of successive increments of volume 1.dx.dz. The work done in lifting each of these elements from

$z = 0$ to z is $\rho.g.z.dx.dz$. Integrating first over all the elements of the column and then over all the columns yields:

$$\Delta F_b = \frac{\rho g}{2} \int\limits_{x=0}^{\infty} z^2 dx \tag{B.12}$$

and then one gets:

$$\Delta F_b = \frac{\sigma_{LV} l_c}{3} [(2 - \sin \theta)(1 + \sin \theta)^{1/2} - 2^{1/2}] \tag{B.13}$$

By combining equations (B.7), (B.11) and (B.13), one obtains the expression (1.21) for the total free energy change (Neumann and Good 1972).

REFERENCE

Neumann, A. W. and Good, R. J. (1972) *J. Colloid and Interface Science*, **38**, 341

Appendix C
Contact angle hysteresis for heterogeneous solid surfaces

Consider a composite solid consisting of alternate strips parallel to the triple line of α and β phases with contact angles $\theta^\alpha = 60°$ and $\theta^\beta = 30°$. If the initial position (1) of TL corresponds to a macroscopic contact angle $\theta_M = 90°$ (horizontal surface of the liquid in Figure C.1.a), the movement of TL from this position up to the position n, corresponding to a line of separation from α to β with $\theta_{(n)} = 60° + \delta\theta$ ($\delta\theta > 0$), will occur spontaneously. Then, TL contacts a β phase with an intrinsic contact angle θ^β lower than $\theta_{(n)}$. Thus, displacement of TL over this β strip up to the position $(n + 1)$ will also be spontaneous. However, at this position, the contact angle is now $\theta_{(n+1)} = 60° - \delta\theta$ which is lower than the intrinsic contact angle θ^α of the next adjacent α strip. As a consequence, TL can be pinned in this position, leading to equation (1.42.a) ($\theta_a(max) = \theta^\alpha$). In practice, TL can move further by jumping over the least wetted α strips and the final position will depend on the vibrational energy of the system and the strip width. As a result, the observed contact angle will lie between $\theta_a(max)$ and the contact angle corresponding to the stable equilibrium for a heterogeneous surface i.e., θ_C (equation (1.41)).

Similarly, if we consider an initial position (1) of TL with $\theta_M \cong 0°$ (Figure C.1.b), the displacement of TL is spontaneous until it reaches a line of separation from α to β (position $(n + 1)$) with $\theta_{(n+1)} = 30° + \delta\theta$. As the next adjacent β strip has an intrinsic contact angle θ^β lower than $\theta_{(n+1)}$, θ_M can increase no more and TL is blocked in position $(n+1)$, a situation described by equation (1.42.b).

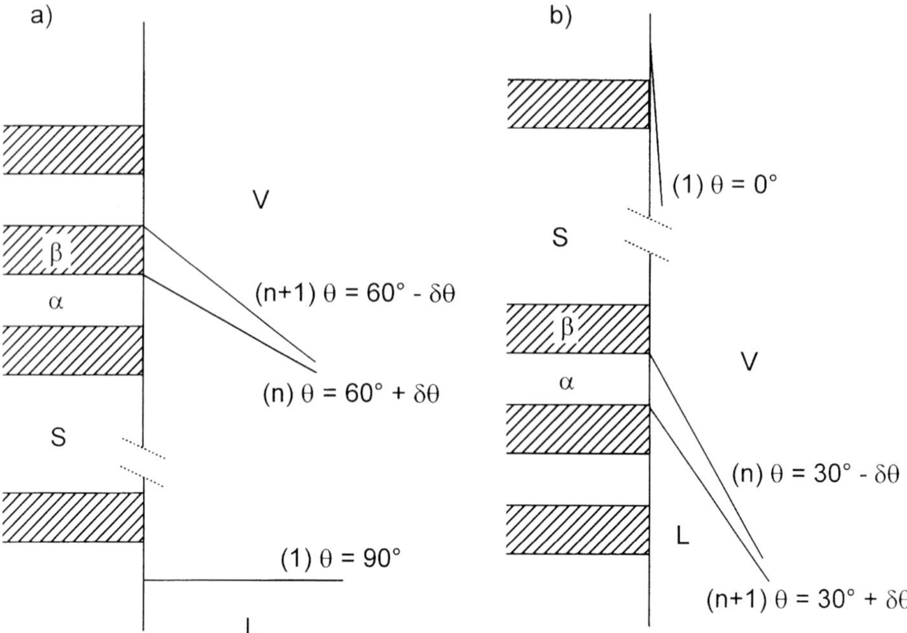

Figure C.1. Advancing (a) and receding (b) contact angles associated with TL positions (n + 1) on heterogeneous surfaces with α and β strips parallel to the triple line ($\theta^\alpha = 60°$, $\theta^\beta = 30°$).

Appendix D
Estimation of the mass of a sessile drop needed for an optimised σ_{LV} measurement

As shown in (Sangiorgi et al. 1982), a value of $\beta > 2$ (see equation (3.8)) is needed to obtain a high-accuracy measurement of σ_{LV}. This requirement can be satisfied using a drop with an optimised mass m_d. To estimate m_d, we need to link it to β through the values of the density ρ, σ_{LV}, θ and the drop volume v:

$$m_d = v^*(\rho g)^{-1/2}\sigma_{LV}^{3/2}\beta^{3/2} \tag{D.1}$$

with $v^* = v/b^3$ where b is the curvature radius at the drop apex. This equation is represented in nomographic form in Figure D.1. The procedure needs the approximate knowledge of the quantities θ, ρ and σ_{LV}. Take for example a metal with $\rho = 8$ g.cm^{-3}, $\sigma_{LV} = 1300$ mJ.m^{-2} and $\theta = 120°$. If the desirable value is $\beta = 3$, $m_d = 3$g.

REFERENCE

Sangiorgi, R., Caracciolo, G. and Passerone, A. (1982) *J. Mater. Sci.*, **17**, 2895

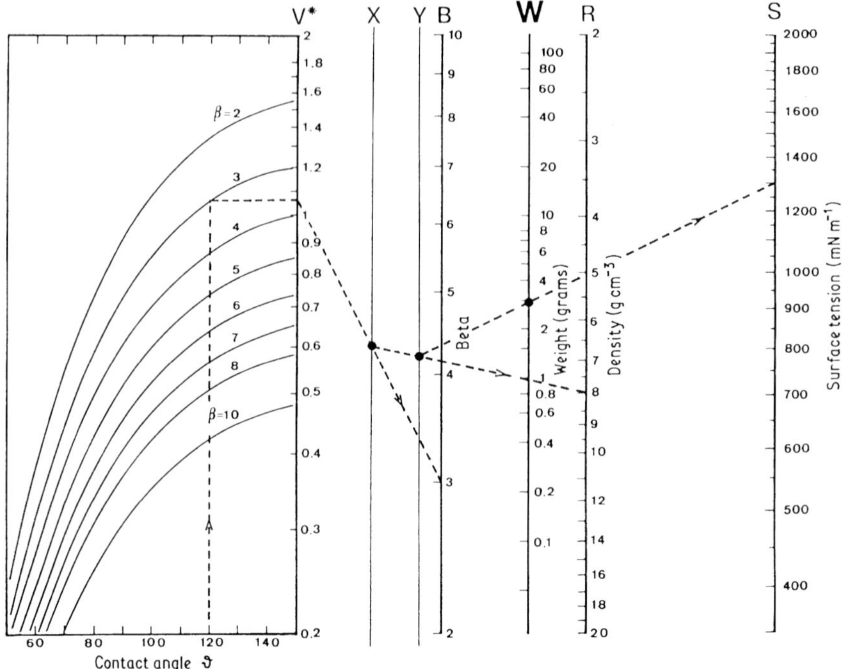

Figure D.1. Nomographic chart for determining the optimum drop mass (noted W on the figure) from a given set of σ_{LV} (noted S), ρ (noted R), θ and β (noted B) values. Reprinted from (Sangiorgi et al. 1982) [2] with kind permission of the authors.

Appendix E
Wetting balance : the case of cylindrical solids

The geometrical quantities used to describe the profile of a meniscus formed on a cylinder of radius r_0 are plotted on Figure E.1. Using the cartesian coordinates, the Laplace equation becomes:

$$\frac{d(x \sin \alpha)}{dx} = -\frac{\rho g}{\gamma_{LV}} zx \qquad (E.1)$$

where ρ is the liquid density and g the acceleration due to gravity.

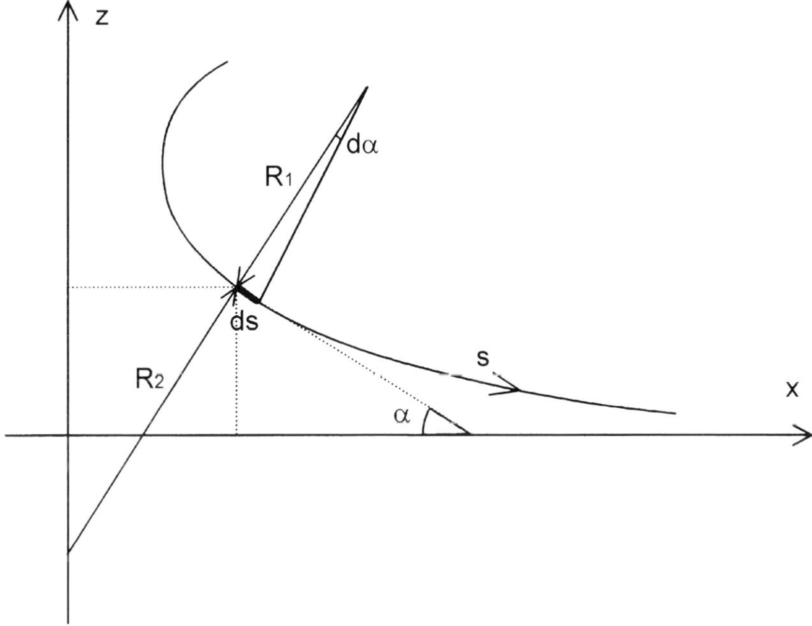

Figure E.1. Geometrical quantities used to describe the profile of a meniscus formed around a circular and constant section cylinder. From (Rivollet et al. 1990) [2].

The coordinates of the joining point of the liquid on the solid are z^*, $x_0 = r_0$ and $\alpha_0 = 90°$ - ϕ where ϕ is the joining angle defined in Figure 3.16. Hartland and

393

Although a rigorous evaluation of the uncertainty on the estimated values of σ_{SV} for Si and diamond cannot be made, it is expected to be less than 50%.

The (111) faces of cubic SiC (as well as basal planes of hexagonal SiC) are polar i.e., can terminate either as a C layer or a Si layer (see Section 7.1.1). However, following the nearest-neighbour model of Skapski, the surface energy of the two faces is the same, and lies between those for Si and diamond. The value of the surface energy of a (111) face of cubic SiC can be estimated from data for Ge using an equation analogous to (F.2). Knowing that $\Delta H_f^{SiC} \cong (L_s^{Si} + L_s^{C} - 2L_s^{SiC})$, the sublimation energy L_s^{SiC} is first calculated from the experimental value of enthalpy of formation of SiC, ΔH_f^{SiC}, and from sublimation energies of Si, L_s^{Si}, and diamond, L_s^{C}. Then equation (F.2) yields $\sigma_{SV}^{SiC} = 1450$ mJ/m^2. As the atomic structure of the (0001) face of hexagonal α-SiC is very similar to that of the (111) face of cubic SiC, the calculated value of σ_{SV}^{SiC} can be used for both faces. The value of 1450 mJ/m^2 is lower but close to the value of 1767 mJ/m^2 estimated by Takai et al. (1985) using Monte-Carlo simulations for a relaxed (111) face of cubic SiC.

REFERENCES

Bardsley, W., Frank, F. C., Green, G. W. and Hurle, D. T. J. (1974) *J. Cryst. Growth*, **23**, 341

Eustathopoulos, N. (1983) *International Metals Reviews*, **28**, 189

Hultgren, R., Desai, P. D., Hawkins, D. T., Gleiser, M., Kelley, K. K. and Wagman, D. D. (1973) *Selected Values of the Thermodynamic Properties of the Elements*, American Society for Metals, Metals Park, Ohio 44073

Lucas, L. D. (1984) in "Densité", Traité Matériaux Métalliques M65, Techniques de l'Ingénieur, Paris

Skapski, A. (1948) *J. Chem. Phys.*, **16**, 386

Takai, T., Halicioglu, T. and Tiller, W. A. (1985) *Surf. Sci.*, **164**, 341

Appendix G
Enthalpy of mixing of binary liquid alloys

Table G.1 reproduces values calculated by Miedema's model (Niessen et al. 1983) for the partial enthalpy of solution at infinite dilution of a liquid metal solute i in a liquid metal solvent j, $\overline{\Delta H}_{i(j)}^{\infty}$ (in kJ/mole). For a i-j alloy, the regular solution parameter λ can be approximated by $[\overline{\Delta H}_{i(j)}^{\infty} + \overline{\Delta H}_{j(i)}^{\infty}]/2$.

Table G.1. Enthalpy of solution (in kJ/mole) at infinite dilution (Niessen et al. 1983).

solvent→ ↓ solute	Ag	Al	Au	Bi	Cd	Cr	Cu	Fe	Ga	Ge	In	Mn
Ag	0	-18	-21	6	-8	119	10	123	-20	-21	-5	57
Al	-17	0	-80	31	12	-43	-34	-48	3	-8	24	-83
Au	-23	-92	0	6	-43	-1	-42	37	-79	-39	-40	-53
Bi	9	48	9	0	5	136	35	146	20	24	-6	16
Cd	-9	14	-41	4	0	81	3	77	3	-7	2	10
Cr	98	-36	-1	74	60	0	51	-6	-2	-23	66	9
Cu	8	-28	-30	19	2	49	0	50	-19	-24	8	15
Fe	102	-41	28	80	58	-6	53	0	-6	-12	63	1
Ga	-21	4	-75	14	3	-3	-26	-8	0	-12	10	-60
Ge	-21	-8	-36	16	-6	-28	-30	-15	-11	0	-3	-84
In	-7	31	-45	-5	2	102	12	95	12	-4	0	13
Mn	46	-70	-39	9	7	8	15	1	-45	-69	8	0
Mo	146	-20	13	123	103	2	82	-9	25	-3	114	22
Na	1	58	-52	-69	-12	329	74	276	21	-39	-21	235
Ni	56	-82	25	31	7	-27	14	-6	-53	-41	5	-33
Pb	13	50	10	0	7	153	41	160	22	23	-3	38
Si	-13	-9	-48	47	13	-87	-40	-75	-1	9	23	-120
Sn	-14	19	-46	5	-1	53	-6	56	4	0	-1	-37
Ta	61	-78	-125	32	35	-30	9	-67	-37	-101	48	-17
Ti	-6	-118	-180	-45	-28	-33	-40	-74	-88	-157	-19	-36
U	1	-132	-177	-53	-33	-13	-33	-53	-101	-170	-27	-25
V	63	-61	-69	33	32	-8	21	-29	-29	-70	40	-3
W	172	-8	44	151	124	4	101	0	41	20	135	28
Zr	-87	-189	-302	-140	-103	-58	-110	-118	-163	-260	-94	-74

Table G.1. (continued)

solvent→ ↓ solute	Mo	Na	Ni	Pb	Si	Sn	Ta	Ti	U	V	W	Zr
Ag	150	1	68	9	-14	-10	60	-6	1	73	171	-78
Al	-20	46	-96	34	-9	14	-75	-119	-114	-69	-8	-164
Au	14	-52	33	8	-57	-38	-135	-203	-171	-87	48	-294
Bi	191	-80	58	0	77	6	47	-69	-69	57	229	-184
Cd	118	-11	9	5	16	-1	39	-33	-34	42	136	-106
Cr	1	220	-27	86	-77	32	-24	-27	-9	-8	4	-41
Cu	67	49	14	23	-34	-3	7	-33	-23	19	80	-78
Fe	-7	195	-6	91	-67	34	-54	-62	-38	-28	0	-85
Ga	28	18	-69	16	-2	3	-40	-98	-96	-36	44	-156
Ge	-3	-31	-51	16	10	0	-99	-159	-148	-80	20	-225
In	148	-20	8	-3	32	-1	61	-25	-31	58	171	-107
Mn	18	152	-33	21	-106	-22	-14	-30	-18	-3	23	-52
Mo	0	311	-32	139	-76	70	-19	-14	8	0	-1	-23
Na	389	0	139	-64	23	-31	392	312	274	342	391	257
Ni	-27	100	0	40	-86	-13	-105	-126	-98	-69	-11	-165
Pb	207	-70	73	0	75	7	73	-38	-40	81	242	-147
Si	-72	18	-98	48	0	21	-149	-190	-175	-128	-54	-239
Sn	97	-32	-21	6	30	0	-13	-101	-99	-4	125	-188
Ta	-20	302	-133	51	-165	-10	0	6	12	-4	-30	10
Ti	-15	225	-154	-26	-202	-74	5	0	-1	-7	-23	-1
U	9	232	-139	-32	-217	-84	14	-1	0	4	5	-13
V	0	231	-75	49	-121	-3	-4	-6	3	0	-3	-13
W	-1	326	-14	166	-58	92	-29	-23	4	-3	0	-34
Zr	-27	208	-236	-117	-299	-161	12	-1	-13	-17	-39	0

REFERENCE

Niessen, A. K., de Boer, F. R., Boom, R., de Châtel, P. F., Mattens, W. C. M. and Miedema, A. R. (1983) *Calphad*, **7**, 51

Appendix H
Secondary wetting

Consider a drop of non-reactive liquid metal on a smooth surface of a polycrystalline metallic solid (Figure H.1). After a first fast spreading, a much slower spreading process is often observed, called "secondary wetting" (Bailey and Watkins 1951-52, Sharps et al. 1981, Weirauch and Krafick 1996, Lequeux et al. 1998) that is generally attributed to enhancement of wetting by grain boundaries. This can be due to an increase of surface roughness (leading to a better wetting for systems with $\theta_Y \ll 90°$, according to the Wenzel equation (1.35)) produced by grain-boundary grooves formed either at the free S/V surface or at the S/L interface (see Figure 4.8).

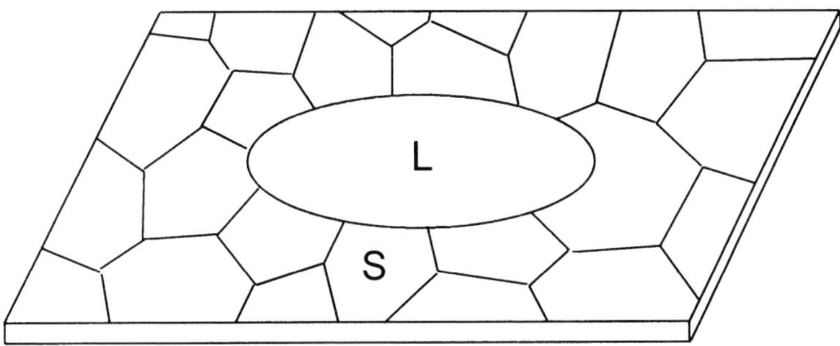

Figure H.1. Sessile drop on a polycrystalline solid.

Grooves formed at the S/V surface create open "capillaries" on the solid surface along grain boundaries ahead of the triple line which can be infiltrated by the liquid (Figure H.2.a). The change in interfacial free energy of the system for a small linear displacement δx of the liquid along the groove, causing a replacement of a S/V surface by an equal area of S/L interface and by a L/V surface of area $(2h\sin(\alpha/2)\theta_Y\delta x)/(\cos(\alpha/2)\sin(\theta_Y))$ (Lorrain 1996) is:

$$\delta F_s = \frac{2h}{\cos(\alpha/2)}\left[\frac{\sin(\alpha/2)}{\sin(\theta_Y)}\theta_Y.\sigma_{LV} + \sigma_{SL} - \sigma_{SV}\right]\delta x \qquad \text{(H.1)}$$

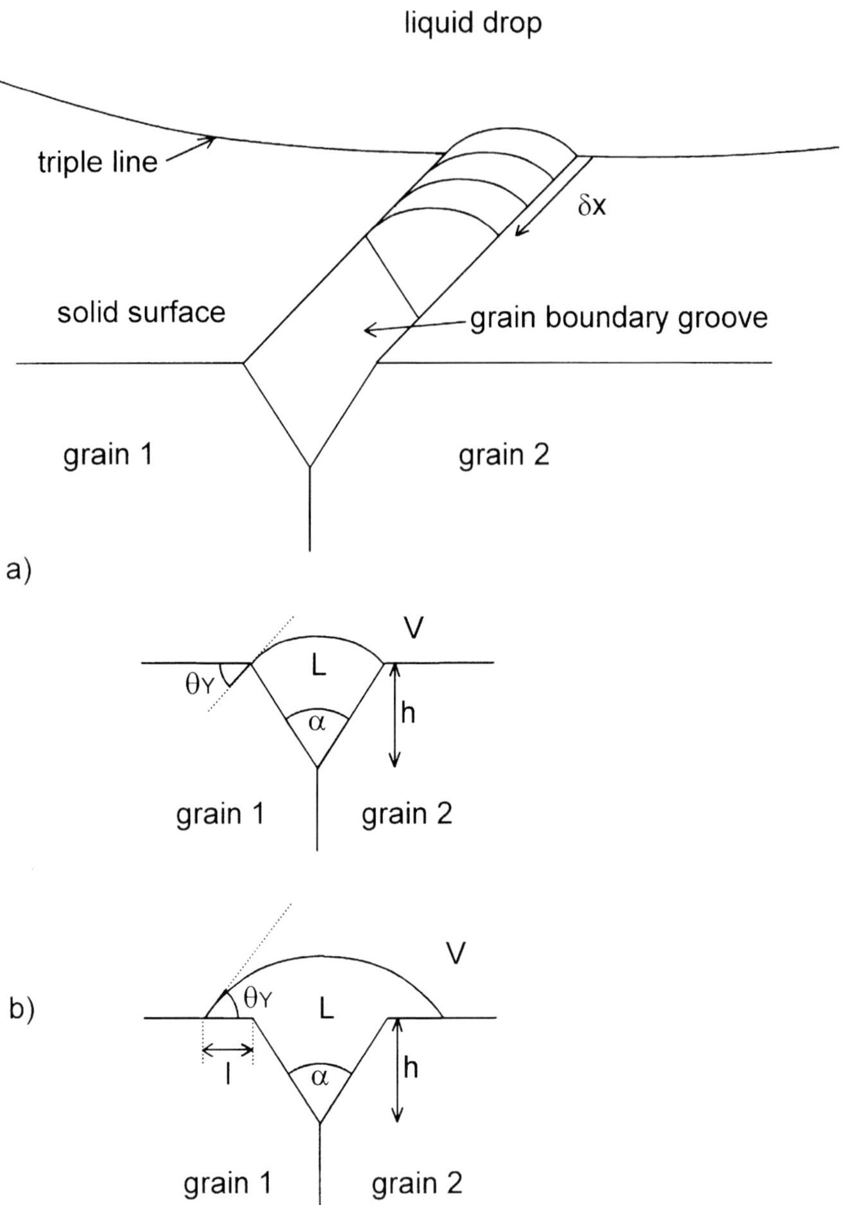

Figure H.2. a) Infiltration of pre-existing grain boundary grooves at S/V surface. b) Extension of the liquid on both sides of the groove.

Introducing the Young equation into (H.1) and setting $d(\delta F_s)/d(\delta x) = 0$, shows that infiltration is possible if the experimental dihedral angle is lower than the value α^* given by:

$$\sin\left(\frac{\alpha^*}{2}\right) = \frac{\cos(\theta_Y)\sin(\theta_Y)}{\theta_Y} \tag{H.2}$$

As $0 \leq \alpha \leq 180°$, this equation cannot be verified for $\theta_Y > 90°$, so that secondary wetting is not expected to occur for non-wetting liquids. For $\theta_Y = 50°$, $\alpha^* = 69°$ and for $\theta_Y = 20°$, $\alpha^* = 134°$. Experimentally, the dihedral angles ψ formed on *thermal* grain boundary grooves are usually higher than $150°$ (see Table 4.4) so that this type of groove cannot affect wetting significantly. However, values of the dihedral angles Φ formed at S/L interfaces lie between 0 and $140°$ (Eustathopoulos 1983) for different systems and temperatures, so the condition $\Phi < \alpha^*$ may hold in many cases. Thus spreading on polycrystalline materials can continue by flow of liquid to fill grooves of grain boundaries that radiate from the nominal triple line (Figure H.3).

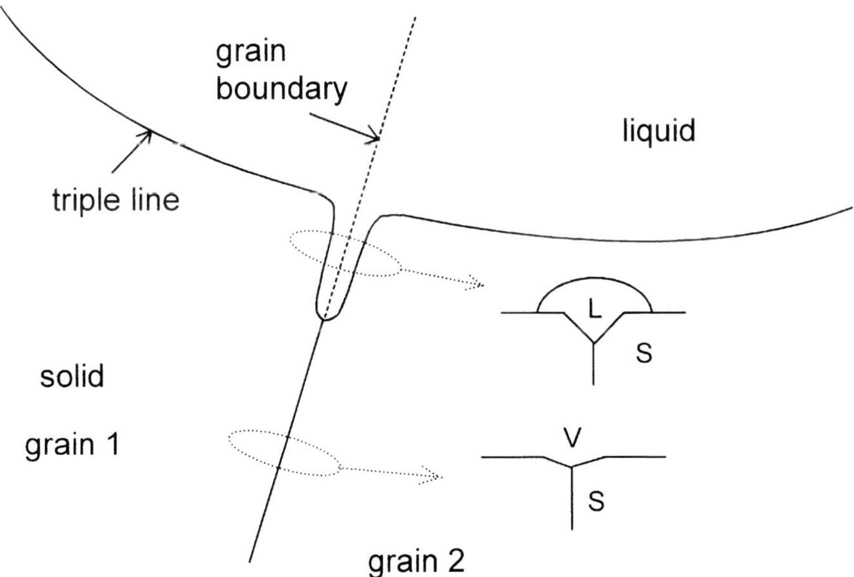

Figure H.3. Formation of a grain boundary groove filled by liquid ahead of the nominal triple line.

Appendix I
Evaluation of the work of adhesion of Ni on SiC

This evaluation will be made using experimental results for the work of immersion of non-reactive Ni-Si alloys on SiC (Section 7.1.1.1) using the atomistic model described in Section 6.5.1. According to this model, the slope of the σ_{SL} versus X_{Ni} curve for $X_{Ni} \to 0$ is related to the adsorption energy of Ni at the Si/SiC interface, $E_{Ni(Si)}^{\infty,SL}$, by:

$$\left. \frac{d\sigma_{SL}}{dX_{Ni}} \right|_{X_{Ni} \to 0} = \frac{RT}{\Omega_m} \left(1 - \exp - \left(\frac{E_{Ni(Si)}^{\infty,SL}}{RT} \right) \right) \tag{I.1}$$

with

$$\begin{aligned} E_{Ni(Si)}^{\infty,SL} &= -(W_a^{Ni} - W_a^{Si})\Omega_m + (\sigma_{LV}^{Ni} - \sigma_{LV}^{Si})\Omega_m - m_l\lambda \\ &= [(\sigma_{SL}^{Ni} - \sigma_{SV}) - (\sigma_{SL}^{Si} - \sigma_{SV})]\Omega_m - m_l\lambda = [W_i^{Ni} - W_i^{Si}]\Omega_m - m_l\lambda \end{aligned} \tag{I.2}$$

In this equation, Ω_m is the molar surface area, m_l is a structural parameter defined in Section 1.1 (see Figure 1.3) and λ is the regular solution parameter of Ni-Si alloy defined by equation (4.3). From the slope of the $\sigma_{SL}(X_{Ni})$ curve for $X_{Ni} \to 0$, the adsorption energy is found to be $E_{Ni(Si)}^{\infty,SL} = -8.2 \text{ kJ/mole}$. Thus, in equations (I.2), all the quantities are known (or can be easily estimated), except W_a^{Ni} and W_i^{Ni} which represent respectively the work of adhesion and the work of immersion of pure liquid Ni in metastable equilibrium with SiC (i.e., for a supposed non-reactive pure Ni/SiC system). The values deduced from equation (I.2) are $W_a^{Ni} = 3.17 \text{ J/m}^2$ and $W_i^{Ni} = -1.35 \text{ J/m}^2$ for pure Ni. They are reported in Figure 7.6 along with the corresponding value of contact angle.

List of symbols

Fundamental constants

g	acceleration due to gravity
k	Boltzmann constant
N_a	Avogadro's number
R	gas constant

Abbreviations

L	liquid
MPE	multiphase equilibrium
P	reaction product
S	solid
TL	triple line
V	vapour
2D(3D)	two (three) dimensions

Common symbols

a_i	thermodynamic activity of component i
b	radius of curvature at sessile drop apex
C	concentration
d	width of a parallel sided gap or joint
D	diffusion coefficient
e	thickness of reaction product at the solid/liquid interface
E	potential energy of a bulk atom
E'	potential energy of a surface atom
E_b	energy barrier
E_c	cohesion energy
E_{diss}	dissipated energy
E_g	gap energy
E_p	potential energy of a spreading drop
E_v	vibrational energy
$E_{i(j)}^{\infty,LV}$	adsorption energy at liquid/vapour surface of solute i at infinite dilution in solvent j

$E_{i(j)}^{\infty,SL}$	adsorption energy at solid/liquid interface of solute i at infinite dilution in solvent j
f	force
f_c	compacity factor
f_d	driving force
f^i	surface fraction of phase i
f_{in}	inertial force
f_v	viscous force
F	free energy
F_b	free energy of bulk phase
F_s	surface free energy
G	Gibbs energy
$\Delta G_{f(i)}^0$	standard Gibbs energy of formation of compound i
$\overline{\Delta G}_{i(j)}^{xs,\infty}$	partial excess Gibbs energy of mixing at infinite dilution of a solute i in a solvent j
h	height of wetting ridge or depth of grain boundary groove
H	sessile drop height
$\Delta H_{f(i)}$	enthalpy of formation of compound i
$\Delta H_{f(i)}^0$	standard enthalpy of formation of compound i
$\overline{\Delta H}_{i(j)}$	partial enthalpy of mixing of component i in component j
$\overline{\Delta H}_{i(j)}^{\infty}$	partial enthalpy of mixing at infinite dilution of a solute i in a solvent j
ΔH_m	enthalpy of mixing of an alloy
K_1	constant first used in equation (2.4)
K_2	constant in equation (2.20)
K_D	equilibrium constant of a dissolution reaction
K_P	equilibrium constant of reaction of formation of a new phase
l	distance
l_c	capillary length
l_e	equilibrium rise or depression of a liquid in a parallel sided gap
L_e	energy or heat of evaporation
L_f	length of a facet on a sawtooth surface
L_m	energy or heat of melting
L_s	energy or heat of sublimation
m_1	for a bulk atom, fraction of nearest neighbour lying in an adjacent plane
m_2	for a bulk atom, fraction of nearest neighbour lying in the same plane
m_d	mass of a drop
n	power term in equation (1.33)
n	coordination number of an oxygen atom dissolved in a metallic phase (equation (6.20))

n_i	number of moles of component i
p	perimeter
P	pressure
P_c	capillary pressure
P_h	hydrostatic pressure
P_i	partial pressure of species i
P_{N2}^d	nitrogen partial pressure for dissociation of a nitride
P_{O2}	oxygen partial pressure
P_{O2}^f	oxygen partial pressure in the furnace atmosphere
P_{O2}^l	equivalent oxygen partial pressure at a solid/liquid interface
$P_{O2}^{d(MO)}$	oxygen partial pressure for decomposition of oxide MO
$P_{O2}^{ox(M)}$	oxygen partial pressure for oxidation of metal M
P_{sat}	partial pressure at saturation of a species
r	radius
r_d	radius of a spherical drop
R	base radius of a sessile drop
R_0	initial sessile drop base radius
R_1, R_2	principal radii of curvature of a surface
R_a	average height of asperities on a surface
R_F	final or stationary drop base radius
s	length of line on a liquid surface
s_r	roughness parameter of Wenzel (equation (1.35))
S_J	joint strength
S_Y	yield strength
S_S	surface excess entropy
$\overline{\Delta S}_{i(j)}^{xs,\infty}$	partial excess entropy of mixing of solute i at infinite dilution in solvent j
t	time
t_0	time corresponding to the initial contact angle θ_0
t_F	time needed to reach a constant drop base radius R_F
T	temperature
T_F	melting temperature
u	fluid velocity
U	triple line velocity
U_0	initial triple line velocity
U_r	rate of reaction
v	volume
v_m	molar volume
w	length of a parallel sided joint

w_m	weight of a meniscus
W_a	work of adhesion
W_a^0	work of adhesion in the absence of adsorption on the solid
W_c	work of cohesion
W_d	work of capillary forces (equation (2.2))
W_i	work of immersion
W_s	work of spreading
x	horizontal coordinate
x_{90}	equatorial radius of a sessile drop
X_i	molar fraction of component i
X_i^j	molar fraction of component i in phase j
X_i^I	molar fraction of component i at the solid/liquid interface
X_i^{eq}	molar fraction of component i for a reaction at equilibrium
X_i^J	molar fraction of component i at point J corresponding to a three phase equilibrium (figure 7.2)
X_O^S	molar fraction of oxygen at the liquid surface
Y_i	molar fraction of component i adsorbed at surface or interface monolayer
Y_i	molar fraction of component i around oxygen atoms dissolved in an alloy i-j (equation (6.19))
z	vertical coordinate
z_{90}	equator-apex height of a sessile drop
z^*	meniscus height
z_{max}^*	maximum (or equilibrium) meniscus height
z_b	depth to which the base of a solid is immersed in a liquid bath
z_e	equilibrium rise or depression of a liquid in a circular cross section capillary
Z	coordination number i.e. number of nearest neighbour atoms
α	angle
α	reduction of the metallic bond energy due to oxygen (equation (6.20))
α	evaporation coefficient
β	slope of surface asperities
β	inclination of a surface to the vertical
β	parameter in equation (3.7)
ε	elastic strain
ε_{ij}	bond energy between i and j atoms
ε_O^i	Wagner interaction parameter between oxygen and solute i
ϕ	joining angle of a meniscus on a solid surface
Φ	dihedral angle of grain boundary groove at solid/liquid interface

Φ_i	dihedral angles at a triple line ($i = 1, 2, 3$)
Φ_O	flow of oxygen
γ_i	activity coefficient of species i
γ_i^∞	activity coefficient of species i at infinite dilution in a solvent
γ_{LV}	liquid/vapour surface tension
γ_{SV}	solid/vapour surface tension (equation (1.3))
Γ_i	absolute adsorption of species i
Γ_i^j	relative adsorption of species i with respect to species j
η	dynamic viscosity
φ	angle between the normal to the liquid surface and the axis of rotation of a sessile drop
κ	curvature
λ	regular solution parameter
$\lambda_{i\text{-}j}$	regular solution parameter of a binary i-j solution
λ_a	average wavelength of asperities on a surface
Λ	Gibbs-Thomson parameter
μ	chemical potential
θ	contact angle
θ_0	initial contact angle or contact angle in the absence of reaction (equation (2.14))
θ_a	advancing contact angle
θ_{app}	apparent contact angle
θ_C	Cassie contact angle on a chemically heterogeneous surface
θ_{eq}	equilibrium contact angle
θ_{equiv}	equivalent contact angle of a small solid drop
θ_F	final or stationary contact angle
θ^i	contact angle on solid phase i
θ_i	contact angle of liquid phase i
θ_M	macroscopic contact angle
θ_P	equilibrium contact angle on a reaction product
θ_r	receding contact angle
θ_W	Wenzel contact angle on a rough surface
θ_Y	Young contact angle
ρ	mass density
σ_{LV}	liquid/vapour surface energy
σ'_{LV}	temperature coefficient of liquid/vapour surface energy
σ_{SV}	solid/vapour surface energy
σ_{SV}^d	van der Waals component of solid/vapour surface energy
σ_{SV}^0	solid/vapour surface energy in the absence of adsorption or reaction

Acknowledgements

[1] Reproduced from Entropie with kind permission.

[2] Reproduced from Journal of Materials Science with kind permission from Kluwer Academic Publishers.

[3] Reproduced from Journal de Chimie Physique with kind permission from EDP Sciences.

[4] Reproduced from Reviews of Modern Physics, Copyright 1985, with kind permission from American Physical Society and P.G. de Gennes.

[5] Reproduced from Journal of Colloid and Interface Science with kind permission from Academic Press Inc. and R. L. Hoffman.

[6] Reproduced from Physics and Chemistry of Glasses with kind permission from Society of Glass Technology.

[7] Reproduced from Acta Materialia, Copyright 1998, with kind permission from Elsevier Science.

[8] Reproduced from Acta Materialia, Copyright 1997, with kind permission from Elsevier Science.

[9] Reproduced from Proceedings of the Second International Conference "High Temperature Capillarity-97" with kind permission from Foundry Research Institute, Cracow, Poland.

[10] Reproduced from Acta Metallurgica, Copyright 1987, with kind permission from Elsevier Science.

[11] Reproduced from Materials Science Forum with kind permission from Trans Tech Publications Ltd.

[12] Reproduced from Journal of Crystal Growth, Copyright 1998, with kind permission from Elsevier Science.

[13] Reproduced from Metallurgical Transactions with kind permission from The Minerals, Metals & Materials Society.

[14] Reproduced from Transactions of the Iron and Steel Institute of Japan with kind permission from the Iron and Steel Institute of Japan.

[15] Reproduced from La Revue de Métallurgie with kind permission.

[16] Reproduced from Annales de Chimie Science des Matériaux with kind permission from Gauthier-Villars/ESME – 23 rue Linois – 75724 Paris cedex 15.

[17] Reproduced from Materials Science and Engineering, Copyright 1998, with kind permission from Elsevier Science.

[18] Reproduced from Transactions of JWRI with kind permission from K. Nogi, Joining and Welding Research Institute, Osaka University.

[19] Reproduced from Acta Metallurgica, Copyright 1972, with kind permission from Elsevier Science.

[20] Reproduced from Journal of Materials Science Letters with kind permission from Kluwer Academic Publishers.

[21] Reproduced from Acta Metallurgica, Copyright 1988, with kind permission from Elsevier Science.

[22] Reproduced from Materials Science and Engineering, Copyright 1996, with kind permission from Elsevier Science.

[23] Reproduced from Acta Materialia, Copyright 1999, with kind permission from Elsevier Science.

[24] Reproduced from Acta Metallurgica et Materialia, Copyright 1993, with kind permission from Elsevier Science.

[25] Reproduced from Solid State Phenomena with kind permission from Scitec Publications Ltd.

[26] Reproduced from Metallurgical and Materials Transactions with kind permission from The Minerals, Metals & Materials Society.

[27] Reproduced from British Ceramic Society Transactions with kind permission from The Institute of Materials.